Lecture Notes in Earth Sciences 70

Springer
Berlin
Heidelberg
New York
Barcelona
Budapest
Hong Kong
London
Milan
Santa Clara
Singapore
Paris
Tokyo

Roger H. Charlier   Christian P. De Meyer

Coastal Erosion

Response and Management

Springer

Authors

Roger H. Charlier, Ph. D., Lit. D. Sc. D.
Professor emeritus
University of Brussels, Bordeaux and Northeastern Illinois
HAECON Inc.
2 avenue du Congo, Bloc 23, B-1050 Brussels, Belgium

Christian P. De Meyer
Civ. Eng., Cand. Sc. D.
President and C.E.O. HAECON Inc.
110 Deinsesteenweg, B-9031 Drongen – Gent, Belgium

"For all Lecture Notes in Earth Sciences published till now please see final pages of the book"

Library of Congress Cataloging-in-Publication Data

Charlier, Roger Henri.
Coastal erosion : response and management / Roger H. Charlier, Christian P. De Meyer.
p. cm. -- (Lecture notes in earth sciences, ISSN 0930-0317 ; 70)
Includes bibliographical references and index.
ISBN 3-540-60022-1 (softcover : alk. paper)
1. Shore protection. 2. Coast changes. 3. Coastal zone management. I. Meyer, Christian P. de, 1942- . II. Title. III. Series.
TC330.C43 1997
627'.58--DC21 97-22225
CIP

ISSN 0930-0317
ISBN 3-540-60022-1 Springer-Verlag Berlin Heidelberg New York

Typesetting: Camera ready by authors
SPIN: 10482929 32/3142-543210 - Printed on acid-free paper

This book has received the support of the University Foundation of Belgium.
Dit boek kreeg de steun van de Universitaire Stichting van België.
Ce livre a bénéficié de l'appui de la Fondation Universitaire de Belgique.

To those who assisted me while writing these pages and to Connie, Jac and Pat
R.H.C.

To my spouse and children
C.P.D.M.

"Lo ! The unbound sea !"

Walt Whitman

PREFACE

Coastal erosion has been discussed in papers published in periodicals (scientific and popular). newspapers and at every congress. symposium and conference addressing the problems of coastal zone study or management. No wonder : at the dawn of the 21st century. human populations are migrating in a steady river-like flow to shores.

In fact. a science of coastal zone management has evolved. and autodidactic managers are being succeeded by professionally trained colleagues. Institutes aiming at preparing workers able to cope with economics. exploitation and protection are multiplying. with the college of the European Federation for Environmental Management Training the most recent (1995). Coastal defense is a major concern of regional managers and national planners alike. Unless solutions are worked out economic disaster will strike some areas. e.g. the Atlantic coasts and the Caribbean. and island countries may be engulfed (Pacific Ocean). The coastal erosion problem has also been treated in several books. mostly as chapters. The present modest volume attempts to cover succinctly the way in which coastal erosion relates to and affects coastal zone management. the causes and mechanics of such erosion. and measures to counteract it. These measures have been placed under the headings of "hard" and "soft" methods. but other recent developments have been included. Considering the interdisciplinary nature of the problems. the material has been approached so as to be of use to managers. economists. earth scientists. and engineers. providing integrated information and guidelines.

While the authors and their coworkers have published several papers dealing with these topics over the past three decades. the book is in part an outgrowth of a three-week workshop on coastal protection and mathematical modeling held in Indonesia. under the auspices of the Belgian government (Bandung, 1991). Mathematical modeling is not included in this book in order to keep it within a reasonable length. Modeling is more "execution" than management. and therefore is more appropriately handled in a separate work.

The authors appreciate the contributions made by engineers E. Blomme. C.E. (Part II). Fransaer and Claeys. members of the workshop teaching team (Part III ; Part IV. *partim* dredging methods). The papers by Van Rijn and others have been adapted in part for use in Part III as they were used in the Bandung Workshop. Papers of other researchers have also been included.

We are grateful to engineer Daniel Decroo. co-author of several papers and parts of the workshop manual. who gave considerable advice. especially regarding some chapters of Part IV.

The workshop was sponsored by and organized for the Belgian Administration for Development Cooperation. A kudo is due to the firms that participated in the Bundung workshop. The manuscript was produced by HAECON and the authors wish to thank its secretarial staff, particularly Mrs. Eveline De Bock, for their patience, efforts and cooperation. Anne-Marie De Wilde provided considerable help in proof-reading.

American English is used in the text. This is due to the fact that the first author spent 35 years teaching at university level in the United States. If the European computer occasionally balked at abiding by such spelling, we apologize.

We hope that this book will prove helpful in guiding decision-makers. It must be kept in mind that, as in medicine where "there are no illnesses, only patients", there are no standard solutions but there are regional situations requiring case-tailored medicine. There is no cure for shoreline retreat but there are steps that can be taken to alleviate and control the situation. Our aim has been to outline, rather comprehensively, what those steps are.

Drongen, Belgium

Prof. Roger H. Charlier
ir. Christian P. De Meyer

TABLE OF CONTENTS

PART I : THE COASTAL ZONE

1. INTRODUCTION 3
2. REFERENCES & NOTES 4

CHAPTER I : COASTAL ZONE MANAGEMENT
1. INTRODUCTION 6
2. THE USE OF THE COASTAL ZONE 6
3. MANAGEMENT 10
4. MANAGEMENT AND ENVIRONMENT 13
5. CONCLUSION 16
6. REFERENCES 18

PART II : COASTAL MANAGEMENT AND COASTAL DEFENSE

INTRODUCTION 21
CHAPTER I : RESPONSE STRATEGIES 22
1. RETREAT 22
2. ACCOMODATION 24
3. PROTECTION 25

CHAPTER II : ENVIRONMENTAL IMPLICATIONS OF ADAPTIVE MEASURES
1. INTRODUCTION 26
2. RETREAT 26
3. ACCOMODATION 27
4. PROTECTION 27

CHAPTER III : ECONOMIC IMPLICATIONS OF ADAPTIVE MEASURES
1. INTRODUCTION 29
2. RETREAT 30
3. ACCOMODATION 30
4. PROTECTION 31

CHAPTER IV : SOCIAL AND CULTURAL IMPLICATIONS OF ADAPTIVE MEASURES
1. INTRODUCTION 35
2. RETREAT 35
3. ACCOMODATION 36
4. PROTECTION 37

CHAPTER V : LEGAL AND INSTITUTIONAL IMPLICATIONS OF ADAPTIVE MEASURES
1. INTRODUCTION 38
2. RETREAT AND ACCOMMODATION 39
3. PROTECTION 40

CHAPTER VI : PRIORITIES FOR ADAPTIVE RESPONSES
1. INTRODUCTION 41
2. PRIORITIES FOR ADAPTATION 42
3. PRIORITIES FOR IMPLEMENTING ADAPTIVE OPTIONS IN DEVELOPING COUNTRIES 44
4. CRITERIA FOR ALLOCATION OF RESOURCES 45
5. REFERENCES 46

PART III : SEDIMENT TRANSPORT

CHAPTER I : BED MATERIAL SUSPENSION AND TRANSPORT IN STEADY UNIFORM CURRENTS
1. BED LOAD TRANSPORT 51
2. SUSPENDED LOAD TRANSPORT 56
3. TOTAL LOAD TRANSPORT 79
4. REFERENCES (IN ALPHABETICAL ORDER) 82

CHAPTER II : BED MATERIAL SUSPENSION AND TRANSPORT IN UNIFORM WAVES
1. INTRODUCTION 85
2. ANALYSIS OF MEASURED CONCENTRATION PROFILES AND TRANSPORT RATES 88
3. COMPUTATION OF TIME-AVERAGED CONCENTRATION PROFILES 97
4. REFERENCES 104

CHAPTER III : DETECTION OF SUSPENDED SEDIMENT
1. PRINCIPLE ... 107
2. REFERENCES ... 111

CHAPTER IV : SEDIMENT TRANSPORT
1. INTRODUCTION ... 112
2. LONGSHORE TRANSPORT ... 113
3. CROSS-SHORE TRANSPORT ... 117

CHAPTER V : COASTAL MORPHOLOGY
1. INTRODUCTION ... 122
2. COASTAL PROFILES ... 122
3. COASTAL MORPHOLOGY MODELS ... 129
4. BEACHES BACKED BY DUNES ... 139
5. MATHEMATICAL MODELING OF BEACHES BACKED BY SEAWALLS ... 141
6. REFERENCES ... 148

PART IV / RETREATING SHORELINES

CHAPTER I : COASTAL EROSION ... 155
1. INTRODUCTION ... 156
2. WAVES ... 159
3. COASTAL PROFILES ... 168
4. COASTAL EROSION, A WORLDWIDE PHENOMENON ... 173

CHAPTER II : COASTAL PROTECTION
1. INTRODUCTION ... 179
2 "HARD" COASTAL DEFENSE WORKS ... 194
3. BEACH NOURISHMENT AS COASTAL PROTECTION ... 222
4. SOCIAL ACTION ... 283
5. REFERENCES ... 285

CHAPTER III : SEDIMENT BUDGET ANALYSIS
1. THE SEDIMENT "BANK ACCOUNT" ... 296
2. BUDGET ANALYSIS ... 297
3. THE CELL THEORY ... 299
4. CONCLUSION ... 299
5. REFERENCES ... 300

CHAPTER IV : EVALUATION OF IMPACTS AND INTERACTIONS
1. INTRODUCTION ... 301
2. IMPACT OF EXISTING STRUCTURES AND SITES ... 304
3. THE BEACH/DUNE SYSTEM ... 305
4. IMPACTS AND INTERACTION ... 309

APPENDICES 311

GLOSSARY 323

SOURCES OF ILLUSTRATIONS 335

INDEX 337

PART I

THE COASTAL ZONE

1 Introduction

1.1 Attraction and Problems

Global changes, global warming, greenhouse effect, sea level rise, are as many terms that have joined the already lengthy list of such in-the-news topics as pollution, environmental impact, ozone layer, sustainability, population explosion. Had the word "catastrophism" not yet been tapped to designate an orogenic theory, it could qualify as a catchword for today's problems. There is more : travel the central spine of Spain, let us say from Zaragoza on, then veer towards Valencia, and one rapidly gets the feeling of being alone. Populations have left. The phenomenon is not unique : everywhere, but more so in already industrialized regions, man is migrating towards cities, in quest of a supposedly higher standard of living, and especially towards shore areas, in search of space, cleaner air, and water[1].

In the United States, for instance, predictions forecast an occupance rate of the coastal fringe that could reach 70 to 80% within the next decade or so. Are there already 50% of populations settled in the coastal zone ? Doubts have been expressed whether these estimates are correct and that as many people will live within 50 miles (90 km) from the coastline by the year 2000 or 2010. But 54% is believed to be realistic, while a 27% increase has occurred in a 20 years span. As a consequence, the coastal zone is under stress, a situation compounded by keen competition among existing and potential users[2-10]. Concern for the future of the coastal zone has sprouted a great number of management plans, ranging from the US Coastal Zone Management Act (CZMA) to legislation and scenarios covering countries in South America, Australia and Asia. While individual countries have put forth some management schemes, Europe lacks a coordinated policy. Present shifts in population and the search for littoral space by a broad array of users, constitute a strong motivation to work out plans for future management.

The concept of coastal zone is not defined with geographical precision ; in fact, it even varies with geographical location, and also with the discipline to which the specialists belong. On one thing, though, they all agree : its crucial importance for mankind.

Taking into consideration the physical reality, and the human activities, a coastal zone encompasses expanses on both sides of the "land-sea boundary", the inner part of the coastal shelf, and a hinterland. Its dynamics, in a delicate equilibrium, are easily affected by natural processes and anthropic activities. Their impact can be felt at considerable distance in time and space ; among man-induced processes are deforestation, river damming, pollution - both domestic and industrial - and marine mining.

Subject to natural evolution trends and several geological hazards, the coastal zone has their effects compounded by human interference. Frequently a close relationship exists between environmental quality and some important economic activities.

Coastlines are generally retreating. Coastal erosions cost can be very high. Remedial action requires familiarization with the genesis and the history of the beach, examination of the current state of the system, and the natural evolution trends.

Hard structures protection is expensive and requires continuous upkeep ; generally, it transfered the erosion problem further down the coast. Beach protection and restoration, can be attained and coast erosion remedied immediately - but frequently such actions dismay both residents and vacationers by numerous prohibitions, dune restoration, beach nourishment with coarse sand, abandon of buildings, setting of a construction boundary line, end of flood control projects on rivers.

With sea-level rise how long is it sensible to try to hold back the sea ? Economic realities notwithstanding the forever stand will have to be abandoned and after each storm consideration will have to be given not to rebuild some homes. How well a community ultimately adapts to sea-level rise will depend largely on the direction it takes when it reaches this crossroads.

Coastal protection management can consider stabilization, let nature take its course, or compromise. In the latter instance land-leasing, rather than selling, and the use of movable buildings seem indicated. A management plan that ignores rising sea-levels, a gnawing erosion, silting access routes, is futile. Disastrous floods do not affect only Bangladesh coasts ; spectacular inundations wrought havoc on the Iberian littorals. Storms can wipe out a beach in a matter of hours, geological subsidence is not uncommon, seismic and volcanic activity affect many a coastal area.

2 References & Notes

1. S.F. Edwards quoted in R.H. Charlier, "Coastal zone : occupance, management and economic competitiveness" *Ocean and Shoreline Manag.* **12,** nos 5/6, 383-402 (1989).

2. Anonymous, "Recent developments in ocean and coastal management" *Ocean and Shoreline Manag.* **12**, 1, 89-100 ; *ibid.* **12,** 3, 363-364 (1989) ;

A. Brandani & E. Schnack, "Conference report. First Latin-American course/seminar on planning and managment of the coastal zone and the Exclusive Economic Zone" *J. Shoreline Man.* **2,** 73-81 (1986) ; J. Mitchell, Coastal zone management, a comparative analysis of national programs. **In** : *Ocean Yearbook III* (E.M. Borgese & N. Ginsburg, eds.) (University of Chicago Press, Chicago, 1982) pp. 258-319 ; J. Sorensen & A. Brandani, "An overview of coastal management efforts in Latin America "*Coast. Z. Man. J.* **15**, No 1, 1-26 (1982) ; L. Van Herwerden & R. Bally, "Shoreline utilization in a rapidly growing metropolitan area : the Cape Peninsula, South Africa "*Ocean & Shoreline Man.* **12**, No 2, 169-178 (1989).

3. R.H. Charlier *op. cit.* note 1 ; *id.*, "Pour une politique européenne de prévention des risques littoraux" *Act. Colloq. Mer et Littoral, Couple à risque* (Biarritz), La Documentation Française, Paris (1987), pp. 409-432 ; P. Koekkebakker & G. Peet, "Coastal zone planning and management in The Netherlands" *Coast. Z. Man. J.* **15**, 121-134 (1987) ; M. Vigneaux, "L'exploitation de la zone côtière : ses perspectives et ses risques pour l'environnement" *Act. Coll. Mer et Littoral, Couple à risque* (Biarritz). La Documentation Française, Paris (1987) pp. 299-310.

4. A. Cendrero & R.H. Charlier, "Resources, use and management in the coastal zone" *Geolis* **III**, No 2, 40-60 (1989).

5. Mitchell & Gosselinck (1986) cited in Cendrero & Charlier note # 4.

6. R.H. Charlier & M. Vigneaux, "Ocean utilization and economic conflicts in the coastal zone" *Int. J. Env. St.* **26**, No 3, 177-189 (1985) ; *ibid.* **26**, No 4, 271-278 (1985).

7. R.H. Charlier, "Marine mineral resources extraction in coastal areas and its impact on the environment and consequences for land use" **In** : *Mineral resources extraction, environmental protection and land-use planning in the industrial and developing countries* (P. Arndt & G. Luettig, eds.) (Schweizerbart Verlag, Stuttgart, 1987) pp. 53-70.

8. R.H. Charlier & C.P. De Meyer, "An environmental purpose artificial island offshore Belgium" *Int. J. Env. St.* **40**, 249-265 (1992).

9. P.R. Burbridge, N. Dankers & J.R. Clark, "Multiple-use assessment for coastal management" *Coast. Z. '89* **I,** 33-45 (1989).

10. J.R. Clark, "*Program Development for Management of Coastal Resources*" (Coast. Man. Coast. Res. No 4, USAIS/NPS Series) Rosenstiel School of Atmosph. Sci., Miami FL, 1989) [112 p.].

Chapter I : Coastal Zone Management

1 Introduction

The concept of "coastal zone" or "coastal fringe" is thus rather vaguely defined. It varies not only according to the geographical area, but also according to the point of view of the different specialists. Here, the term is understood to include the inner part of the continental shelf, the coastline and a hinterland a few kilometres in width. No attempt is made here to establish precise limits.

The coastal zone is the interface between the atmosphere, hydrosphere and lithosphere. It constitutes a high energy zone with very active and complex dynamics, in which the Quaternary record shows clearly a history of considerable instability. It is logical to assume that this instability is going to continue, probably even to increase because of human interference.

The dynamics of the coastal zone are in many cases a situation of delicate equilibrium, easily affected by processes, natural or man-induced[1].

Therefore, solving land-use problems in the coastal zone requires, more than in other regions, careful planning[2]. This needs prior knowledge of the area to be exploited. Thus, research and inventory must precede land- and resource-use planning, so that sensible exploitation of resources and conservation of the environment are ensured[3].

2 The Use of the Coastal Zone

Undoubtedly, an economic conflict exists between those who champion the total utilization of the coast for industrial, mining and commercial expansion, and the partisans of a "balanced" occupance between, on the one hand industry, and on the other tourism and habitat interests[4].

2.1 Types of Economic Activities

Some long established activities, are now being displaced by new ones : fisheries, artisanal or industrial, and mariculture, port-related, commercial and industrial complexes, energy production, the exploitation of mineral resources, to name a few. Mariculture is a broad term covering "new aquaculture", such as fish husbandry in sea farms, conchyliculture and algae harvesting.

Almost everywhere, industrial activities develop, sometimes linked to maritime transport - naval construction and repairs - but embracing diverse demands such as steel and iron works, refineries and petrochemical installations, agrifood industries and others.

Electrical power stations on the coast are fairly common on sites near harbors and estuaries, because of the increased demand for energy in these areas. The construction of nuclear power plants reflects the need for cooling water. In fact, the coastal zone is often suitable for the construction of tidal-, wave- and ocean thermal electrical plants.

Tourism activity flourishes over an ever-increasing mileage. Coastal zones in Third World countries are no exception. The steadily growing demand for tourist facilities triggers an important development of coastal accommodations (villas and other buildings, camping grounds, pleasure harbors) entraining an increase in environmental quality problems.

Coastal tourism adds new dimensions to coastal development. If investors are willing to risk making substantial investments, one can only conclude that customers are willing to pay steep prices.

Exploitation of biological resources, once an artisanal endeavor, now industrialised in many developed nations, will soon be in the developing countries, requiring construction of more sophisticated harbor facilities, ships and electronic equipment repair shops, new factories and cold storage warehouses. The exploitation of mineral resources, offshore and in the open ocean, for its part will call for the construction of treatment facilities, refineries, oil and gas pipeline terminals[5,6].

Mining operations in the coastal zone could, and in spots already, include coal (extracted underneath the sea-floor since the sixteenth century), petroleum, natural gas, sulphur, tin, sand and gravel, ... The economic importance of sand, gravel and coquina has steadily increased due to the demand for building materials where land supplies are depleted or in short supply, (Charlier 1990,1992)[7]. Mining must be carefully monitored since it may severely alter benthic life on the seabed, while equilibrium may be endangered. Gravel extraction is more problematic than sand extraction, because layers of 10-50 m thickness frequently overlie the gravels. Sea-shell deposits (often exploited for cement) are an important source of calcium ; they can be abundant in bays, lagoons and estuaries.

Nearshore mining of sand or other minerals can disrupt beach stability and endanger barrier islands accelerating erosion processes by removing the sand that naturally replenishes beaches. The matter was examined several decades ago in connection with operations along the coasts of Lebanon and Israel[8].

The removal treatment and tailings rejections of all such materials can seriously influence the benthos, filterers, beach stability and water clarity.

Seawater itself is an important resource. Desalination costs remain high, and such operations are economic only where water supplies on land are limited, as in some petroleum-rich countries.

Relatively shallow continental shelves are the habitat of abundant biological resources, allowing the exploitation of fish, crustaceans and other invertebrates. Consumer fisheries share the markets with industrial fisheries used for animal feed and food industries, ink, glue, and paint manufacture. Marine plants are harvested. Coastal zone management needs to protect such economically sound activities.

The currently modest exploitation of ocean energy is potentially expandable. Small plants of regional or insular significance may replace or supplement those burning costly fossil fuels, a heavy financial burden for non-petroleum producers.

2.2 Conflicting Interests

Beyond matters physical and geological, exploitation of a coastal zone triggers off many economic and social consequences. Whether an activity is coastal, offshore or open ocean sited, such activities as processing, transportation and marketing of marine products will have socio-economic and political ramifications. Economic and environmental conflicts between various types of users of the coastal zone are often inevitable, and deep-seated. Priorities always create a dilemma for coastal zone management.

An area used for touristic purposes may be ideally suited for the establishment of a harbor, a petroleum terminal, or an unsightly but sorely needed artificial island with a waste processing factory.

This raises the dilemna of choosing between creation of new long term, and well-paid, jobs, and seasonal work for the fisherman or for service jobs during the holiday season.

Among the economic aspects of a mining operation in the coastal zone, one must consider the nature of risks involved, alternative solutions of resource allocation and indemnities to be paid to the resource owner, types of benefits to be drawn from mining, and the economic rent. Furthermore, responsible parties must scrutinize resource conservation problems and ensure that the resource be extracted at an acceptable level. This means a tailored mineral extraction plan as a function of its environmental location and the extraction technology, a considerable task requiring a high level of competency.

To reiterate the underlying situation, the problem to solve is to establish how economic development of the marine milieu can be reconciled, and how to simultaneously protect the environment while avoiding, or at least reducing, conflicts between various competing ocean-user groups. The concept of *sustainability* has evolved of these environmental considerations.

3 Management

Any human activity, albeit residential, industrial or commercial will more or less directly impact on the coastal zone (onshore, nearshore and offshore), and on the coastal shelf in general. The exploitation of this region must be carefully planned, its consequences duly examined, and its environmental and social impacts assessed. Management of such region requires an in-depth study of the best resources to exploit, decision-making as to choice, and constant vigilance and monitoring, within environmental imposed constraints[9].

3.1 Adaptation, Planning and Land-Use

Planning for coastal zone management must consider proposed utilization, whether single or multiple. Conflicts inevitably ensue as urbanization, commercial and industrial development, tourism and traditional and, even artisanal, occupations vie for space. A general occupation scheme is required rather than a step by step policy or one of individual freedom. It must consider the exploitation at the chosen site of the proposed activities, which requires a study of human factors, adaptation to the local geographical conditions, and efforts to satisfy the space requests, while paying attention to aesthetics, conservation and environmental protection of both the coastal zone and hinterland. From the economics viewpoint, planning will include a search for financing sources and an assessment of the comparative benefit-cost ratios.

It is self-evident, yet important, that the nature and the stability of a coast will determine to a high degree the action to be taken and the scenario to be followed.

The coastal zone ought also to be managed with a view for its touristic and recreational potential, its transport, communication and even waste dumping possibilities, though the latter are severely legislated, restricted and foreseeably may be completely prohibited. The proportion of beaches in use for recreation is steadily expanding as a result of improved public transportation, the "shortening" of distances by air travel, and the attractiveness of far-away places - all corollaries of social change following the two world wars. Yet, one may deplore the anarchy which marked the post- World War II touristic occupation of the coastal zone[10].

If a touristic use is to be assigned to a coastal area, so that industrial and service jobs be created, the planning process should include site reconnaissance, in-depth environmental study, exploitation, and management. As far as developing countries are concerned, coastal zone use decisions should be based on multidisciplinary scientific studies by a team made up of specialists from academe and industry. Only at a later stage should exploitation be considered. The expansion of coastal tourism, the use of the littoral fringe for recreation and parcelling of the dunal belt have all helped to accelerate erosion by destroying dunes and by slowing down beach restoration after its Winter regression.

3.2 Management and the Multidisciplinary Approach

Any attempt at economic development that is based on adaptation to, and conservation, of the littoral fringe can only be effectively carried out if a policy of fundamental and pluridisciplinary research is concurrently pursued. Such a global approach will allow an understanding of the especially complex and varied processes which operate in the coastal zone, and will also provide the basic information needed for a scientific and technical foundation for coherent management of the littoral fringe.

Management and use of the coastal zone require pluridisciplinary investigations which, although the disciplines involved all belong to the domain of oceanology "*sensu largo*", necessitate expansion of the field of study generally considered as "marine science" (Charlier and Vigneaux 1973).

The coastal fringe is the scene of many diverse and interactive processes, and it has to be borne in mind that any coastal project will produce perturbations in the environment.

In view of its complexity, the coastal environment obviously constitutes a unique milieu, wherein the various scientific disciplines able to help in improving the knowledge base cannot be separated. Any problem pertaining to a coastal zone must necessarily be studied through the efforts of teams made up of physicists, biologists, geologists, chemists, marine science specialists, geographers, economists and engineers.

Such collaboration often involves the pooling of techniques and data, depending on the magnitude of the actions to be deployed, and may call for cooperation on an international scale.

Rational management of the coastal region requires, at the outset, a detailed study of the marine topography and the compilation of surface geological maps. Knowledge of the sediments, and of sedimentary processes is needed ; among them the study of the benthic boundary layer, the carbonate cycle, geochemistry and sediment diagnosis, and the role played by organic substances. Equally important are facies analysis and paleo-oceanography inclusive of a biostratigraphic study.

3.3 A Bird's Eye View of Legislation

A management program is usually launched by a government agency reacting to resource degradation, exposure to major hazards, utilization conflicts, or to the need for social-economic development. Such a program should be carried out on a continuous basis rather than being a one shot undertaking. If implemented, a policy of decision allocation must be established. Programs will be designed on the basis of, and follow, a pluri-sector approach, taking into account the reciprocity between the coastal environment systems and public services. The program will have flexible landward limits, which may vary with time. Three decades ago, the Belgian "littoral" was a strip 1 -2 km wide, rapidly grading into pastoral landscapes shielded by the dunal barrier. Present-day development has removed farms, destroyed traditional occupations, and impacted the hinterland, in places affecting cities situated far inland.

Planning may involve a single district, or, if there are several of them, may still consider them individually neglecting the overall approach. “At present, programs to study coastal processes ... tend to set up independent surveillance activities that ignore the interactions and so overlook the biological processes. This is an area where greater interdisciplinary interaction is badly needed ... : there are broad zonal patterns in ecosystems that transcend ... local variations.”

In South America, for instance, all countries, with the exception of Surinam, established management programs for coastal fisheries, and nearly all provided for the preservation of parks and nature reserves, but without coordination between these two sectors. In practice, therefore , only Brazil, Ecuador, Colombia and Costa Rica have "real" programs.

North Sea riparian states have paid small heed to the management of their coastal zones ; however, a determined approach is gaining momentum to protect and control the future development of tourism, habitat, industry, and to insure coast protection. Coastal activities in each country are often uncoordinated, and the matter is left to the national government, although area utilization falls within the municipal jurisdiction.

The United States' Coastal Zone Management Act (CZMA) mirrors a national concern to harmonize the demands of urbanization, recreation, industry and energy development in the littoral fringe, and recognizes the recurrent incompatibility of these uses. The Act underscores the nefarious role of anarchic development generated by repeated development demands. More than 56 million hectares (138 million acres) are affected by this legislation ; it provides a financial inducement to regional jurisdictions to set up rules aiming to solve opposing requests involving the coastal zone, to establish a priority order for its use, and to determine by whom and how final decisions will be made.

4 Management and Environment

Conflicts between the various users of the coastal zone make if difficult for planners and ecologists to make the right decisions. Further conflict arises in trying to reconcile development and the environment. This is reflected in the philosophy of the (US) Coastal Zone Management Act which aims not so much at saving the *status quo* as in harmonizing environmental protection and development.

4.1 Preservation, Protection and Environmental Equilibrium

4.1.1 Pollution Management

Even though pollution management should be on an international scale, regional management may of itself protect and improve existing conditions. Drawing chemical, physical and biological baselines, determining distribution and dispersion mechanisms, monitoring pollution levels, and gathering valid prognoses about the effect, upon the environment, of new activities and of the introduction of new technologies in the coastal zone are necessary.

Several international and national organizations, both governmental and non-governmental concern themselves with ocean pollution. The United States took measures as long ago as 1899, when coastal pollution first became serious (note the Rivers and Harbors Act). Management of the coastal zone has to face both occasional pollution incidents and more continuous pollutants emissions.

4.1.2 Coastal Erosion Management

To properly monitor the coast, profiling, at frequent intervals during each year, has to be undertaken. The profiles should be taken close to protective structures. Aerial photographs, at a precise scale, can be taken to survey changes in the shoreline. Observations made from satellites hold a great deal of promise.

Besides the "do-nothing" and the "retreat" approaches, now favored in many cases, coastal protection may involve building of hard defenses such as groins, breakwaters, seawalls, the application of soft methods, i.e. artificial nourishment, or (often) a combination of several methods.

4.2 Environmental Impact

Some fundamental questions must be answered in devizing a protective plan:

- what is the genesis and history of the area ?
- what is the present state of the system ?
- what are the natural evolution trends ?
- what will the impact of the proposed action be ?

Once the answers are known, the limits of the natural milieu tolerance, in relation to the planned installations, must be determined. The follow-up stage, once a project is completed, monitors the environment's adaptation and reaction.

The understanding of the spatial scale of phenomena requires a site study. The site's evolution must be explained through the study of a series of parameters which quite often go beyond the immediate geographic zone. Hence, answers to a particular problem frequently come from parameters and characteristics geographically some distance from the site itself. In a similar vein, the environmental impact may be felt, favorably or otherwise, some distance away from the site in question.

The timescale involved in environmental responses (physical and biological) to natural and man-made events may vary considerably. A catastrophic pollution event can lead to a rapid and almost total destruction of the flora and fauna in the contaminated zone. Again, excessive exploitation of biological resources, exceeding the natural capacity of the species to reproduce, may result in a gradual extinction of these species. The limits of sustainability have then been exceeded. At sea, discharges of toxic products, even in relatively small quantities, can (over a period of years) have equally catastrophic consequences, particularly when toxic substances enter foodchains and concentrate to high, and increasingly toxic, levels. In coastal sedimentology, beach evolution provides an excellent example of the diversity of time scales. Indeed, sediment transport phenomena on the littoral fringe may be recognized, operating at different scales from near-instantaneous transport such as sediments placed in suspension due to waves, to a millennium variation of the shoreline, with yearly and seasonal changes in between.

A first step is the formulation of a "health report" for the coast, in which three types of zones can be identified :

a) zones that are still intact and able to undergo some development in the near future ;
b) endangered zones, because they are particularly fragile, such as estuaries, deltas, lagoons, or sandy shores undergoing continual evolution ;
c) already damaged zones, because of chemical and/or biological pollution, or because of insensitive development (e.g. sites of camping grounds or construction at the sea's edge liable to erosion).

Based on this identification, two kinds of action can be undertaken : continuous and individual. So far, France, for example, has achieved :

Continuous actions :

- type "Réseau National d'Observation de la Qualité du Milieu Marin" : survey and control of water quality in clearly bounded zones ;
- type "Surveillance Ecologique du Blayais" : establishment of an ecological reference state prior to development, followed by monitoring of the milieu's evolution after development (e.g. a nuclear power station near Blaye);
- the case of an unconsolidated littoral in continual evolution, with on-going surveillance in the form of a sedimentary dynamic study (collection of beach profiles at regular intervals, determination of volumes of eroded or accreted sediments, and so on).

Individual actions :

- impact study prior to implementation of any new development ;
- impact assessment of various discharges upon the quality of water which could lead to ecosystems modification, even though the marine environment can "absorb" waste far better than its land counterpart through dilution, dispersion and degradation ;
- assessment of impacts of new constructions near the edge of the sea on the shoreline's evolution ;
- managing pollution ; limitation of discharges, purification stations, ...
- managing coastal erosion : provision of defense structures.
- Finally, the development of protected areas is necessary. One must define "control zones" and monitor their integrity, in order to safeguard a part of the natural patrimony, so that it can be conserved for future generations, and also to preserve the equilibria that are necessary, e.g. in fish spawning grounds.

5 Conclusion

The coastal zone is subject to such pressures, not counting anthropication, as erosion, sea-level rise, hazards, but thus also to exploitation and increased migration. While any management scheme must provide for relief of pollution, natural and otherwise, it must face the reality of sea-level rise and of retreat of the shoreline.

Subsidence and anthropogenic activities such as oil-, gas-, sand extraction, and barrage construction, contribute to beach retreat, further aggravated by offshore airport-runway or artificial islands construction, coral reef exploitation, mining and tourism depredations.

Multiple-use of coastal zones has been practiced in many geographical areas, and usually flourished harmoniously. But contemporary competitiveness differs greatly from ancestral arrangements. Management systems often fail to harness the resource potential offered by natural systems.

Management of processes may be more important than management of uses, which may be better based on complementarity. The coastal zone management concept has gained acceptance in many non-industrialized countries. To achieve significant results, the natural functions of complex coastal resources systems, e.g. wetlands, should be assessed correctly, goods and services appraised, and practical techniques to promote system conservation and sustainable resource use developed. It has been shown that single purpose approaches and the exercise of unharnessed private property rights are frequently detrimental to good management.

Planning is facilitated by carving up the coastal zone in survey units, thereby permitting a.o. possible impact and users' reciprocal influence assessment. It must consider relationships, for instance the interaction of a beach and an adjoining touristic complex. Complementarity or parallelism of uses could be strived for. Beneficial multi-use strategies should be planned.

The cost of coastal erosion can be very high. Remedial action requires familiarization with the genesis and the history of the beach, examination of the current state of the system, and the natural evolution trend. Economic realities notwithstanding, the "forever" defense stand will have to be abandoned and after each storm consideration will have to be given to rebuild or not some homes. How well a community adapts to such dilemma will depend largely on the direction it takes when it reaches this crossroads.

Coastal protection management can consider stabilization, let nature take its course, or compromise. Land-leasing rather than selling, and use of movable buildings may be indicated. A management plan that ignores rising sea-levels, a gnawing erosion, silting access routes is futile. Storms can wipe out a beach in a matter of hours, geological subsidence is not uncommon, seismic and volcanic events affect many a coastal area.

Economic potential maps would constitute a working dossier for planners and allow weighing the benefits-disadvantages balance of major undertakings. Thus consequences were not projected when constructing the Volta River barrage in Ghana : the drastic reduction of sediment apport near the Volta's mouth led to the dissapearance of the cities of Ada and Keta. The harbor extension and construction in Cotonou (Benin) and neighbouring Lomé (Togo) led to disquieting erosion.

Major coastal zone processes should be monitored continuously. The casual approach to its occupation and use must bow to the imperatives of the coming decades : scientific approach, careful planning, resistance to selfish private interests, a "global" approach, synchronized legislation. Research should a.o. specifically aim to provide long- and short-term solutions to planning, development and management.

6 References

1. Bird, E.C.F., 1985, *Coastline changes : a Global Review* : Somerset, N.J., J. Wiley.

2. Charlier R.H., 1989, Coastal zone : occupance, management and economic competitiveness : *Ocean and Shoreline Management* 12, 5/6, 383-402

3. Charlier R.H., 1989, Land-use problems, planning and management in the coastal zone : An Introduction *: Ocean And Shoreline Management* 12, 5/6, 403-417

4. Charlier R.H., 1975, An assessment of the oceanic, geologic and economic significance of the littoral fringe : *3rd Int. Ocean Devel. Conf.* (Tokyo, 1975), *Proc.*.

5. Charlier, R.H., 1992, Mining potential of the inner continental shelf. In : Cendrero, A. *et al. (edit), Lecture Notes in Earth Sciences*, vol. 42 : Berlin/Heidelberg, Springer Verlag ; pp. 331-370.

6. Charlier, R.H. and Charlier, C.C.P., 1992, Ocean non-living resources : historical perspective on exploitation, economics and environmental impact : *Int. J. Env. St.*, 40, 123-184.

7. Charlier, R.H. and Charlier, C.C.P., 1992, Environmental, economic and social aspects of marine aggregates exploitation : *Env. Conserv.* 19, 1, 29-38.

8. Charlier, R.H., De Meyer, C.F. and De Croo, D. 1989, Soft beach protection and restoration In : Borgese, E.M. Ginsburg, N. and Morgan, J.R. (edits.), *Ocean Yearbook 8*, Chicago, University of Chicago Press, pp. 289-328.

9. Charlier, R.H. et Vigneaux, M., 1973, Frange littorale : appréhension géologique et stratégie d'aménagement : *Bull. Inst. Géol. Bass. Aquit.* 14, 181-192.

10. United Nations, Ocean Economics and Technology Branch, 1982, *Coastal area management and development : United Nations.*

PART II

COASTAL MANAGEMENT AND COASTAL DEFENSE

Introduction

The selection and timing of adaptive measures in response to coastal erosion and sea-level rise depends on the physical, social, economic, political and environmental characteristics of the affected areas[1]. Although such measures can be implemented on a case by case basis, growing population pressures and conflicting demands in many of the world's coastal areas favor implementation of comprehensive and systematic coastal management programs.

The three principal objectives of coastal management are to :

a) avoid development in areas that are vulnerable to flooding ;
b) ensure that critical natural systems continue to function ;
c) protect human lives, essential properties and economic activities against the ravages of the seas.

Accordingly, such programs should give full consideration to ecological, cultural, historic and aesthetic values and to the needs for human safety and economic development[2].

Coastal management programs usually include governmental controls and private sector incentives. Vulnerable areas are managed to minimize loss of life and property through such means as setback lines, limits on population densities, minimum building elevations and coastal hazard insurance requirements. Resilient natural protective features, such as beaches, sand dunes, mangroves, wetlands and coral reefs, are conserved and enhanced, which also maintains biological diversity, aesthetic values and recreation potential.

Comprehensive plans for protecting existing economic activities help to ensure that defense measures are consistent with other coastal management objectives. Policies that specify which activities and development are permitted in new areas promote efficient private land-use with the least risk of exposure to coastal hazards.

Successful coastal management programs require public education to gain broad based support and public participation to ensure equal representation of interests.

Chapter I : Response Strategies

Responses strategies fall into three broad categories :

Retreat :
abandonment of land and structures in vulnerable areas and resettlement of inhabitants ;

Accommodation :
continued occupancy and use of vulnerable areas ;

Protection :
defense of vulnerable areas, especially population centers, economic activities and natural resources.

1 Retreat

Options for retreat include :

a) Preventing development in areas near the coast ;
b) Allowing development to take place on the condition that it will be abandoned if necessary (*planned phase-out*) ;
c) No direct government role other than through withdrawal of subsidies and provision of information about associated risks.

Governmental efforts to limit development generally invoice land acquisition, land-use restrictions, prohibited reconstruction of property damaged by storms and reductions of subsidies and incentives for development in vulnerable areas. Many nations have purchased large areas on the coast and designated them as nature reserves (national and regional parks). Preventing development can reduce future expenditures for adaptation. India, Western countries, Australia and the United States already require new buildings be set back from the sea. These regulations could be modified to consider the future impacts from a rising sea level, but most nations would require payment of compensation to coastal property owners[3].

The second option gives the government a more limited role in that it lays out the "rules of the game" (the eventual transgression of the sea)[4]. Investors are accustomed to evaluating uncertainty and can determine whether development should proceed given the constraint.

This approach can be implemented through :

- regulations that prohibit private construction of protective structures,
- conversion of land ownership to long term or conditional leases which expire when the sea reaches a particular level or when the property owner dies.

The third option would be to depend on the workings of the private market. Productive crop and timber lands may be left to slowly and progressively deteriorate as a result of salt intrusion into the groundwater or by surface flooding. Wells and surface water exposed to salt water intrusion would gradually be abandoned. Natural resources, such as mangroves, marshes and coral reefs, would be left to their natural processes as sea level rises. However, salt intrusion with consequent loss of wells use, does not necessarily result in habitant departure. On Long Island, N.Y., water was simply provided from outside sources.

Under this option, governments could take the more limited role of ensuring that all participants in potentially vulnerable areas have full knowledge about the expected sea-level rise and its associated uncertainties. Development would presumably not occur if developers, lenders and insurers were not willing to accept the risks. However, if people continue to build in vulnerable areas, governments must be prepared to take the necessary actions to ensure public safety.

For small island states, retreat does not offer a broadly applicable alternative. There would be little or no land for resettlement, in addition to loss of heritage and cultural upheaval. This is particularly true for Pacific Ocean island countries which have hardly any elevation above present sea level.

2 Accommodation

The strategy of accommodation, like that of retreat, requires advanced planning and acceptance that some coastal zone values could be lost. Many coastal structures, particularly residential and small commercial buildings, could be elevated on pilings for protection from floods. To counter surging water and high winds, building codes could specify minimum floor elevations and piling depths, as well as structural bracing. Drainage could be modified. Storm warning and preparedness plans could be instituted to protect the affected population from extreme events. Where salt water damages agricultural lands and traditional crops, salt tolerant crops may be a feasible alternative. Fundamental changes in land-use may be desirable, such as the conversion of some agricultural lands to aquacultural uses. So may cattle raising be considered.

Human activities that destroy the natural protection values of coastal resources can be forbidden. Perhaps the most important controls would be to prohibit filling wetlands, damming rivers, mining coral and beach sands and cutting mangroves. Rebuilding of wetlands has been undertaken in some countries[22]. Undeveloped land with sufficient elevation and slope can be set aside to accommodate natural re-establishment of wetlands and mangroves. Within deltaic areas, natural processes can be maintained by diverting water and sediment. In response to salinity intrusion into groundwater aquifers, management controls can be implemented to regulate pumping and withdrawal practices.

Requiring private insurance coverage in vulnerable areas is an important method to compensate injuries and damages caused by natural disasters. It forces people to consider whether risks are worth taking and provides the necessary funds to repair damages and compensate innocent victims.

3 Protection

This strategy involves defensive measures and other actions to protect areas against inundation, tidal flooding, effects of waves on infrastructure, shore erosion, salinity intrusion and the loss of natural resources[5]. The measures may be drawn from an array of "hard" and "soft" structural solutions[6]. They can be applied alone or in combination, depending on the specific conditions of the site. There is no single or generic "best solution", as each situation must be evaluated and treated on its particular merits. However, there are some basic steps in the selection of measures likely to produce the highest economic returns. First, those charged with planning, design or management responsibilities in the coastal zones should be cognisant of the potential for future sea-level rise. Moreover, proposed plans should leave options open for the most appropriate future response. For example, many protection structures can be planned and designed with features that allow for future incremental additions which, if needed, could accommodate increased water levels and wave action. This can often be done without significant additional costs to the initial investment.

Yet, the capital costs associated with the "hard" set of options may prove a barrier to consideration of this option by developing countries and small island states.

Chapter II : Environmental Implications of Adaptive Measures

1 Introduction

Most coastal areas contain habitats that are important to fish, shellfish, sea turtles, sea birds and marine mammals. These areas also have high recreational, cultural and aesthetic values for many people. Working Group II of I.P.C.C. concluded that a large net wetland loss would result as sea level rises, because the area onto which new wetlands might expand is less than the area of wetlands at risk[1].

2 Retreat

Enabling wetlands to migrate inland is one possible motivation for a retreat strategy. Coastal wetlands can be found along most of the world's coastal margins, notably in the tropical and subtropical regions. From a global perspective, there is presently a large scale *de facto* retreat in process, given that most of the world's coastal wetlands border on land areas with low population densities and little major development[7]. Nevertheless, these areas may be developed in the future. Governments should focus attention on their wetland areas and where appropriate, establish zones to which wetlands will be allowed to retreat using the measures outlined in the previous section. However, even a retreat cannot prevent a large net loss of coastal wetlands.

Developed nations with large land masses, such as the United States and Australia, have implemented retreat strategies along sections of their coasts in the interest of allowing coastal ecosystems, particularly tidal wetlands, to adjust to increased levels of the sea through a slow landward migration. By contrast, on small islands the lack of land for inland migration would restrict the applicability of this option. In the case of atolls, many ecosystems could be completely lost.

3 Accommodation

The implications of this option would be a compromise between retreat and protection. However, resource exploitation practices would change. For example, people may harvest mangrove wood for use as pilings to elevate houses. Flood control efforts might alter water flow patterns which could adversely affect the coastal environment.

4 Protection

This strategy is most relevant for areas having relatively large populations and an important infrastructure. These conditions inherently alter the environments. However, the structural measures related to a protection strategy can impose additional alterations not only to the immediate environment but also to the unaltered coastal ecosystems beyond the area of protection. Therefore, environmental impact assessments (E.I.A.) are particularly important when protective measures are envisaged. A subsidiary social impact assessment (S.I.A.) may be considered (cf. 4.1.).

4.1 Hard Structures

Along ocean coasts, a sea-wall constructed landward of the shoreline would have little immediate impact on the beach systems. However, an eroding shore would eventually reach the sea-wall and result in a loss of the natural beach[8]. This impact can be avoided by means of beach nourishment. Similarly, structures could block the inland migration of coastal wetlands. For example, in the United States, the loss from a one meter rise would be 29-66 percent under the retreat option, but 50-82 percent if shores are protected with bulkheads[9].

Groins trap sediment moving along the shore. However, protection of one area is generally at the expense of increased erosion downdrift from the protected area. Because these structures do not increase the total sediment available to beaches and barrier islands, their long term impact is primarily a geographic shift of the erosion. Detached breakwaters often have similar effects, although they allow for nearshore habitat shifts in some cases and often provide desirable fish habitats ("artificial reefs") in much the same fashion as natural reefs.

Dams and salt water intrusion barriers can protect water supplies and fresh water habitats. On the negative side, these structures can retain sediments which in turn can increase erosion of coastal headlands and impair the ability of deltaic wetlands to keep pace with sea level rise.

In deltaic areas, levees might be constructed along rivers to prevent flooding due to sea-level rise. The resulting "channelizing" of rivers could, in some cases, prevent annual river floods from providing sediment and nutrients necessary to enable deltas to keep pace with sea-level rise and maintain the fertility of agricultural lands[10].

4.2 Soft Structures

Soft structures have a less severe impact than hard structures, since they usually consist of simulated natural features, such as beaches and wetlands[11].

The most common "soft engineering" approach is beach nourishment which involves dredging sand from back bays, navigation channels, or offshore, or excavating material from a land-based source and placing it on the beach. Because beach ecosystems are already adapted to annual erosion/accretion cycles, the placement of sand on the beach generally has little impact on beach ecosystems. Impact of "profile" feeding is still under study. By contrast, the dredging itself can seriously disrupt shallow water ecosystems and wetland habitats, both due to the direct effects of removing material and the resulting increase in turbidity.

Chapter III : Economic Implications of Adaptive Measures

1 Introduction

The potential economic implications of responses to sea-level and temperature rise over periods of 50 to 100 years are extremely difficult to quantify. The variables to be considered include both the cost of the strategies themselves and the effects of those costs on national economies. Thus far, only the cost of protecting against floods and erosion has been estimated worldwide. Considerably more research needs to be done.

The cost of an adaptive response is site-specific. The nationwide impact of such costs will be greater on rural, subsistence economies, often found in coastal areas in developing countries. Losses of resources such as biologically productive wetland areas and important mangrove stands and their products would compound such hardship. Reduction in the productivity of fisheries and the loss of land, resources and jobs are a further consideration. Significant costs can also be associated with the establishment and operation of the institutional mechanisms necessary to implement retreat or accommodation strategies. Finally, and especially if structural response options were exercised, operation and maintenance costs are a factor.

A fundamental element in the decision-making process is a cost-benefit assessment to weigh the life cycle costs and economic returns of the various alternatives[12]. Not all of the important factors are totally quantifiable in monetary terms, however. This is particularly so for cultural, environmental and social factors. Nevertheless, these non quantifiable aspects must be evaluated and given due consideration in an equitable trade of analysis in order to formulate and implement an acceptable adaptive response.

2 Retreat

In densely populated and productive areas, retreat may prove to be the least economically viable response option because of nearly irretrievable losses involved, or, in the case of small islands, the lack of land on which people can resettle. Inundation of fertile coastal agricultural land and frequent flooding of industrial sites and urban centres would threaten the value of past investments and drastically limit future growth. In such cases, it is highly unlikely that the economic benefits of retreat would exceed the costs.

Large scale resettlements could severely tax the planning, infrastructural and distributive capabilities of most countries, especially for developing countries. In particular, small island nations would face the most serious economic implications of retreat. At its most extreme, it would involve resettlement of the populations of entire nations.

The slow (albeit increasing) rate of sea-level rise permits appropriate planning and incremental implementation of retreat options and this may reduce costs[13]. However, in the case of arable lands, the inability to produce an adequate food supply may cause further national hardship through both unemployment and loss of exports.

3 Accommodation

Accommodation provides opportunities for inundated land to be used for new purposes. Thus, some compensatory economic benefits could be derived from accommodation or adaptation to inundation and flooding. For example, agricultural land may, in some instances, be found suitable for aquaculture ; salt resistant crops may be grown in areas previously dependent on freshwater. Nonetheless, considerable costs may be involved in the planning and restructuring of land-use. The necessary expenditure may place significant stress on national budgets, especially in developing countries. In the case of an increase in extreme events induced by climate change, such as tropical storms, altered wave regimes and storm surge frequencies, significant expenditures would be involved in disaster planning and preventing loss of life. Responding to such events would require considerable national planning and might involve compensation.

4 Protection

The economic benefits accruing from protection depend on the values of the land being protected. Benefit categories, as measured against taking no action include :

a) prevention of physical damage to property as a result of waves and flooding ;
b) prevention of loss of [economic] production and income ;
c) prevention of land loss through erosion ;
d) the prevention of loss of natural resources (environmental and recreational).

Costs include capital, operation and maintenance of the protective measures, as well as any cultural, environmental and social changes that may result. For example, some hard structural protection works may cause beaches to disappear. For economies heavily dependent on tourism (e.g. Caribbean islands) this may have serious adverse consequences. As already said, the non quantifiable aspects of cultural, environmental and social impacts must be considered when selecting any response strategy. Options may be restricted for some developing countries because of costs or lack of technology.

Cost Estimates for Protection

While the potential economic developments in the next few decades are difficult to predict, an approximation of basic implementation costs is possible. Although any such calculations are only rough approximations, they provide a useful first estimate and a guide for future data collection efforts. Table 1 illustrates estimates based on a sea-level rise scenario of one meter in 100 years, for 181 countries and territories with a marine coast. These estimates show that preventing inundation alone would cost, at a minimum some US $ 500 billion (none discounted) in today's dollars.

This value only reflects the marginal or added costs of protecting against the effects of a one meter rise in water level over the next 100 years. Costs associated with basic coastal protection in place or necessary to meet present coastal defense needs are not included. The estimate does not include either the value of the unprotected dry land or ecosystems that would be lost, nor does it consider the costs of responding to salt water intrusion nor the impacts of increase storm frequency. For Indonesia for instance, the cost for adaptive measures based on a sea-level rise scenario of one meter in 100 years, would be as follow (in millions of 1990 US $) :

a) low coast : (low lying coastal zones, with a population density exceeding 10 per km^2, 26 per sq. mi) : 5,904 for 19,680 km (12,269 mi) of coastline ;
b) cities : (length of sea defense for coastal cities) : 1,176 for 98 km (57 mi) ;
c) harbors : 488 for 27,1 km^2 (10.46 sq. mi) ;
d) beaches : (recreational beaches in a sandy shoreline), 1,080 for 200 km (124.3 mi).

Total costs per capita : US $ 51.
Annual costs as % of GNP : 0.10 %

Table 1. Estimate of marginal costs involved in protecting countries, worldwide, against the effects of a 1 m sea-level rise in 100 years (billion US $)

	Region	Total Prot. Costs bill. (US $)	Total Costs per capita (US $)	Annual Prot. Costs as % of GNP
1.	North America	106.2	306	0.03
2.	Central America	3.0	117	0.12
3.	Caribbean Islands	11.1	360	0.20
4.	South America Atlantic Ocean coast	37.6	173	0.09
5.	South America Pacific Ocean coast	1.7	41	0.04
6.	Atlantic Ocean small islands	0.2	333	0.12
7.	North and West Europe	49.8	190	0.02
8.	Baltic Sea coast	28.9	429	0.07
9.	Northern Mediterranean	21.0	167	0.04
10.	Southern Mediterranean	13.5	87	0.06
11.	Africa Atlantic Ocean coast	22.8	99	0.17
12.	Africa Indian Ocean coast	17.4	98	0.17
13.	Gulf States	9.1	115	0.02
14.	Asia Indian Ocean coast	35.9	34	0.14
15.	Indian Ocean small islands	3.1	1333	0.91
16.	South-East Asia	25.3	69	0.11
17.	East Asia	37.6	38	0.02
18.	Pacific Ocean large islands	35.0	1550	0.17
19.	Pacific Ocean small islands	3.9	1809	0.75
20.	(Former) USSR	25.0	89	0.01
	TOTAL	488.1	103	0.04

Source:
"A world wide cost estimate of basic coastal protection measures", paper by Dutch Delegation (Rijkswaterstaat / Delft Hydraulics, Note n° H1068 to IPCC-CZM Workshop, Perth, Australia, February 1990.

Remarks

The above figures are based on a United Nations worldwide study and are only meant as a first rough estimation of the needed responsive measures to the assumed sea-level rise. They call for corrections based on a more refined and detailed study at the country level.

Protection works consist of civil engineering structures, such as beach replenishment, dikes, revetments, closure dams and elevation of very low areas ;

As to the design conditions of these civil engineering works, the following description serves as a basic concept. Starting point for the protection measures considered here is maintaining the present protection level of human beings and infrastructure in the coastal zone at risk. (Coastal zone " hazards" were the topic of a French Ministry of the Environment colloquium ; cf. Ministère de l'Environnement-Délégation aux Risques Majeurs, 1988, Actes du Colloque *Mer et Littoral - Couple à Risques* ; Biarritz, Sept. 1987 : Paris, La Documentation Française (565 pp.). As an example, the above formulated starting point implies for a region protected by dikes, assuming that other conditions remain the same, a raising of dikes approximately equal to the rise of sea level. This starting point often does not coincide with optimal coastal protection and may even leave coastal regions with too low a protection level. In particular this is the case for coastal zones which at present are insufficiently protected due, for example, to recent land occupation or subsidence. These situations ask for a solution, but in principle this is independent of sea level-rise.

A reduction in the scope is that only civil engineering aspects of these protection measures are considered. Institutional, legal and technical aspects, all essential to maintain infrastructural integrity, are not considered, nor are their costs, which might be substantial.

Cultural and environmental aspects, such as the impact of sea-level rise on coastal wetlands, should be considered as well. Socio-economic aspects are, in the present study, reduced to the costs for construction of the civil engineering works.

Sociological Impact Assessments (SIA) are commonly part of an impact report.

Chapter IV : Social and Cultural Implications of Adaptive Measures

1 Introduction

The social and cultural implications of adaptive response measures may affect hundreds of million people living in coastal zones with average width of 50 kilometres[14]. In addition to the general population move towards the coasts, in some coastal areas, inhabitants are highly concentrated in a narrow coastal belt (e.g. Java, India and China).

The lifestyles of many people are tied directly to the coast and its predominant local characteristics. The coast also features strongly in the mythology of many cultures. Numerous places of particular cultural significance are situated on the coast and many people in developed and developing nations view the sea, coasts, reefs and beaches as central to their lives.

Social and cultural implications of adaptive options are likely to vary considerably from country to country and from site to site. Options that are socially and culturally beneficial in some situations may be less desirable in others. Affected communities must be consulted and participate in the decisions to adopt particular options. This is probably one of the best means available to identify the social and cultural implications for particular cases.

2 Retreat

Retreat, as an option, may imply a partial, incremental process or a sudden large scale event. In some circumstances, there may be a need to relocate inhabitants, or even entire communities, which could have major financial and social implications in developing countries. The loss of the traditional environment that normally sustains economies and cultures and provides for recreational needs, could severely disrupt family life and create social instability, with a resulting adverse impact on the entire community, especially on the young and the elderly. In addition, places of great cultural significance, for example, burial grounds, historic places, of religious centers, could also be lost if retreat occurs.

All retreat options have been identified as having potentially significant implications both socially and culturally. This is particularly the case with abandonment and the resulting need to resettle whole populations. Even though migration is relatively common in some areas, for example, the South Pacific, a need for social adjustment remains. Situations where an individual's or a community's identity is closely associated with a particular piece of land or access to particular resources, as in most subsistence economies, can have implications which are difficult to resolve.

The greatest implications of retreat may lie in being denied access to the original coast. A well planned retreat that provides for access to alternative resources could minimize some of these impacts[15]. An associated issue is that of the social implications for the host people at the place of relocation (s). There exists a potential for conflict, and existing social services may be heavily taxed in the host area if relocation is not well planned and managed. People may choose not to abandon even vulnerable coastal areas in anticipation of climate change impacts, if there is strong population pressure in adjacent areas.

3 Accommodation

The social and cultural implications of accommodation, while not as severe as those of retreat, may still be significant. A change in the economic activity of an area, for instance, from farming to aquaculture, will change lifestyles. Accommodating change may lead to living conditions being less desirable, for example, if properties are subject to periodic flooding, or if problems with sewage disposal occur. Public safety and health will thus be adversely affected by this option.

Accommodation is a more socially desirable option when applied in areas where there is a tradition of adapting to water, for instance, if people live in houses on stilts or in house-boats.

4 Protection

Protection has fewer identified social and cultural implications. However, hard structures are likely to have less aesthetic value than the original environment and access to the shore may also be restricted by some protective options. Beach losses could impair recreation, while loss of wetlands may affect fish stocks. If protective options involve non local labor, there may be social and cultural friction which could lead to community disruption. Options that can be implemented by communities themselves are less likely to have social and cultural implications than those which require outside labor.

If the protection structures cause alteration to places of cultural significance there could be opposition to their construction. The loss of any biological resources resulting from protection activities could also be of cultural significance. In some areas, for example, if a significant species is seriously threatened it may no longer be available for ritualistic or economic purposes.

Chapter V : Legal and Institutional Implications of Adaptive Measures

1 Introduction

Existing .institutions and legal frameworks may be inadequate to plan and implement adaptive responses. New institutions and legal authorities may be needed in many coastal states. National legislation and institutions for coastal zone management can provide the needed planning. In addition, legal structures to require advance consideration of likely impacts, such as environmental impact review by those planning new projects, can encourage needed foresight.

One matter to consider is that virtually any adaptive option involves the use of "private" land. In some nations such use by individuals may be prohibited by law, while in other states the government may not have authority to use the land without the consent of the land owner. The government in some states may have the authority to use private land, but only upon providing compensation to the land owner (right of eminent domain)[16].

An accelerated rate of increase in the global sea level also raises the possibility of legal issues pertaining to maritime boundaries and jurisdiction and transboundary matters. These issues may require a review of existing international arbitration procedures. An example of the first issue would be if a nation loses maritime boundary base points and therefore a legal claim to sea territory, or if beach nourishment measures are required in the vicinity of national borders. An example of this second issue occurs when protective measures interrupt or impede the longshore sediment transport benefitting an adjoining coastal state. In the worst case, sea-level rise may result in the total land loss of an island nation ; the resulting legal implications are difficult to assess or even predict[17].

2 Retreat and Accommodation

The resettlement option could raise significant transboundary implications. The legal authority and institutional capability to manage or direct a relocation on a temporary or permanent basis must be clearly established. Authority to facilitate and encourage relocation from vulnerable areas and to subsequently deal with the use of abandoned lands may be needed. In extreme cases when individuals will not leave areas subject to great risk, authorities for condemnation of land and facilities may also be necessary.

Whether relocation is on a temporary or a permanent basis, accommodation for displaced inhabitants must be provided. In some coastal states, relocation could involve tens of millions of inhabitants. The relocation may be further complicated by the lack of land within small coastal or island states. If relocation outside such states is required, then the assistance of regional or international institutions may be needed.

The first option for both measures is to discourage growth in population, or additional development in vulnerable areas that would increase either the risks of losses, or would increase to unacceptable levels the costs of later retreat. In order to implement this option, the coastal state must have the institutional facility for identifying vulnerable coastal zones. Many developing countries do not have the institutional structures and will require assistance to develop a national plan for management of coastal resources and coastal development. Alternatively, a state might choose to encourage private retreat and accommodation actions through non-regulatory measures, such as providing information to the affected population.

In addition to an institutional structure to plan and manage coastal development, legal authorities are needed to enforce restrictions or conditions on coastal development if a coastal or island state chooses to take an active regulatory role for implementing those strategies. Legal authorities may also be needed both to ensure the integrity of natural coastal protection systems and to avoid placing coastal populations and developments in jeopardy from sea-level rise. For example, a coastal or island state may need new authority to restrict access or activities to certain areas in order to protect natural systems (such as from the use of mangrove for firewood) as well as have the authorities to restrict residential and commercial development (such as new settlements on deltas).

3 Protection

An important implication of selecting an option to protect against sea level-rise is liability for the failure of public protection structures. Structures to protect against sea level rise enable commercial, agricultural and residential activities to continue in protected areas. Therefore, people and economic resources will be attracted to and concentrate in areas so protected. Should the structure fail, significant loss of life and property could result. Each type of structure is different and each requires some type of maintenance in order to perform as designed. Where the entity responsible for maintenance is different from the entity which designed and/or constructed the structure, it may be difficult to assess any liability for damage resulting from a failure of the structure. Some public or private entity within each nation, therefore, must have responsibility for maintenance of the physical integrity of these structures.

Chapter VI : Priorities for Adaptive Responses

1 Introduction

The projected rise in sea level warrants urgent policy responses in many coastal states, particularly those with populated coral atolls and deltas, or those with estuary dependent fisheries. Such actions must focus on human safety and on sustainable development of coastal resources.

Even though sea level rise is predicted to be a relatively gradual phenomenon, adaptive strategies may require lead times in the order of 50 to 100 years, to tailor them to the unique physical, social, economic, environmental and cultural considerations of a particular coastal area. Moreover, even though there may be no need to begin building dikes that are not needed for 50 years, it is appropriate to begin planning now to avoid actions that could increase vulnerability to the impacts of sea-level rise. It will take ten years to implement plans, in view of the time required for the necessary analyses, training the people, developing the plans and mobilizing public and political awareness and support. Therefore, the process should started without delay. Protection from coastal impacts of sea-level rise and other impacts of global climate change include both capital investment in defense structures and maintenance costs. Moreover if the sea continues to rise, these structures may have to be augmented or replaced. Similarly non-structural options to reduce vulnerability to impacts of sea-level rise, such as land use plannings, may require actions to implement and enforce them.

It is important to recognize that decisions today on planning for coastal development will greatly influence cost for later adaptation to impacts of sea-level rise. Charlotte (South Carolina), Venice, Shanghai, New-Orleans and Lagos are all vulnerable because of decisions made 200-2,000 years ago. It is therefore necessary, to establish some immediate priorities for planning and management of coastal resources and for technical and financial assistance to developing countries to facilitate their planning.

Developing countries should be provided with the technical and financial assistance required to plan for coastal development in order to reduce vulnerability to impacts of sea-level rise. There is also a need to estimate the future long term funding requirements for developing countries that may be required if protection options are needed[18].

Finally, the success of strategies to limit climate change is a factor to be considered. Limitation measures will be likely to reduce the costs of adaptation to the coast impacts of sea-level rise. It is likely, however, that some adaptation to sea-level rise will be required regardless of the limitation strategies eventually implemented[19].

2 Priorities for Adaptation

2.1 Scientific Knowledge and Monitoring

There is still considerable speculation regarding sea-level rise and other impacts of global climate change. This makes the selection of adaptation options extremely difficult. In particular, regional, national and site specific data needed to make decisions on adaptive options is lacking.

For example, a system to monitor, detect and predict sea-level rise is needed to determine the need for construction of protective structures or relocation of coastal inhabitants. Information on other impacts, such as changes in tropical storms, in order to plan for natural emergencies is equally required[20].

2.2 Information

Those areas that are most vulnerable to the impacts of sea-level rise must be identified. The identification should concentrate on densely populated low lying areas, deltas and small atoll islands.

The need for clearing house arrangements to facilitate exchange of information and international databases accessible to all nations has also been established.

Development of models and assessment techniques to support coastal planning needs to be undertaken in order to provide decision makers insight into the complicated interactions and conflicting interests that are involved in coastal zone management. Equally important is the transfer to developing countries of existing coastal adaptation technologies and the provision for training in coastal zone management, engineering and environmental monitoring. Such training might also include technology research centers, extension services, technology advisory committees, technology research and development, technology conferences and pilot projects to enhance technology transfer.

2.3 Planning

Many priorities have been identified within the broad area of planning. These include :

Emergency management planning to reduce vulnerability of inhabitants in areas exposed to extreme weather events ;
Coastal management planning to reduce impacts on development structures and on natural resources of the highest priority. Technical and financial assistance to developing countries may be required to develop and implement national plans for management of coastal development.

2.4 Education and Community Participation

Public education and education of decision-makers regarding the impacts of sea-level rise and the impacts of ongoing activities is essential, so that everyone understands the risks of development in a coastal area[21]. The involvement of members of the local communities in selecting and implementing response options is essential for the success of adaptive responses.

3 Priorities for Implementing Adaptive Options in Developing Countries

3.1 Retreat

Technical assistance to developing countries is required for timely planning for resettlement and emergency management pending resettlement. Financial assistance also may be needed to facilitate the resettlement.

Assessments of potential relocation sites should be made to minimize dislocation difficulties such as linguistic diversity, cultural differences and long term viability.

3.2 Accommodation

Education, technical assistance and training are required for developing countries so that their populations can understand the risks of development in coastal areas in order to reduce vulnerability to impacts of possible sea-level rise.

Technical assistance on alternative economic activities, for example mariculture instead of agriculture, is required to mitigate the social, cultural and economic implications of various options. Experience in this field exists in several developing countries and should be shared. The same holds true for alternatives to current coastal development activities. For example, using coastal areas for tourism rather than for industrial or residential activities, may also provide a solution.

3.3 Protection

Protection options involving structures in most developing countries is likely to require external assistance. For example, the building of hard structures could require assistance such as transfer of skills and/or capital. There may also be a need for transfer of planning skills to support the choice of appropriate options.

4 Criteria for Allocation of Resources

In addition to identifying priorities for adaptive responses it is also appropriate to identify the priorities that might be used to allocate resources. As the necessary resources vary considerably depending on the adaptive option and the coastal area, allocation criteria must include consideration of both the options and the area. A list of sample criteria are provided in table 2. It is not suggested that any one criterion should be pre-eminent. Some may be more significant in some situations while other criteria may be more important.

Table 2. : Criteria for Allocation

Related to the coastal area :

1. The contribution of current activities within the coastal area that contribute to its vulnerability to sea-level rise ;
2. The importance of the coastal area in terms of :
   - urgency of risk,
   - proportion of national land area,
   - population affected,
   - environmental importance,
   - social and cultural importance, and
   - regional importance.
3. The national ability to finance the response option ;
4. The institutional and political ability to realize implementation.

Related to the adaptive response option :

1. The cost of the option ;
2. The effectiveness of the option ;
3. Cost effectiveness ;
4. The economic, environmental, social, cultural, legal and institutional implications of the adaptive option ;
5. The vulnerability of the option to the impacts of an accelerated sea level rise ;
6. Performance under uncertainty ;
7. Equity.

5 References

1. Anonymous, 1990, Strategies for adaptation to sea-level rise. Intergovernmental Panel on Climate Change (I.P.C.C.) - Working Group I. *Report of the Coastal Zone Management Subgroup.*

2. Jansen, M., 1990, Changing Climate and the Coast, *op. cit.* (1)

3. Leatherman, S.P. 1990. Environmental Implications of Shore Protection Strategies along Open Coasts (with a Focus on the United States). In : *Changing Climate and the Coast, op. cit.* (1)

4. Titus, J.G., 1990, Strategies for Adapting to the Greenhouse Effect : *Journal of the American Planning Association* ; J.G. Titus, Greenhouse Effect and Coastal Wetland Policy : *Environmental Management* (in press).

5. Pope, J.J. and T.A. Chisholm, 1990, Coastal Engineering Options by which Hypothetical Community might adapt to Changing Climate. In : *Changing Climate and the Coast, op. cit.* ; and Sorensen, R.M., R.N. Weisman and G.P. Lennon, 1984. Control of Erosion. Inundation and Salinity Intrusion. In : Barth and Titus (eds). Greenhouse effect and sea-level rise : a challenge for this generation : New York, Van Nostrand Reinhold.

6. U.S. Army Corps of Engineers, Coastal Engineering Research Center, 1977. Shore Protection Manual : Fort Belvoir, Virginia, U.S.A, Coastal Engineering Research Center,.

7. Misdorp, R., 1990, Strategies for Adapting to the Greenhouse Effect. A Global Survey of Coastal Wetlands : Delft, The Netherlands, Rijkswaterstaat, Tidal Waters Division, Note GWWS-90.008.

8. Howard, J.D., O.J. Pilkey and A. Kaufman, 1985, Strategy for Beach Preservation Proposed. *Geotimes* 30, 12, 15-19.

9. Titus, J.G., R. Part and S. Leatherman, 1990, The Cost of holding back the Sea. *Coastal Management* (in press).

10. Park, R.A., 1990, Implications of Response Strategies for Water Quality. In : *Changing Climate and the Coast, op. cit.*

11. Leatherman, S.P., 1990, Environmental Impacts of Sea Level Response Strategies. In : *Changing Climate and the Coast, op. cit.*

12. Moser, D.A., E.Z., Stakhiv and L. Vailianos, 1990, Risk-Cost Aspects of Sea-Level Rise and Climate Change in the Evaluation of Coastal Protection Projects. In : *Changing Climate and the Coast, op. cit.*

13. Yohe, G.W., 1990, Toward an Analysis of Policy, Timing and the Value of Information in the Face of Uncertain Greenhouse-Induced Sea-Level Rise. In : *Changing Climate and the Coast, op. cit.*

14. Charlier, R. H., 1987. Planning for Coastal Areas. In : Ecology for Environmental Planning. In : Wollf, F.C. ed., Trondheim, Norway, Norges Geologiske Underskølse, pp. 2-17 ; Charlier, R.H., 1987, Marine mineral resources extraction in coastal areas and its impact on the environment and its consequences for land-use. In *Mineral Resources Extraction, Environmental Protection and Land Use Planning in the Industrial and Developing Countries* (P. Arndt & G. Lüttig, eds) (Schweizerbart'sche Verlag und Buchhandlung, Stuttgart) pp. 53-70 ; Cendrero U., A & R. H. Charlier, 1989, Resources and use and management in the coastal zone, *Geolis III,* 32, 40-60.

15. Yohe, G.W., 1990, cf. (13).

16. Fishman, R.L., and L. St. Amand, 1990, Preserving coastal wetlands and sea-level rises : Legal opportunities and constraints". In : *Changing Climate and the Coast, op. cit.*

17. Shihab, H. (ed.), n.d., *Proceedings of the Small States Conference on Sea Level Rise*, Environment Section. Male, Republic of the Maldives, The Conference.

18. Campbell , J., 1990. Funding Implications for Coastal Adaptations to Climate Change. Some Preliminary Considerations. In : *Changing Climate and the Coast, op. cit.*

19. Warrick, R.A., and J. Oerlemans, 1990, IPCC-Working Group I : Chapter 9 : *Sea-Level Rise (see (1) above).*

20. Intergovernmental Oceanographic Commission of UNESCO, 1990, *Global Sea Level Observing System (GLOSS). Implementation Plan.* UNESCO/IOC Secretariat 7, place de Fontenoy, Paris, France-75700.

21. Maroukian, K., (e.d.) 1990 ; Implications of Sea-Level Rise for Greece ; Erol O., Impacts of Sea-Level Rise on Turkey ; Muehe, D. and C.F. Neves, Potential Impacts of Sea-Level Rise on the Coast of Brazil ; Andrade B. and C. Castro, Impacts of and Reponses to Sea Level Rise in Chile ; Adam, K.S., Implications of Sea-Level Rise for Togo and Benin ; and Ibe, A.C., Adjustments to the impact of Sea-Level Rise Along the West and Central African Coast. In : *Changing Climate and the Coast, op. cit.*

22. Charlier, R.H., and Charlier, C.C.P., 1995, Sustainable multiple-use and management in the coastal zone : *Env. Man. and Health* 6, 1, 14-24 ; F.R. Burbridge *et al.,* 1989, Multiple use assessment for coastal management : *Coastal zone '89* I, 330-345 ; S. Sakai *et al.*, 1992 Challenge for co-existence of the fishery and marine recreation in Japan ; *Proc. PIANC Meeting* (Surabaya, Indonesia) 243-260 ; Anonymous, 1989, Recent developments in ocean and coastal management" *Ocean & Shoreline Man.* 12, 1, 89-100 ; ibid. 3, 363-369.

PART III

SEDIMENT TRANSPORT

Chapter I : Bed Material Suspension and Transport in Steady Uniform Currents

1 Bed Load Transport

1.1 Introduction

Usually, the transport of particles by rolling, sliding and saltating is called the bed-load transport. For example, Bagnold (1956) defines the *bed-load* transport as that in which the successive contacts of the particles with the bed are strictly limited by the effect of gravity, while the *suspended load* transport is defined as that in which the excess weight of the particles is supported wholly by a random succession of upward impulses imported [and impacted] by turbulent eddies.

Einstein (1950), however, had a somewhat different approach. Einstein defined the bed-load transport as the transport of sediment particles in a thin layer of 2 particle diameters thick just above the bed by sliding, rolling and, sometimes, by making jumps with a longitudinal distance of a few particle diameters. The bed layer is considered as a layer in which the mixing due to the turbulence is so small that it cannot influence the sediment particles, and therefore suspension of particles is impossible in the bed-load layer. Further, Einstein assumed that the average distance traveled by any bed-load particle is a constant distance of 100 particles, independent of the flow conditions, transport rate and the bed composition. In Einstein's view, saltating particles belong to the suspension mode of transport, because the jump lengths of saltating particles are considerably longer than a few grain diameters.

Many formulas to predict the bed-load transport rate are described in the literature. The earliest equation is that of Du Boys (1879), who assumed that sediment particles move along the bottom in layers of progressively decreasing velocities in a vertical downward direction. The first empirical formula was presented by Meyer-Peter and Müller (1948). They performed flume experiments with uniform particles and with particle mixtures. Based on data fitting, a still frequently used and relatively simple formula has been obtained. Kalinske (1947) and Einstein (1950) introduced statistical methods to represent the turbulent behavior of the flow. Kalinske assumed a normal distribution for the instantaneous fluid velocity at grain level. Einstein gave a detailed but complicated statistical description of the particle motion in which the exchange probability of a particle is related to the hydrodynamic lift force and particle weight. Einstein proposed the d_{35} as the effective diameter for particle mixtures and the d_{65} as the effective grain roughness diameter.

Frijlinck's (1952) very practical approach made a simple fit of the Meyer-Peter-Müller and Einstein formulas. Bagnold (1966) introduced an energy concept and related the sediment transport rate to the work done by the fluid. Van Rijn (1984) solved the equations of motion of an individual bed-load particle and computed the saltation characteristics and the particle velocity as a function of the flow conditions and particle diameter.

1.2 Meyer-Peter and Müller (1948) Prediction Method

Extensive experimental work has been carried out by Meyer-Peter and Müller at the "Eidgenössische Technische Hochschule" (ETH) in Switzerland. The experiments were performed in a laboratory flume with a cross-section of 2 x 2 m^2 (2 x 21.5 sq. ft.) and a length of 50 m (164 ft.). Uniform bed material as well as particle mixtures were used in the experiments.

The bed-load transport rate is expressed as (cf. Fig. 1) :

$$q_{b,c} = 8\,[(S-1)\,g]^{0.5}\,d_{50}^{15}\,(\mu\theta - 0.047)^{15} \qquad (1)$$

in which :

$q_{b,c}$ = bed-load transport by current (m^2/s)

$\theta = \dfrac{\tau_{b,c}}{(\rho s - \rho) g d_{50}}$ = effective particle mobility parameter (-)

μ = $(C/C')^{1.5}$ = bed-form factor (-)

$\tau_{b,c}$ = $\rho\,g\,h\,I$ = current-related bed-shear stress (N/m^2)

d = water depth (m)

I = energy gradient (-)

C = 18 log ($12d/k_{s,c}$) = overall Chézy-coefficient ($\sqrt{m/s}$)

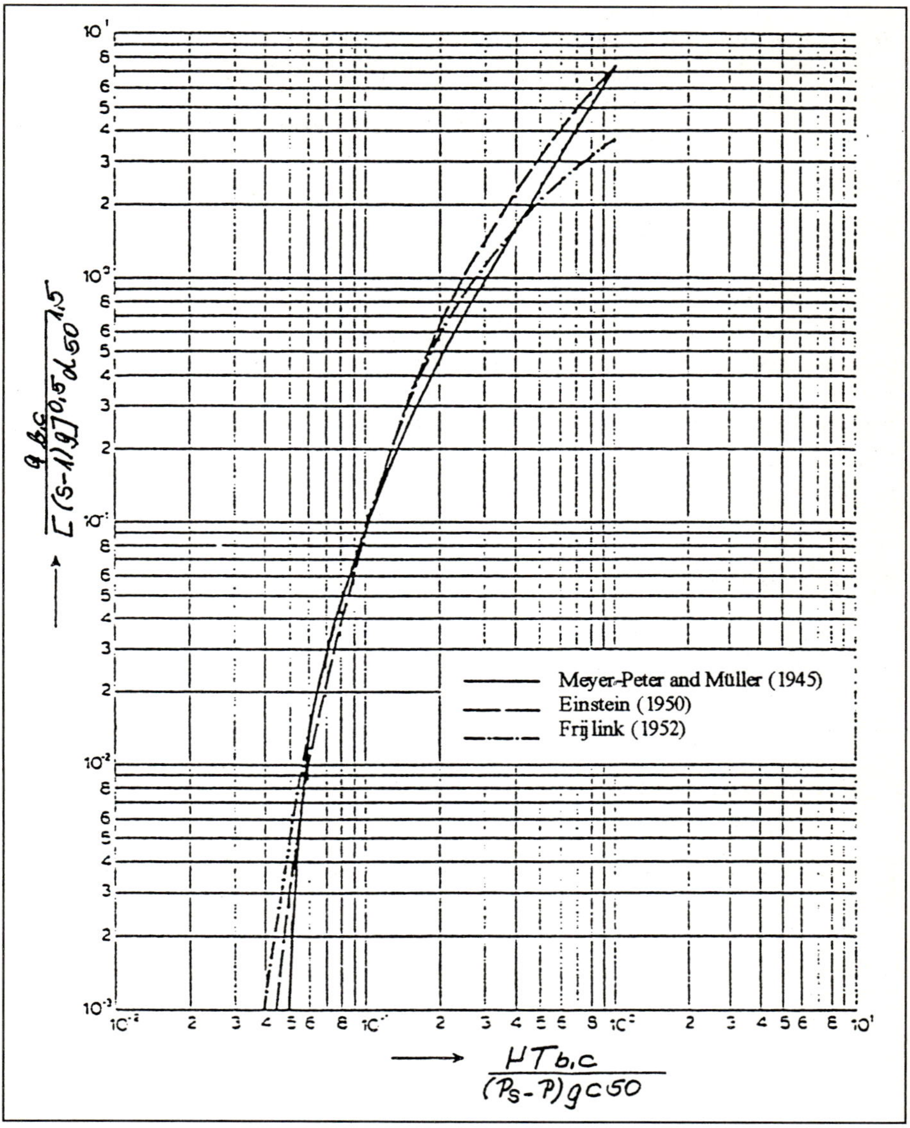

Fig. 1 : Formulas of Meyer-Peter Müller, Einstein and Frijlink

C'	= 18 log (12d/d_{90}) = grain-related Chézy-coefficient	(m/s)
$k_{s,c}$	= effective current-related bed roughness	(m)
d_{50}, d_{90}	= particle diameters	(m)
$s = \rho_s/\rho$	= specific density	(-)

Eq. (1) is based on particle diameters in the range 3.17 to 28.6 mm. The d_{50} is used as the effective diameter for particle mixtures, while the d_{90} is used as the effective diameter for grain roughness.

1.3 Prediction Method of Frijlink (1952)

The formula of Frijlink is essentially an approximation of the formulas of Meyer-Peter and Müller (1948) and of Einstein (1950), as shown in Fig. 1.

The formula, which is because of its simplicity given here, reads :

$$\rho_{b,c} = 5 \sqrt{\mu} u*_{,c} \, d_{50} e^{0.27/(\mu\Theta)} \quad (2)$$

with : $u*_{,c}$ = bed shear velocity.

The parameters are identical to those given in 1.2.

1.4 Prediction method of Bagnold (1966)

Bagnold (1966) introduced an energy balance concept and related the bed-load transport to the work done by the fluid (product of bed-shear stress and depth-averaged velocity). He explicitly incorporated the local bed slope in his bed-load formula :

$$q_{b,c} = \frac{e_b}{(\rho_s - \rho)g(\tan d\phi - \tan\beta)} \tau_b \bar{u} \quad (3)$$

The bed-load transport rate is related to the overall bed-shear stress and not to the grain-shear stress. Thus, the effect of the bed-forms is not taken into account.

1.5 Prediction Method of Van Rijn (1984)

Van Rijn followed the approach of Bagnold (1954) assuming that the motion of the bed load particles is dominated by particle saltations under the influence of hydrodynamic fluid forces and gravity forces. Van Rijn (1984) computed that the bed-load transport rate for particles in the range of 200 - 2000 μm as :

$$q_{b,c} = 0.053\ [(s - 1)g]^{0.5}\ d_{50}^{1.5}\ D_*^{-0.3}\ T^{2.1} \qquad (4)$$

in which :

$q_{b,c}$	= bed load transport	(m^2/s)
D_*	= $d_{50}\ [(s - 1)\ g/\nu^2]^{1/3}$ = particle diameter	(-)
T	= $(\tau'_{b,c}-\tau_{b,cr})/\tau_{b,cr}$ = bed-shear stress parameter	(-)
$\tau'_{b,c}$	= $\mu\tau_{b,c} = \rho g\ [\bar{u}\ /\ C\ ]^2$ = effective bed-shear stress	(N/m^2)
C'	= 18 log ($12d/3d_{90}$) = grain-related Chézy-coefficient	($\sqrt{m/s}$)
d	= water depth	(m)
s	= ρ_s/ρ = specific density	(-)
$\bar{u}$	= depth-averaged velocity	(m/s)
d_{50}, d_{90}	= particle diameters of bed material	(m)
ν	= kinematic viscosity coefficient	(m^2/s)

$\tau_{b,cr} = (\rho_s - \rho)\ g\ d_{50}\ \theta_{cr}$ = critical bed-shear stress according to Shields

2 Suspended Load Transport

2.1 Introduction

The depth-integrated suspended load transport is defined as the depth-integration of the product of velocity (u) and concentration (c), as follows (Fig.3) :

$$q_{s,c} = \int_a^h ucdz \tag{5}$$

Application of Eq. (5) requires information of the velocity profile, the concentration profile and a reference concentration (c_a). This latter variable generally is specified at a reference level (z=a) close to the bed. Three prediction methods based on this approach are described : Einstein (1950), Bijker (1971) and Van Rijn (1984), while the formula of Bagnold (1966) is also given.

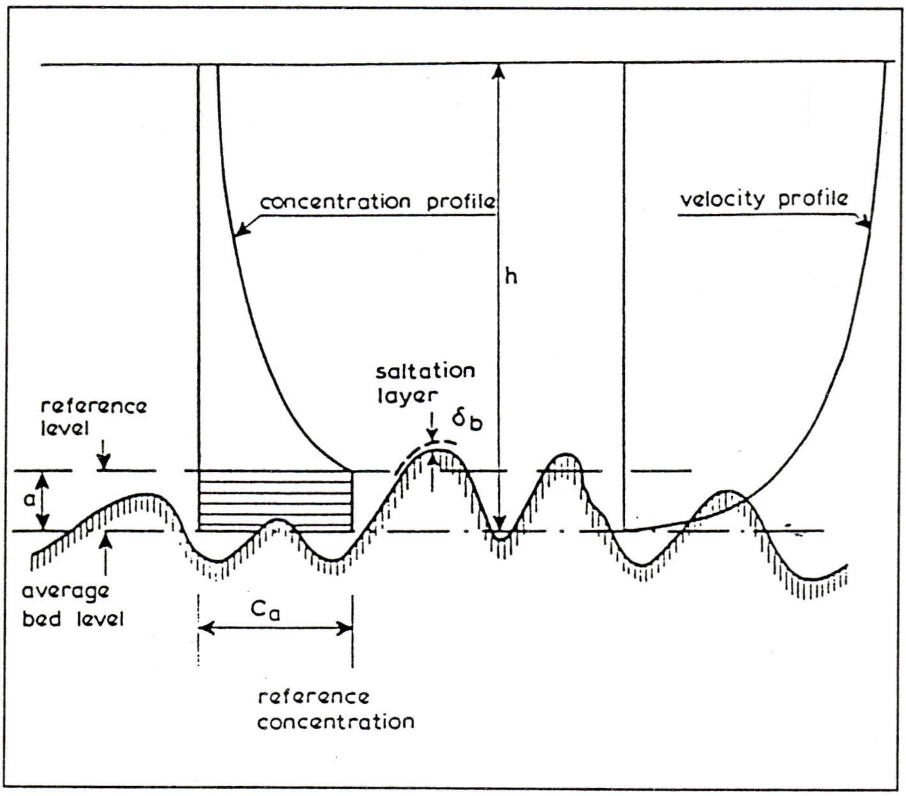

Fig. 2 : Schematization velocity and concentration profile.

2.2 Mass - Balance Equation for Concentration

In a steady uniform flow the vertical equilibrium of the upward turbulence as related and downward gravity-related transport can be expressed as :

$$c w_{s,m} + \varepsilon_{s,c} \frac{dc}{dz} = 0 \tag{6}$$

in which :

c is sediment concentration at height z above bed

$W_{s,m}$ is particle fall velocity in a fluid-sediment mixture

$\varepsilon_{s,c}$ is sediment mixing coefficient related to current

2.3 Sediment Mixing Coefficient

Various distributions of the fluid mixing coefficient (E_f) can be found in the literature. Herein the following distributions are given (see Fig. 3) :

$$\text{constant} : \varepsilon_{f,c} = \frac{1}{\alpha_1} \kappa u_{*,c} d \tag{7}$$

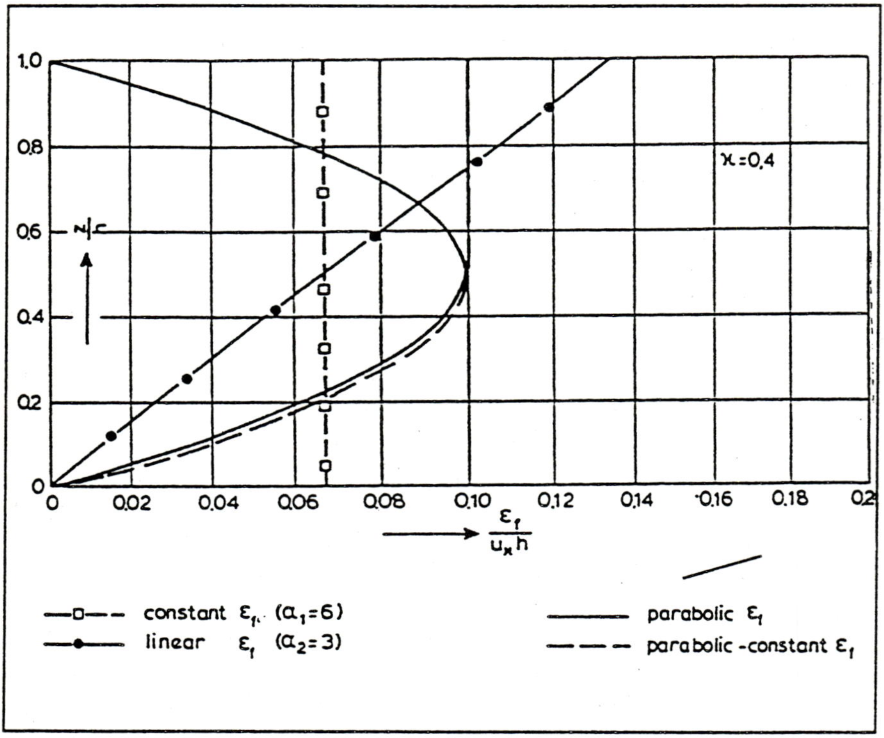

Fig. 3 : Fluid mixing coefficients.

linear : $$\varepsilon_{f,c} = \frac{1}{\alpha_1} \kappa u_{*,c} d \frac{z}{c} \tag{8}$$

parabolic : $$\varepsilon_{f,c} = \kappa u_{*c,} d \frac{z}{d}(1 - \frac{z}{d}) \tag{9}$$

parabolic-constant :
$$\begin{aligned} \varepsilon_{f,c} &= \kappa u_{*,c} d \frac{z}{d}(1 - \frac{z}{d}) \quad \text{for } \frac{z}{hd} < 0.5 \\ \varepsilon_{f,c} &= 0.25 \kappa u_{*,c} d \quad \text{for } \frac{z}{d} \geq 0.5 \end{aligned} \tag{10}$$

in which :

$u_{*,c}$ = current-related bed-shear velocity

d = water depth

z = vertical coordinate

κ = von Karman constant (0.4)

α_1, α_2 = constant

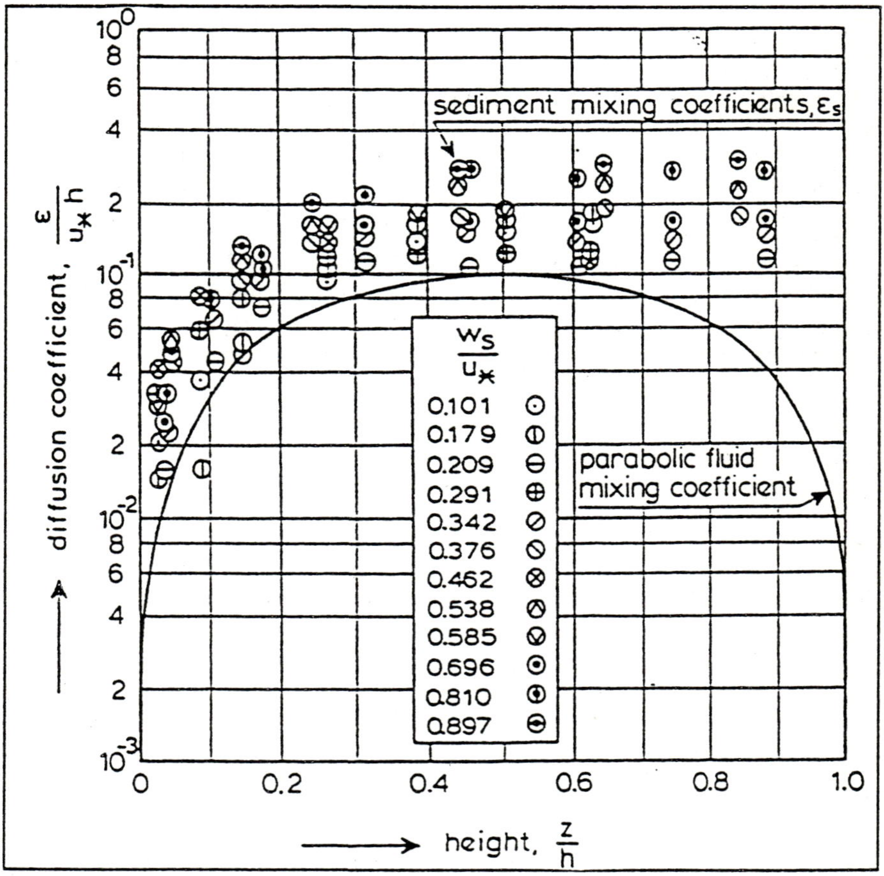

Fig. 4 : Sediment mixing coefficient distribution (Coleman, 1970).

The parabolic distribution is most satisfactory in a physical sense because it is based on a linear shear stress distribution and a logarithmic velocity profile. A disadvantage of the parabolic distribution is that it yields a zero-concentration at the water surface. Sediment mixing coefficients based on the analysis of measured concentration profiles (Coleman, 1970) indicate a parabolic-constant distribution rather than a parabolic one, as shown in Fig. 4. Usually, the mixing or diffusion of the sediment particles is related to the fluid mixing coefficient for a clear fluid (ε_f) as follows :

$$\varepsilon_{s,c} = \beta\theta\varepsilon_{f,c} \tag{11}$$

2.3.1 β-factor

The β-factor describes the difference in the diffusion of a fluid "particle" (or small coherent fluid structure) and a discrete sediment particle. Herein, the β-factor is assumed to be constant over the flow depth. Information of the β-factor can be obtained from a study carried out by Coleman (1970). Based on eq. (6), Coleman computed the sediment mixing coefficients (see Fig. 4). We will use Coleman's results to determine the β-factor, defined as $\beta = \varepsilon_{S,max} / \varepsilon_{f,max}$. The $\varepsilon_{f,max}$-value was computed from Eq. (2.151) for z/h = 0.5. The $\varepsilon_{S,max}$-value was determined as the average value of the ε_S-values in the upper half of the flow (as given by Coleman, Fig. 4) where the concentrations are relatively small. The β-factors can be represented by the following function :

$$\beta = 1+2\left(\frac{W_S}{u_{*,c}}\right)^2 \quad \text{for } 0.1 < \frac{W_S}{u_{*,c}} < 1 \tag{12}$$

Equation (12) specifies a value larger than unity, indicating a dominant influence of the centrifugal forces which cause the particles to be thrown to the outside of the eddies with a consequent increase of the effective mixing length. Given the limited knowledge of the physical processes involved, it is not advisable to use a β-factor larger than 2.

2.3.2 Φ-factor

The Φ-factor expresses the influence of the sediment particles on the turbulence structure of the fluid (damping effects). Usually the damping effect is taken into account by reducing the constant of Van Karman (κ).

Several investigators have observed that the constant of Van Karman falls below the value 0.4 (clear flow) in the case of a heavy sediment-laden flow over a rigid, flat bed. It has also been observed that the flow velocities in a layer close to the bed are reduced, while in the remaining part of the flow there are larger flow velocities. Apparently, the mixing is reduced by the presence of a large amount of sediment particles.

According to Einstein [1950], and Chien [1954], who determined the amount of energy needed to keep the particles in suspension, the constant of Von Karman is a function of the depth-averaged concentration, the particle fall velocity and the bed-shear velocity.

Although Ippen [1971] supposed that the constant of Van Karman is primarily a function of some concentration near the bed, an investigation of Einstein and Abdel-Aal [1972] showed only a weak correlation between the near-bed concentration and the constant of Von Karman.

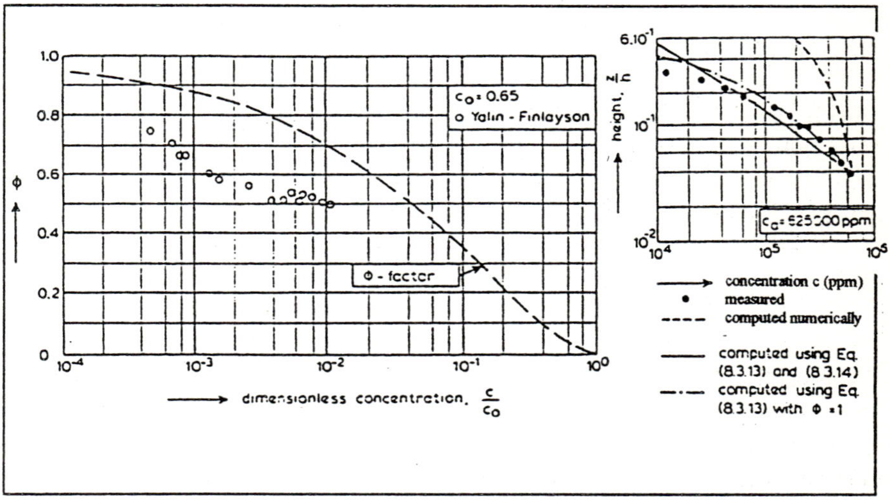

Fig. 5 : (a) θ-factor
(b) Concentration profile for experiment of Einstein-Ning Chien (1950, 1954)

Coleman (1980] questioned the influence of the sediment particles on the constant of von Kármán. He re-analyzed the original data of Einstein-Chien and Vanoni-Brooks and concluded that they used an erroneous method to determine the constant of Von Karman.

These contradictions allow to question whether the concept of an overall constant of von Karman for the entire velocity is correct for a heavy sediment-laden flow.

Van Rijn (1984) presumed a concentration-dependent Φ-factor and found by calibration using measured concentration profiles, the following expression (Fig . 5) :

$$\Phi = 1 + \left(\frac{c}{c_0}\right)^{0.8} - 2\left(\frac{c}{c_0}\right)^{0.4} \qquad (13)$$

in which :

c = local concentration (volume)

c_0 = maximum concentration = 0.65

Fig. 5 shows one of the calibration results (experiment of Einstein and Ning Chien, *op cit.*). As can be observed, the Φ-factor according to eq. (13) yields good agreement in the near-bed region, but a lesser agreement in the upper layer. With Φ = 1, computed concentrations are an order of magnitude too large compared with measured values (Fig. 5).

Finally, it is noted that the Φ-factor is only important for concentrations larger than about 0.001 (- 2.5 kg/m^3 or 0.16 lb/ft.^{3}), (Van Rijn, 1984) .

2.4 Concentration Profiles

Expressions for the sediment concentration profiles can be obtained by integration of the convection-diffusion equation (6) using relevant expressions for the mixing coefficient (see eqs. 7 - 11 with Φ = 1). As bed-boundary condition a reference concentration (c_a) at a small distance (a) above the mean bed is applied (known as the integration constant).

2.4.1 Low Concentrations (c < 0.001)

Hindered settling effects and turbulence damping effects can be neglected. Integration of eq. (6) yields :

constant α_1:
$$\frac{c}{c_a} = e^{-\alpha(Z/h)(z-a)} \tag{14}$$

linear ε_f:
$$\frac{c}{c_a} = \left(\frac{a}{z}\right)^{\alpha_2 Z} \tag{15}$$

parabolic ε_f:
$$\frac{c}{c_a} = \left(\frac{d-z}{z}\cdot\frac{a}{d-a}\right)^Z \tag{16}$$

for parabolic-constant ε_f:
$$\frac{c}{c_a} = \left(\frac{d-z}{z}\cdot\frac{a}{d-a}\right)^Z \quad \text{for } \frac{z}{d} < 0.5 \tag{17}$$

$$\frac{c}{c_a} = \left(\frac{a}{d-a}\right)^Z e^{-4Z(z/d-0.5)} \quad \text{for } \frac{Z}{f} \geq 0.5$$

in which :

c = concentration at height z above the mean bed
c_a = reference concentration at height z = a above bed
d = water depth
$Z = w_s/(\beta\kappa u_{*,c})$ = suspension number
w_s = fall velocity in clear water

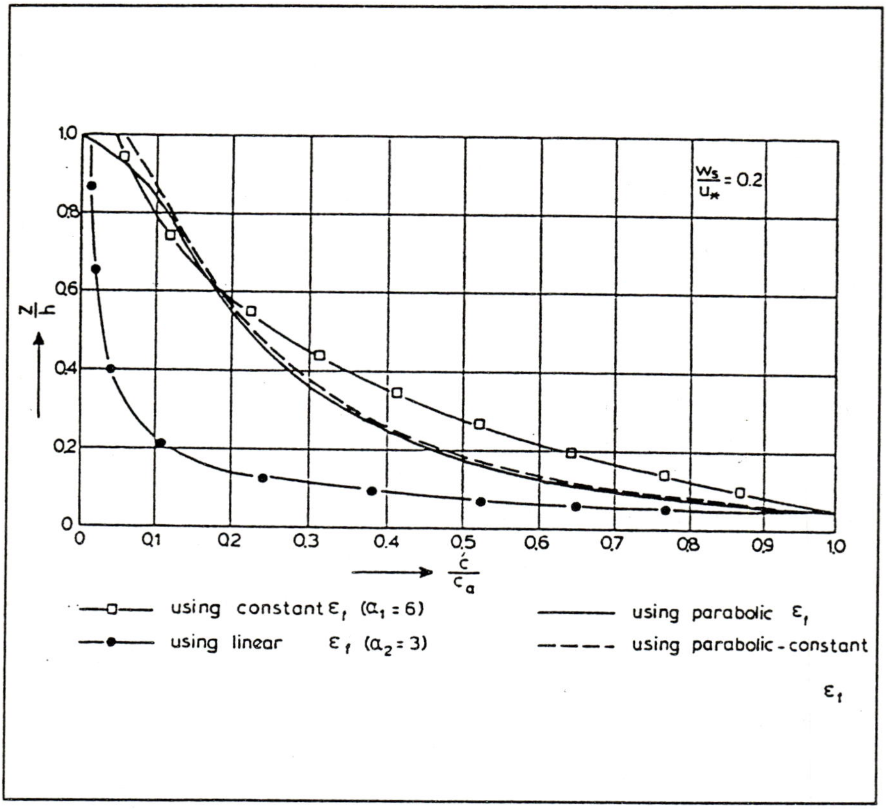

Fig. 6. Concentration profiles for a uniform flow.

Eq. (16), first derived by Rouse in 1936, is known as the Rouse equation.

Fig. 6 shows the above-given concentration profiles for $w_s/u_{*,c} = 0.2$, $\beta = 1, \kappa = 0.4$ giving $Z = 0.5$.

Equations (16) and (17) yield the best agreement with measured concentrations. Fig. 7 shows eq. (16) for different values of Z in comparison with measured concentrations (Vanoni, 1946). Equation (17) yields a finite concentration at the water surface, while eq. (16) gives a less realistic zero concentration at the water surface.

Chien (1954) applied eq. (16) to determine the Z-parameter from measured concentration profiles. The results show smaller measured Z-values. This can be interpreted as a β-factor larger than 1 ($\beta > 1$).

2.4.2 High concentrations (c > 0.001)

In case of flow in the upper regime the sediment concentration profile should be described by taking into account the hindered settling effect and the turbulence damping effect (Φ-factor). Modelling of both effects requires the application of numerical solution methods to solve eq. (5). The hindered settling effects tend to increase the concentration because the fall velocity is reduced, while the turbulence damping effect tends to decrease the concentration because the sediment mixing coefficient is reduced.

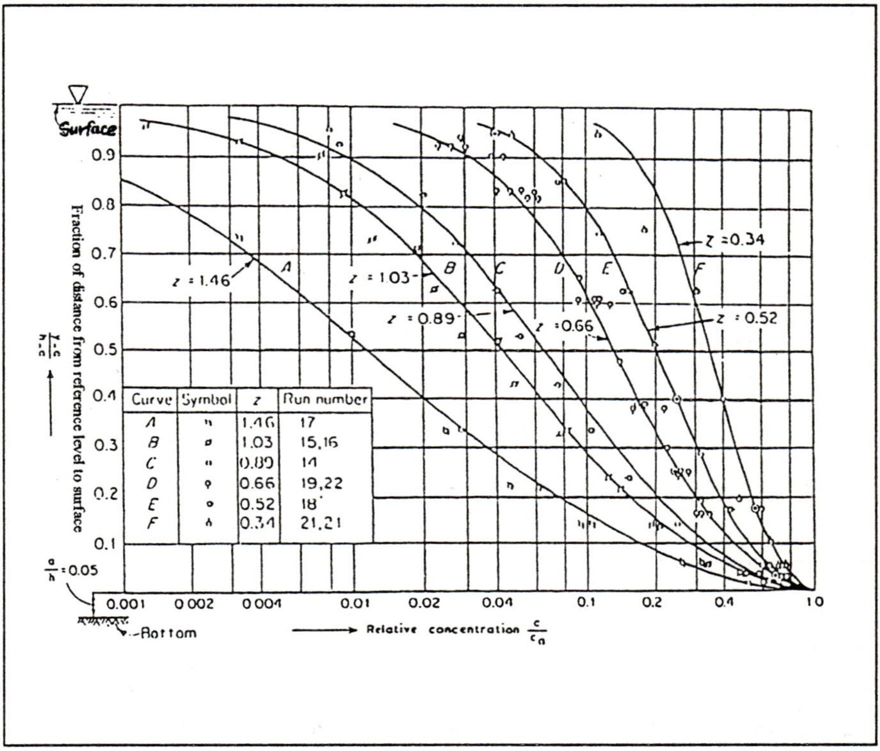

Fig. 7 : Concentration profiles (Vanoni, 1946)

According to Winterwerp *et al.* (1989), turbulence damping due to the presence of large vertical concentration (density) gradients occurs for concentrations up to c = 0.2. For increasing concentrations (c > 0.2) the concentration profile tends to become uniform due to dominating hindered settling effects and turbulence damping disappears because the concentration gradients disappear (Fig. 8). Equation (16) in combination with eq. (A) representing hindered settling effects shows reasonable agreement with measured values ($c_a \approx 0.35$).

Van Rijn (1984) presented a simplified method based on the application of eq. (17) to describe the concentration profile in combination with a modified suspension number Z' defined as :

$$w_{s,m} = (1 - C)^{\gamma}\, w_s \qquad \text{(A)}$$

in which :

$w_{s,m}$ = particle fall velocity in a suspension
w_s = particle fall velocity in a clear fluid
c = volumetric sediment concentration (-)
γ = coefficient (-)

$$Z' = Z + \Psi \qquad (18)$$

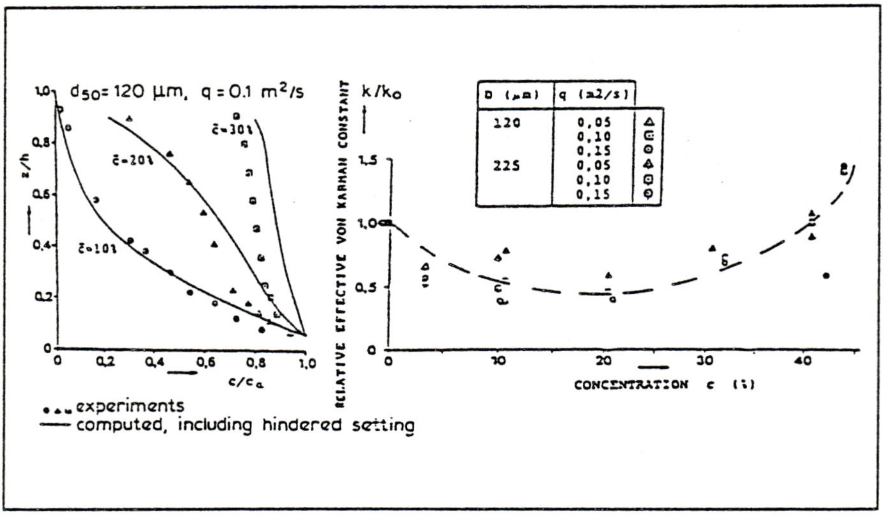

Fig. 8 : (a) Measured Concentration profiles (Winterwerp et al, 1989)
(b) Von Karman coefficient as a function of concentration.

in which :

Z = suspension number
Ψ = correction factor representing hindered settling effects and turbulence damping effects.

The Ψ-factor can be determined as follows. First, the concentration profile is computed numerically solving eq. (6) and applying eq. (A) to represent the fall velocity $w_{s,m}$ and eqs. (10), (11), (13) to represent the sediment mixing coefficient. Second, the concentration profile is computed from eqs. (17) and (18), applying a ψ -factor that yields a concentration profile similar to that obtained by the first (numerical) method.

This procedure has been repeated for a range of conditions. Analysis of all results gave the following relationship :

$$\psi = 2.5 \left(\frac{w_s}{u_{*,c}} \right)^{0.8} \left(\frac{c_a}{c_0} \right)^{0.4} \quad \text{for } 0.01 \leq 1 \tag{19}$$

in which :
c_0 = maximum volume concentration = 0.65

Fig. 5b shows (based on eq. 17) computed and measured (eq. 18) concentrations for the high-concentration experiment of Ning Chien. Good agreement can be observed.

2.5 Velocity Profile

In case of low sediment concentrations ($c < 0.001$) and rough hydraulic conditions, the logarithmic velocity profile can be applied, as follows :

$$u = \frac{u_{*,c}}{\kappa} \ln \left(\frac{z}{z_0} \right) \tag{20}$$

in which :

$u_{*,c}$ = current-related bed-shear velocity
z_0 = $0.033\, k_s$ = zero-velocity level
κ = constant of Von Karman ($\kappa = 0.4$).

Eq. (19) is the well-known Prandtl-Von Karman Equation.

Velocity profiles in high (hyper) concentration flow have been studied by many researchers (Einstein-Ning Chien (1955), Coleman (1980) and Winterwerp *et al.* (1989).

Einstein-Ning Chien found reduced velocities in the near-bed region, which was represented as a reduced κ-value.

Coleman also observed reduced velocities near the bed. He introduced an additional turbulence-related parameter to represent the effect of the sediment concentration on the velocities, the κ-parameter was kept constant.

Winterwerp *et al.* did experiments with depth-averaged volume concentrations ranging from $c = 0$ to $c \approx 0.4$. They found a reduction of the κ-parameter from $\kappa = 0.4$ to $\kappa \approx 0.25$ for concentrations increasing from $c = 0$ to $c \approx 0.1$, a nearly constant κ-value of about 0.25 for concentrations in the range of $c \approx 0.1$ to $c \approx 0.2$ and an increase of the κ value to approximately $\kappa \approx 0.4$ for concentrations increasing to $c \approx 0.4$. (Fig. 8b). According to Winterwerp *et al.*, turbulence damping due to the presence of large vertical concentration gradients occurs for concentrations up to $c = 0.2$. For increasing concentrations ($c > 0.2$) the concentration profile tends to become uniform due to dominating hindered settling effects and turbulence damping disappears because the concentrations gradients disappear (κ nears 0.4, Fig. 8a,b).

2.6 Reference Concentration and Reference Level

2.6.1 Reference Level in Case of a Flat Bed

The most logic assumption for the location of the boundary (or reference level) for sediment concentrations is the upper edge of the bed-load layer. For a perfectly flat bed a good estimate of the thickness of the bed-load layer can be obtained by taking the maximum saltation height of the bed-load particles (Van Rijn, 1984). This leads to a value of the order of $10d_{50}$ for flat bed flow in the upper regime. The concentrations close to the bed will be rather large and hence the hindered settling effect and the turbulence damping effect must be taken into account. However, at the present stage of research, the knowledge of these processes is rather limited which may easily result in large errors of the predicted concentrations as will be shown by a computation example for a flat mobile bed. The concentration profile has been computed numerically by applying eqs. (15) and (A) in combination with a parabolic-constant mixing coefficient. The suspension number $Z = W_S/(\beta \kappa u_{*,c})$ is taken as 0.625. Because of errors in w_s, β, κ and $u_{*,c}$ the inaccuracy of the suspension number will be rather large (say at least + 25%). Fig. 9 shows concentration profiles for $Z = 0.625 + 0.175$ (+ 25%). As can be observed, an error of about 25% in the Z-parameter leads to a concentration error of a factor 3 at a level of $z = 0.5$ h (mid depth) and an error of a factor 2 at a level of $z = 0.1$ h. Based on these results, it is evident that the application of a reference level at the upper edge of the bed-load layer is not very attractive for reasons of accuracy.

2.6.2 Reference Level in Case of Bedforms

Basically, the flow over a boundary covered with bedforms (dunes) is nonuniform showing acceleration effects at the upsloping sections and deceleration effects at the downsloping sections. The bed-load particles are transported close to the bed by rolling, sliding and saltating at the upsloping sections of the dunes. Arriving at the top of the dune, the majority of the bed- load particles jump over the edge and roll down the leeside slope of the dune towards the dune trough where they are buried waiting for a new [transport] cycle. The suspended load particles are entrained from the bed-load layer developing at the upsloping section. The entrainment rate is maximum at the dune top (large velocities). Above the dune trough the suspended particles are transported further upwards by turbulence mixing produced in the shear layers of the vortex generated in the dune trough. A detailed mathematical representation of the afore-mentioned processes requires the application of a sophisticated model, e.g. the K-E model of Rodi (1980), in combination with a convection-diffusion model for the suspended sediment particles. The horizontal grid size should be much smaller than the dune length ($\Delta x \ll \lambda$), while the reference level should be applied at the upper edge of the bed-load layer ($a \ll \Delta$). To get a first understanding of the phenomena involved, a relatively simple two-dimensional vertical model of the Delft Hydraulics Laboratory (Van Rijn, 1985) was used to compute the concentration profiles for the flow over dunes.

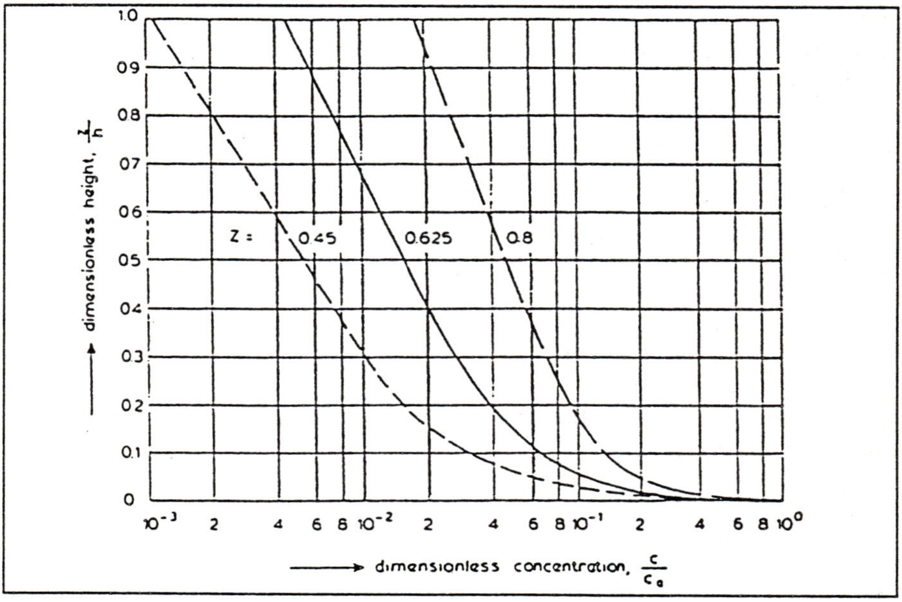

Fig. 9 : Influence of errors in hydraulic parameters on concentration profile.

Generally, the application of mathematical models for estuarine and coastal problems requires a schematization on a macro-scale (limited computer facilities) applying a horizontal grid size much larger than the bedform length. Consequently, a detailed representation of the concentration profiles and also the velocity profiles is not feasible. The effect of the dunes on the flow field is, usually, represented by introducing an effective roughness parameter (k_S) as shown in Fig. 10.

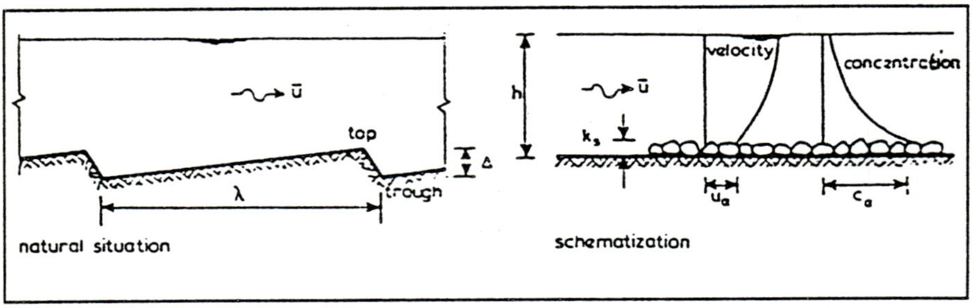

Fig. 10 : Effective roughness parameter k_S

To represent the effect of the dunes on the concentration field, it is proposed :

- to represent the concentration profiles by a schematized concentration profile being an estimate of the spatially-averaged (over the dune length) concentration profiles,
- to specify an effective concentration (c_a) at a level through the top of the dunes ($a = \frac{1}{2}\Delta$ or $a = k_S$ when the dune dimensions are unknown, $a_{minimum}$ = 0.01 h for reasons of accuracy).
- to represent the transport of particles below the reference level as bed-load transport using a simple formula.

2.6.3 Reference Concentration

Based on the bed-load transport given in Section 2.1.5., Van Rijn (1984) has proposed a function for the reference concentration (Fig. 2) :

$$c_a = 0.015 \frac{d_{50}}{a} \frac{T^{1.5}}{d*^{0.3}} \tag{21}$$

in which :

c_a = reference concentration (volume)
a = reference level above the mean bed
D* = dimensionless particle parameter
T = dimensionless bedshear stress parameter

The constant in eq. (21) is based on calibration using flume and field data with velocities in the range of 0.4 to 1.6 m/s, depths in the range of 0.1 to 25 m and mean particle size in the range of 180 to 700 μm (Van Rijn, 1984). Voogt *et al.* (1989) showed that eq. (21) is valid for velocities upto 3 m/s. The reference level is assumed to be equal to half the bedform height ($a = \frac{1}{2}\Delta$) or equal to the effective bed roughness ($a = k_S$) when the bedform height is unknown. The reference level should not be chosen smaller than $a = 0.01$ h for reasons of accuracy.

2.7 Influence of Non-Uniform Bed Material

Under natural conditions the bed material is non-uniform. Observations in flume and field conditions have shown that the bed material and the suspended sediment material have different particle size distributions. Usually, the suspended sediment particles are considerably smaller than the bed material particles. Basically, it is possible to compute the suspended load for any known type of bed material and flow condition by dividing the bed material into a number of size fractions and assuming that the size fractions do not influence each other.

Van Rijn (1984) has applied the size fraction method to determine a representative suspended sediment diameter (d_S) which accounts for the non-uniformity effects. Using the size fraction method (as proposed by Einstein), the total suspended load has been computed for various conditions, whereafter the representative (suspended) particle diameter (d_S) was determined by trial and error gave the same value for the suspended load than according to the size-fraction method.

In all, six computations have been done using two types of bed material with a geometric standard deviation $\sigma_S = (d_{84}/d_{50} + d_{16}/d_{50}) = 1.5$ and 2.5. The d_{50} of the bed material was equal to 250 μm. The mean flow velocities were 0.5 and 1.5 m/s (1.6 and 4.92 ft./s). The flow depth was assumed to be 10 m (32.81 ft.). The concentration profile was computed by applying eq. (17) with $\beta = 1$, $= 1$ and $\kappa = 0.4$. The reference concentration according eq. (21) was applied at a level of $z = 0.05$ h. The flow velocity profile was assumed to be logarithmic, eq. (20). The computed d_S-values have been related to σ_S-parameter and to the dimensionless shear-stress parameter T, as follows :

$$\frac{d_S}{d_{50}} = 1 + 0.011(\sigma_s - 1) \quad (T - 25) \tag{22}$$

Equation (22) has been compared with experimental data of Guy *et al.* (1966). The scatter of the data is too large to detect a clear influence of the σ_S- parameter, but, in an average sense the agreement between measured and computed values ($\sigma_S = 2.5$) is quite good.

2.8. Prediction Method of Einstein (1950)

Einstein's method is based on a parabolic distribution of the fluid mixing coefficient and a logarithmic distribution for the velocity.

The suspended sediment transport rate can be expressed as :

$$q_{s,c} = 11.6 u'_{*,c} c_a a \left[I_2 + I_1 \ln\left(30.2 eh / d_{65}\right)\right] \tag{23}$$

$$I_1 = 0.216 \frac{A^{Z-1}}{(1-A)^Z} \int_A^1 \left(\frac{1 - z'Z}{z'} dz' \right) \tag{24}$$

$$I_2 = 0.216 \frac{A^{Z-1}}{(1-A)Z} \int_A^1 \left(\frac{1 - z'Z}{z'} \ln(Z') dz' \right) \tag{25}$$

in which :

$q_{s,c}$	= current-related suspended load transport	(m^2/s)
$u_{*,c}$	= current-related bedshear velocity due to the grains	(m/s)
c_a	= reference concentration (volume) = $q_b/(11.6\, u'_{*,c} a)$	(-)
a	= reference level (= 2 d)	
h	= water depth	(m)
d	= particle diameter	(m)
A	= a/h = dimensionless reference level	(-)
z'	= z/h = dimensionless vertical coordinate	(-)
Z	= $W_S/(\kappa u'_{*,c})$ = suspension number	(-)
e	= correction factor	

According to Einstein, the suspended load transport is related to the grainshear velocity (u_*) and not to the overall shear velocity (u_*).

The reference concentration is determined from the bed load transport, assuming a bed load layer thickness equal to 2 particle diameters. The I_1 and I_2 -integrals can be determined graphically (cf. Einstein, 1950) or numerically.

The method of Einstein is not given in full detail here, because of its complexity. Furthermore, an extensive verification study of White *et al.* (1973) has shown that the predicting ability of the Einstein method is much less than other (and simpler) methods.

2.9 Prediction Method of Bagnold (1966)

Bagnold (1966) introduced an energy balance concept and related the suspended load transport to the work done by the fluid. The suspended load transport is expressed as :

$$\rho_{s,c} = \frac{e_s(1-e_b)}{(\rho_s-\rho)g(w_s/\bar{u}-\tan\beta)}\tau_b\bar{u} \qquad (26)$$

in which :

$q_{s,c}$	= current related suspended load transport	(m^2/s)
e_s	= efficiency factor (- 0.02)	(-)
e_b	= efficiency factor (- 0.1)	(-)
τ_b	= bed-shear stress	(N/m^2)
$\bar{u}$	= depth averaged velocity	(m/s)
w_s	= particle fail velocity	(m/s)
β	= angle of local bottom slope	(-)

Kachel and Sternberg (1971) have shown that the efficiency factors (e_s and e_b) are not constant, but are strongly related to the bedshear stress and the particle diameter. Basically, these results show that the validity of the Bagnold concept is questionable.

2.10 Prediction Method of Bijker (1971)

Based on the concept of Einstein (1950), Bijker (1971) proposed :

$$q_{s,c} = 1.83\, q_{b,c} \left[I_2 + I_1 \quad 33h \quad k_s\right]$$

in which :

$q_{s,c}$	= suspended sediment transport rate	(m^2/S)
$q_{b,c}$	= bed-load transport rate	(m^2/S)
I_1	= integral according to eq. (33)	(-)
I_2	= integral according to eq. (34)	(-)
a	= k_S = reference level	(m)

The bedload transport rate (m^2/S) is expressed as :

$$q_{b,c} = b u_{*,c}\, d_{50}\, e^{-0.27}\, (\mu\Theta) \tag{28}$$

in which :

$u_{*,c}$	= overall bed shear velocity
Θ	= mobility parameter
μ	= $(C/C')^{1.5}$ = bed form factor (see eq. (5))
C	= overall Chézy coefficient
C'	= grain-related Chézy coefficient (see eq. (5))
b	= coefficient ($\approx$ 5)

Because the reference level is assumed to be equal to the bed roughness ($a=k_S$), the ratio $q_{s,c}/q_{b,c}$ can be expressed as a function of the parameter Z and k_S/h .

2.11 Prediction Method of Van Rijn (1984)

Applying eq. (17), (18) and (20), the depth-integrated suspended sediment transport rate can be computed from eq. (6), yielding :

$$q_{s,c} = \frac{u_{*,c} c_a}{\kappa}\left(\frac{a}{d-a}\right)^{Z'}\left[\int_a^{0.5d}\left(\frac{(d-z)}{z}\right)\ln\left(\frac{z}{z_0}\right) + \int_{0.5d}^{d} e^{-4Z'(z/d-0.5)}\ln\left(\frac{z}{z_0}dz\right)\right] \tag{29}$$

Equation (29) cannot be solved analytically. An approximate solution accurate to about 25% for $0.3 < Z' < 3$ and $0.01 < a/d < 0.1$ is given by :

$$q_{s,c} = F\,\overline{u}\,d c_a \tag{30}$$

$$F = \frac{(a/d)^{Z'} - (a/d)^{1.2}}{(1-(a/d))^{Z'}(1.2 - Z')} \tag{31}$$

in which :

$q_{s,c}$	= current-related suspended sediment transport	(m^2/s)
$\bar{u}$	= depth-averaged flow velocity	(m/s)
d	= water depth	(m)
F	= dimensionless shape factor	(-)
c_a	= reference concentration	(volume)

Summarizing, the complete method to compute the suspended load transport per unit width, is as follows :

$$q_{s,c} = F\,\bar{u}\,dc_a \qquad (32)$$

in which :

$D_* = d_{50}\left(\frac{(s-1)g}{v^2}\right)^{1/3}$	= particle parameter	(-)
$T = \frac{\tau'_{b,c} - \tau_{b,cr}}{\tau_{b,cr}}$	= bedshear stress parameter	(-)
$c_a = 0.015\frac{d_{50}}{a} \cdot \frac{T^{1.5}}{D_*^{0.3}}$	= reference concentration	(-)
$F = \frac{\left(\frac{a}{h}\right)^{Z'} - \left(\frac{a}{h}\right)^{1.2}}{\left(1-\frac{a}{h}\right)^{Z'}(1.2 - Z')}$	= shape factor	
$Z' = Z + \Psi$	= suspension number	(-)
$Z = \frac{w_s}{\beta\kappa u_{*,c}}$	= suspension number	(-)
$\Psi = 2.5\left(\frac{w_s}{u_{*,c}''}\right)^{0.8}\left(\frac{c_a}{c_0}\right)^{0.4}$	= correction factor related to turbulence damping and hindered settling	(-)

$w_s = \frac{10\nu}{d_s}\left[\left(1+\frac{0.01(s-1)gd^3_s}{\nu^2}\right)05-1\right]$ = particle fall velocity (m/s)

$d_s =$ (m)

$\left[1+0.011(\sigma_S-1)(T-25)\right]d_{50}$ =- representative particle

(size of suspended sediment, $d_5 = d_{50}$ for T>25)

$\beta = 1+2\left(\frac{w_s}{u_{*,c}}\right)^2$ = ratio of sediment and fluid mixing coefficient (β_{max}=2) (-)

$\sigma_S = \frac{1}{2}\left(\frac{d_{84}}{d_{50}}+\frac{d_{50}}{d_{16}}\right)$ = gradation coëfficiënt of bed material (-)

$\tau_{b,cr}=(\rho_S - \rho)\ g\ d_{50}\ \Theta_{cr}$ = critical bedshear stress according to Shields (N/m²)

$\Theta_{cr} = 0.24\ D_*^{-1}$ if $1 < D_* \leq 4$

$\Theta_{cr} = 0.14\ D*^{-0.64}$ if $4 < D_* \leq 10$

$\Theta_{cr} = 0.04\ D*^{-1}$ if $10 < D_* \leq 20$ Shields curve (-)

$\Theta cr = 0.013\ D*^{029}$ if $20 < D_* \leq 150$

$\Theta cr = 0.055$ if $1 < D_* \leq 150$

$\tau_{b,c} = \rho g\left(\frac{\bar{u}}{C'}\right)^2$ = current-related effective bedshear stress (N/m²)

$u_{*,c} = \frac{\sqrt{g}}{C}\bar{u}$ = current-related overall bed-shear velocity (m/s)

$C' = 18\log\left(\frac{12d}{3d_{90}}\right)$ = grain-related Chézy coefficient (m/s)

$C = 18\log\left(\frac{12d}{k_s}\right)$ = overall Chézy coefficient (m/s)

$q_{s,c}$	= current-related suspended load transport	(m^2/s)
$\overline{u}$	= depth-averaged velocity	(m)
d_{50}	= median particle diameter of bed material	(m)
d_{50}, d_{84}, d_{90}	= diameters of bed material	(m)
d	= water depth	(m)
k_s	= overall bed roughness height	(m)
g	= acceleration due to gravity	(m/s^2)
a	= reference level ($a = \frac{1}{2}\Delta$, or $a = k_s$, $a_{minimum} = 0.01\ d$)	(m)
Δ	= bedform height	(m)
c_0	= maximum concentration = 0.65	(1)
s	= specific density = 2.65	(-)
ν	= kinematic viscosity coefficient	(m^2/S)
κ	= Von Karman constant = 0.4	(-)

When the bed-load transport and the suspended load transport are known, the total load transport of bed material can be determined by addition ($q_t = q_b + q_s$).

The ratio of the suspended and total load transport can be expressed as :

$$\frac{q_s}{q_t} = \frac{q_s}{q_b + q_s} = \frac{1}{q_b / q_s + 1} \tag{33}$$

For reasons of simplicity the bed load transport is here defined as :

$$q_b = a c_a \overline{u}_a \tag{34}$$

in which :

c_a = reference concentration

$\overline{u}_a$ = effective transport velocity of bed-load particles

a = reference level (= bed-load layer thickness)

Introduction of eq. (32) and (34) in eq. (33) yields :

$$\frac{q_s}{q_t} = \frac{1}{\left[\left(F^{-1}\right)\left(\overline{u_a} / \overline{u}\right)(a / h)\right] + 1} \tag{35}$$

The ratio $\overline{u_a} / \overline{u}$ may be identified as the ratio of the average transport velocity of the bed-load and suspended load particles, which varies from about 0.4 for large, steep bedforms in the lower flow regime to about 0.8 for flat bed conditions in the upper flow regime. Fig. 11 shows the ratio of the suspended load and the total load as a function of the ratio of the bed-shear velocity and particle fall velocity for different values of $\overline{u_a} / \overline{u}$ with $\kappa = 0.4$, $\beta = 1$ and a/h=0.05. An empirical relationship given by Laursen (1958) and data from Guy *et al.* (1966) are also shown.

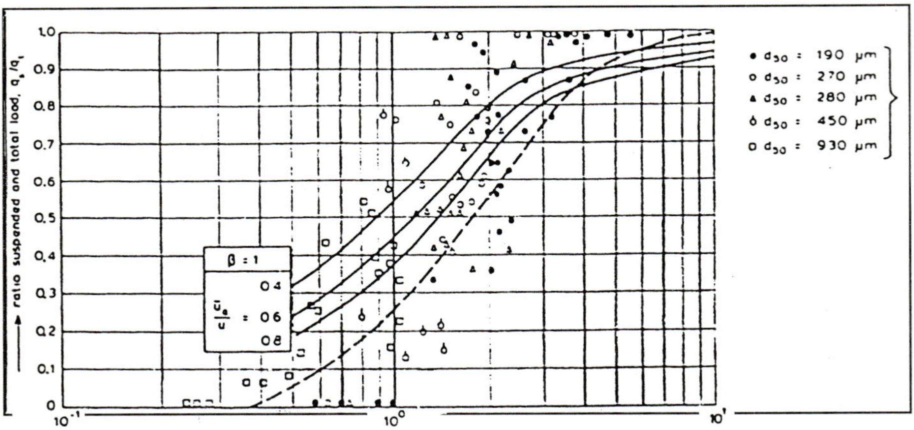

Fig. 11 : Ratio of suspended and total transport.

——— Van Rijn (1984)

– – – Laursen (1958)

} measurements Guy et al. (1955) n:0,1-0,3m

3 Total Load Transport

3.1 Prediction Methods of Einstein, Bagnold, Bijker and Van Rijn

The total transport of bed material particles according to the methods of Einstein (1950), Bagnold (1966), Bijker (1971) and Van Rijn (1984) can be obtained by addition of the bed load and suspended load transport :

$$q_t = q_b + q_s \tag{36}$$

Two typical total load formulas are those of Engelund-Hansen (1967) and Ackers-White (1973). Both methods have a good predicting ability compared with measured data, as shown by a verification study of White *et al.* (1973).

3.2 Prediction Method of Engelund-Hansen (1967)

Based on energy considerations and calibration using about 100 flume data, the following formula was proposed :

$$q_{t,c} = \frac{0.05\left(\overline{u}\right)^5}{(s-1)^2\sqrt{g}\,d_{50}C_3} \tag{37}$$

in which :

$q_{t,c}$ = total bed material transport (m^2/s)

$\overline{u}$ = depth-averaged velocity (m/s)

C = overall Chézy-coefficient (m /s)

d_{50} = median particle diameter of bed material (m)

s = $\rho_{s/s}$ = specific density (= 2.65) (-)

g = acceleration of gravity (m/s^2)

3.3 Prediction Method of Ackers-White (1973)

Based on analysis of 925 sets of flume and field data, the following empirical formula was proposed :

$$q_{t,c} = K\bar{u}\ d_{35}\left(\frac{\bar{u}}{u_{*,c}}\right)^{n}\left(\frac{Y - Y_{cr}}{Y_{cr}}\right)^{m} \qquad (38)$$

in which :

$q_{t,c}$ = total bed material transport (m^2/s)

$\bar{u}$ = depth-averaged velocity (m/s)

$u_{*,c}$ = current-related bed shear velocity (m/s)

Y = particle mobility parameter (-)

Y_{cr} = critical particle mobility parameter (-)

n,m,K = coefficients (-)

ν = kinematic viscosity coefficient (m^2/s)

s = specific density (= 2.65) (-)

$K = \exp[2.86 \ln (D_*) - 0.434 (\ln D_*)^2 - 8.13]$ for $D_* < 60$

$n = 1 - 0.56 \log (D_*)$ for $D_* < 60$

$$m = \frac{9.66}{D_*} + 134 \qquad \text{for } D_* < 60$$

$$Y_{cr} = \frac{0.23}{D_*^{0.5}} + 0.14 \qquad \text{for } D_* < 60$$

$K = 0.025$, $n = 0$, $m = 1 .5$, $Y_{cr} - 0.17$ for $D_* > 60$

$$Y = \left(\frac{u_{*,c}^{n}}{\sqrt{(s-1)gd_{35}}}\right)\left(\frac{\bar{u}}{5.66\log(10d / d_{35})}\right)^{1-n}$$

$$D_* = d_{35}\left(\frac{(s-1)g}{\nu^2}\right)^{1/3}$$

3.4 Comparison of Prediction Methods

Van Rijn (1984) used 486 sets of river data to verify the methods of Engelund-Hansen (1967), Ackers-White (1973) and Van Rijn (1984). Bed material sizes were in the range of 100 to 400 μm. Flow velocities were in the range of 0.4 to 2.4 m/s (1.28 to 76.8 ft./s). The results have been expressed in terms of a discrepancy ratio (r) defined as the ratio of predicted and measured transport rate. The table below shows the percentage of r-values of all data falling in the range of $0.5 < r \leq 2$. The method of Van Rijn yields the best results with 76 % of the predicted transport rates within a factor 2 of the measured values. Voogt et al (1989) carried out large-scale flume experiments with bed material of 200 μm and velocities in the range of 1 to 3m/s (3.2 to 9.6 ft./s). Comparison of predicted and measured transport rates showed good results for all three methods. Voogt *et al.* also compared predicted rates with 120 sets of estuary data. Bed material sizes were in the range of 200 to 300 μm. Flow velocities were in the range of 1 to 2 m/s (3.2 to 9.6 ft./s). The results are given in the table below. The method of Van Rijn yields the best results with about 90 % of the predicted transport rates within a factor 2 of measured values. The results of the other two methods are not as good. Both methods gave a considerable overprediction, particularly the Ackers-White method.

Method	River data (486)	Estuary data (120)
Engelund-Hansen (1967)	64%	33%
Ackers-White (1973)	63%	26%
Van Rijn (1984)	76%	89%

White *et al.* (1973) examined various transport formulas using about 1000 flume data and 260 field data.

Froude numbers greater than 0.8 were excluded. The results of the Ackers-White (1973), Engelund-Hansen (1967), Einstein (1950) and Bagnold (1966) formulas are given in the following table.

Formula	Flume and field data (1260)
Ackers-White (1973)	68%
Engelund-Hansen (1967)	63%
Einstein (1950)	46%
Bagnold (1966)	22%

The Ackers-White method yields the best results with 68% of the predicted transport ranges within a factor 2 of measured values. The method of Bagnold yields poor results with a score of 22%.

4 References

Ackers, P. and White, W.R., 1973, Sediment Transport : New Approach on Analysis, *Journal of the Hydraulics Division,* ASCE, No. HY 11, USA.

Bagnold, R.A., 1956. The Flow of Cohesionless Grains in Fluids, *Proc., Royal Soc. Philos. Trans., London*, Vol. 249.

Bagnold, R.A., 1966, An Approach to the Sediment Transport Problem from General Physics, *Geological Survey Prof. Paper 422*-I, Washington.

Bijker, E.W., 1971, Longshore Transport Computations, *Journal of the Waterways, Harbors and Coastal Engineering Division*, Vol. 97, No. WW4.

Chien, N., 1954. The Present Status of Research on Sediment Transport, *Proc. ASCE*, Vol. 80.

Coleman, N.L., 1970, Flume Studies of the Sediment Transfer Coefficient, *Water Resources Research*, Vol. 6, No. 3, USA.

Coleman, N.L., 1980. Velocity Profiles with Suspended Sediment, *Journal of Hydraulic research*, (The Netherlands), Vol. 19, No. 3.

Einstein, H.A., 1950. The Bed-Load Function for Sediment Transportation. In : *Open Channel Flow Technical Bulletin No. 1026*, U.S. Dep. of Agriculture, Washington, D.C.

Einstein, H.A. and Abdell-Aal, F.M., 1972, Einstein Bed-Load Function at High Sediment Rate, *Journal of Hydraulics Division,* ASCE, Vol. 98, No. HY 1 , USA.

Engelund, F. and Hansen, E., 1967, A Monograph on Sediment Transport in Alluvial Streams, *Nordic Hydrology* 7.

Frijlink, H.C., 1952, Discussion of Bed-Load Movement, *Formulas Report No. X2344/LV*, Delft Hydraulics, Delft, The Netherlands.

Guy, H.P., Simons, D.B. and Richardson, E.V., 1966, Summary of Alluvial Channel Data from Flume Experiments, 1956-1961, *Geol. Survey Prof. Paper, 462*-I, Washington.

Ippen, A.F., 1971, A New Look at Sedimentation in Turbulent Streams, *Journal of the Boston Soc. of Civ. Engrs.,* USA, Vol. 58, No. 3.

Kachel, N.B. and Sternberg, R.W., 1974, Transport of Bed Load as Ripples during an Ebb Current, *Marine Geology*, 19, p 229-244.

Kalinske, A.A., 1947, Movement of Sediment as Bed Load in Rivers, *Trans.American Geophysical*, Vol. 28, No. 4.

Laursen, E.M,. 1958, The Total Sediment Load of Streams, *Journal of the Hydraulics Division,* ASCE, No. HYI.

Meyer-Peter, E. and Müller, R., 1948, Formulas for Bed-Load Transport, *Sec. Int. IAHR Congress*, Stockholm, Sweden.

Vanoni, V.A., 1946, Transportation of Suspended Sediment by Water, *Trans.ASCE*, Vol. 111.

Vanoni, V.A. and Brooks, N.H., 1957, Laboratory Studies of the Roughness and Suspended Load of Alluvial Streams Sedimentation Laboratory California Institute of Technology, Pasadena, *Report E-68.*

Van Rijn, L.C., 1984a, Sediment Transport, Part : Bed-Load Transport, *Journal of Hydraulic Engineering*, Vol. 110, No. 10.

Van Rijn, L.C., 1984b, Sediment Transport, Part I : Suspended Load Transport, *Journal of Hydraulic Engineering*, Vol. 110, No. 11.

Van Rijn, L.C., 1984c, Sediment Transport, Part II : Bed-Forms and Alluvial Roughness, *Journal of Hydraulic Engineering*, Vol. 110, No. 12.

Van Rijn, L.C., 1985, *Mathematical Models for Sediment Concentration Profiles in Steady Flow,* Euromech 192, Transport of Suspended Solids in Open Channels, Munich, F.R. Germany.

White, B.R. and Schulz, J.C., 1977, Magnus Effect in Saltation, *Journal of Fluid Mechanics*, Vol. 81.

White, W.R., Milli, H. and Crabbe, A.D., 1973, Sediment Transport, An Appraisal of Available Methods, *Int. Report 119*, Wallingford, England.

Winterwerp, H., De Groot, M.B., Mastbergen, D. and Verwoert, H., 1989, Hyperconcentrated Sand-Water Mixture Flows, Part I : Flow over a Flat Bed, *Journal of Hydraulic Engineering* (in press).

Chapter II : Bed Material Suspension and Transport in Uniform Waves

1 Introduction

Wave motion over a movable fine sand bed (50 to 500 μm) can generate a sediment suspension with relatively large sediment concentrations in the near-bed region as shown by Nakato *et al.* (1977) and by Bosman (1982) for the ripple regime, and by Horikawa *et al.* (1982) and by Staub *et al.* (1984) for the plane bed (sheet flow) regime.

The variation of the *instantaneous* sediment concentrations in space and time which is relevant for the cross-shore transport processes in non-breaking waves, appears to be a complicated physical process particularly in the ripple regime. This process seems hardly predictable due to the random character of the phenomena involved. Cross-shore transport processes in non-breaking waves are caused by various mechanisms :

- the asymmetry of the wave motion yielding larger onshore peak velocities under the wave crests than offshore peak velocities under the wave troughs which may result in a net offshore-directed transport in case of a ripple-covered bed or in a net onshore directed transport in case of a plane bed (sheet flow),
- the generation of net mean (Eulerian) onshore-directed velocities in the near-bed region which may result in a net onshore-directed transport (Fig. 12),
- the generation of forced long waves due to a mean water surface decrease ("trough") under large-amplitude wave groups resulting in secondary offshore-directed orbital velocities under the "trough" and a net offshore-directed transport because the sediment concentrations are largest under large-amplitude waves (Fig. 12).

Time-averaged sediment concentrations which are relevant for cross-shore and longshore transport processes due to breaking waves are less complicated to describe and have been studied by many researchers. Transport processes in breaking waves are caused by the following mechanisms :

- the generation of net mean offshore-directed velocities in the lower layers (undertow, Fig. 12),
- the generation of large-scale horizontal circulation cells with longshore currents and with offshore-directed rip currents (Fig. 12).

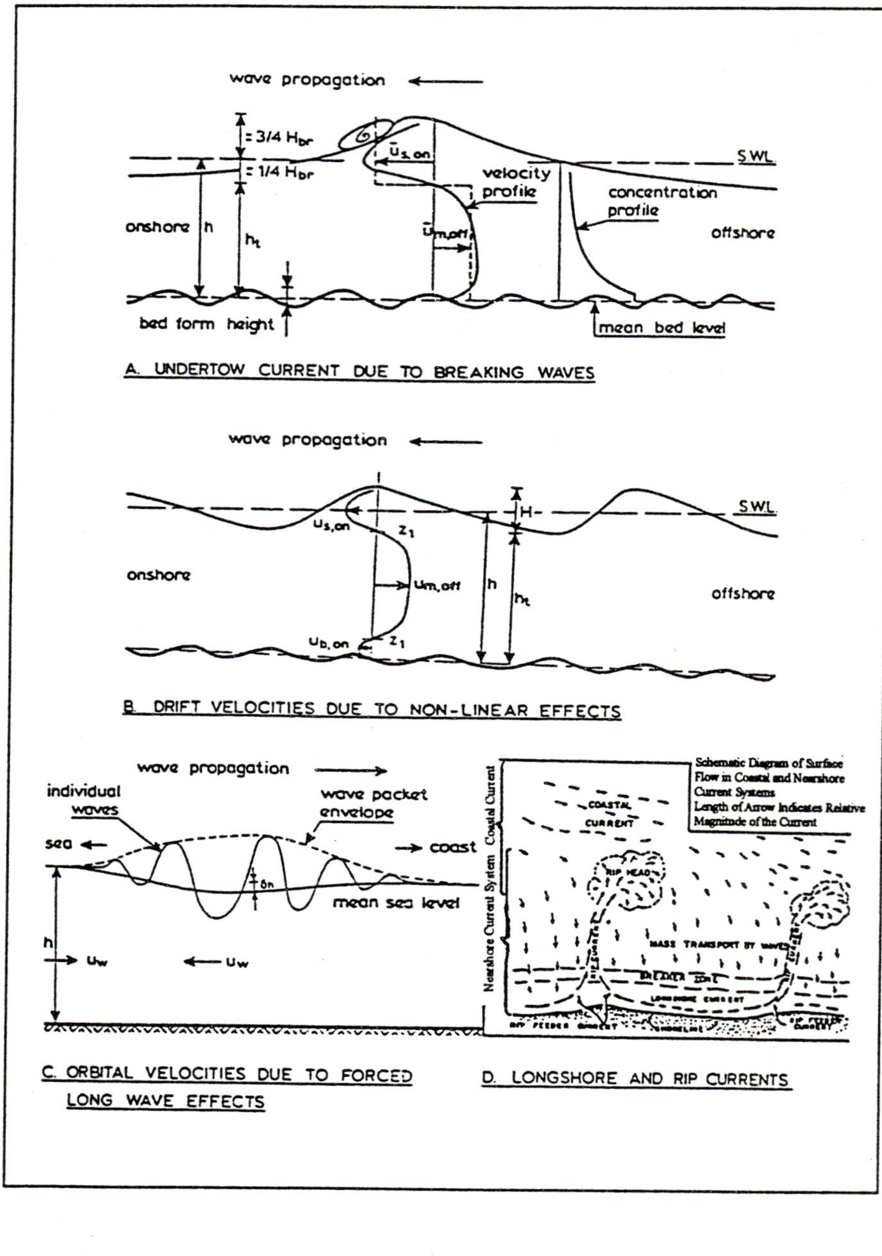

Fig. 12 : Wave-induced velocities and currents

The total sediment transport rate (q_t) can be computed from the vertical distribution of fluid velocities and sediment concentrations, as follows :

$$q_t = \int_0^{d+n} UCdz \tag{39}$$

in which :

U	= local instantaneous fluid velocity at height z above bed
C	= local instantaneous sediment concentration at height z above bed
d	= water depth to mean surface level
η	= water surface elevation

Defining : $U = u + \tilde{u}$ and $C = c + \tilde{c}$ (40)

in which :

u	= time and space-averaged fluid velocity at height z
c	= time and space-averaged concentration at height z
$\tilde{u}$	= oscillating fluid component (including turbulent component)
$\tilde{c}$	= oscillating concentration component (including turbulent component)

By introducing eq. (40) in eq. (39) and averaging over time and space, we get :

$$\overline{q_t} = \int_0^d ucdz + \int_0^d \overline{\tilde{u}\tilde{c}}dz \tag{41}$$

in which :

$\overline{q_c} = \int_0^d ucdz$ = current-related sediment transport rate

$\overline{q_w} = \int_0^d \overline{\tilde{u}\tilde{c}}dz$ = wave-related sediment transport rate

Current-related sediment transport is defined as the transport of sediment particles by the time-averaged (mean) current velocities (longshore currents, rip currents, undertow currents). Both the current velocities and the sediment concentrations are affected by the wave motion. It is known that the wave motion reduces the current velocities near the bed, but strongly increase the near-bed concentrations due to its stirring action. Wave-related sediment transport is defined as the transport of sediment particles by the oscillating fluid components (cross-shore orbital motion). The oscillating components ($\tilde{u}$ and $\tilde{c}$) may also be affected by the current velocities.

In this section the attention is focused on the wave-related sediment concentrations and transport processes, presenting the following subjects :

- analysis of measured concentrations and transport processes,
- computation of time-averaged concentration profiles,
- computation of transport rates.

2 Analysis of Measured Concentration Profiles and Transport Rates

2.1 Instantaneous concentrations

In section 1 it has been shown that instantaneous concentrations are especially important with respect to the wave-related transport processes

2.1.1 Ripple regime

Fig. 13 shows ensemble mean concentrations within a wave period in a wave tunnel measured by Bosman (1982). A sinusoidal oscillatory motion with a period of T = 1 sec and a velocity amplitude of U_δ = 0.3 m/s (0.96 ft./s) was generated over a sand bed (d_{50} = 200 μm, d_{90} = 240 μm). The bed was covered with almost perfectly two-dimensional ripples (length = 0.055 m or 0.176 ft., height = 0.01 m or 0.032 ft.). An optical instrument was used to measure the concentration values above the ripple crest and trough. The exact measuring locations are shown in Fig. 13. The ripple crest measurements show ensemble mean concentrations and standard deviations based on 100 periods. About 70% of all measurements fall within the standard deviation lines. As regards the trough measurements, only ensemble mean values are shown. The following phenomena can be observed above the crest :

- the (random) scatter is quite large (roughly + 50%),
- two large concentration peaks just after flow reversal and probably generated by leeside eddy-velocities,
- two smaller concentration peaks at the moment of maximum flow, probably generated by stoss-side velocities,
- asymmetrical-concentration distribution (water motion being symmetrical).

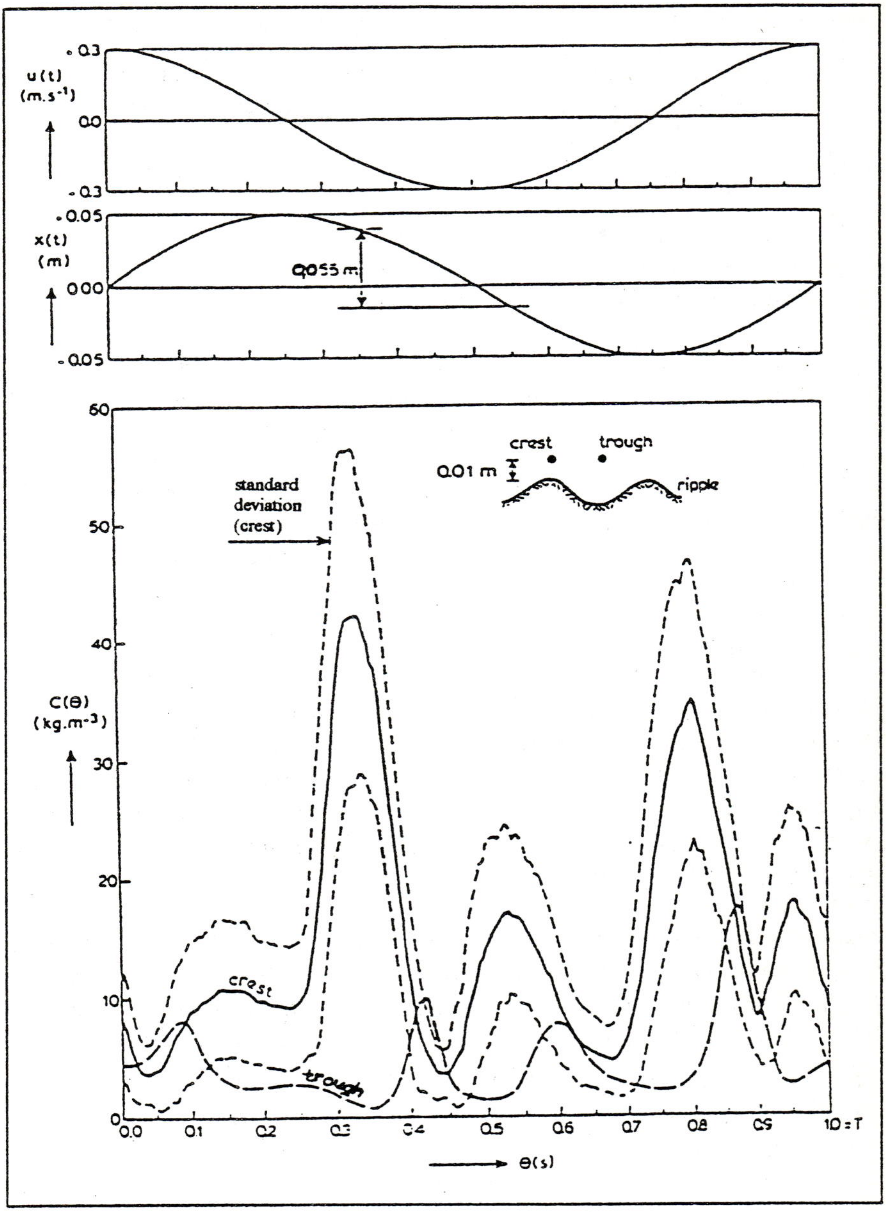

Fig. 13 : Ensemble mean and total standard deviation of concentration in single period T.

The phenomena above the trough are :

- the (random) scatter is also quite large (not shown),
- two larger concentration peaks after flow reversal and probably generated by leeside eddy-velocities (time lag is larger compared with concentration measurements at the crest),
- two smaller concentration peaks after maximum flow and probably generated by stoss-side velocities,
- asymmetrical concentration distribution,
- the peaks above the trough are smaller than those above the crest due to dispersion and settling of sediment particles.

According to Bosman, the scatter is mainly caused by (slight) local ripple changes resulting in small differences in the local velocities and hence concentrations. Based on this, it seems very difficult to relate the local instantaneous sediment concentration to a local instantaneous fluid velocity. The measurements of Nakato *et al.* (1977), with periods in the range of 1 to 3 seconds show similar results as those of Bosman. Mathematical models to compute the instantaneous sediment concentrations in the ripple regime are not yet available.

2.1.2 Sheet flow regime

Instantaneous concentrations generated by non-breaking waves in the sheet flow regime have been measured by Horikawa *et al.* (1982) in oscillatory flow over a sand bed (d_{50} = 200 μm) in a wave tunnel using an electro-resistance concentration meter. The maximum concentrations are generated at the moment of maximum velocities (Θ = 90°). The concentration is minimum at the moment of minimum velocities (Θ= 0°).

The thickness of the concentration layer is about 0.02 m (0.064 ft.). At this level the time-averaged concentration was about 0.5 kg/m^3 (0.03 lb/ft.^{3}). The velocities have been determined by applying a camera technique focusing on the interior particles to eliminate the wall effect. Analysis of the instantaneous sand transport rates show that most of the transport occurs below the initial bed level (Θ = 30°, 60°, 90°) .

2.2 Time-Averaged Concentrations

Time-averaged concentrations (averaged over many waves) are relevant for the current-related sediment transport. Time-averaging is necessary to eliminate the large random scatter of the instantaneous concentrations. Time-averaged concentrations have been measured by various researchers. Some of the experimental results are analyzed here focusing on the following cases :

1. concentrations by non-breaking waves over a rippled bed,
2. concentrations by breaking waves,
3. concentrations in the sheet flow layer.

2.2.1 Concentrations by Non-Breaking Waves over a Rippled-Bed

In case of oscillatory flow over a rippled bed the instantaneous sediment concentrations are varying in time and space, as shown in Fig. 13. To eliminate these variations, it is necessary to average the concentrations over time (many waves) and over space (many bedforms). In laboratory conditions this can be done quite simply by moving the measuring instruments forwards and backwards over a certain distance during the sampling period.

Nieuwjaar-Van Der Kaaij (1987) observed the following phenomena :

- concentrations increase for increasing wave height H_s
- negligible small concentrations for $z/h > 0.3$
- constant concentration gradient at each wave height H_s
- steeper concentration profiles for increasing wave heights
- ripple height increases slightly for increasing wave height

Van Rijn (1987) observed the following phenomena :

- increasing concentrations for increasing wave height upto $H_s = 0.69$ m (2.21 ft.) and decreasing concentrations for $H_s = 1.1$ m (3.52 ft.) in the near-bed layer,
- two layer concentration profiles with large gradients in the near-bed layer of $z/h < 0.05$ (3 ripple heights) and small gradients for $z/h > 0.05$,
- decrease of ripple height from $\Delta_r = 0.02$ m (0.8 in.) to 0.001 m (0.04 in.) for wave height increasing from $H_s = 0.69$ m (2.21 ft.) to $H_s = 1.1$ m (3.52 ft.).

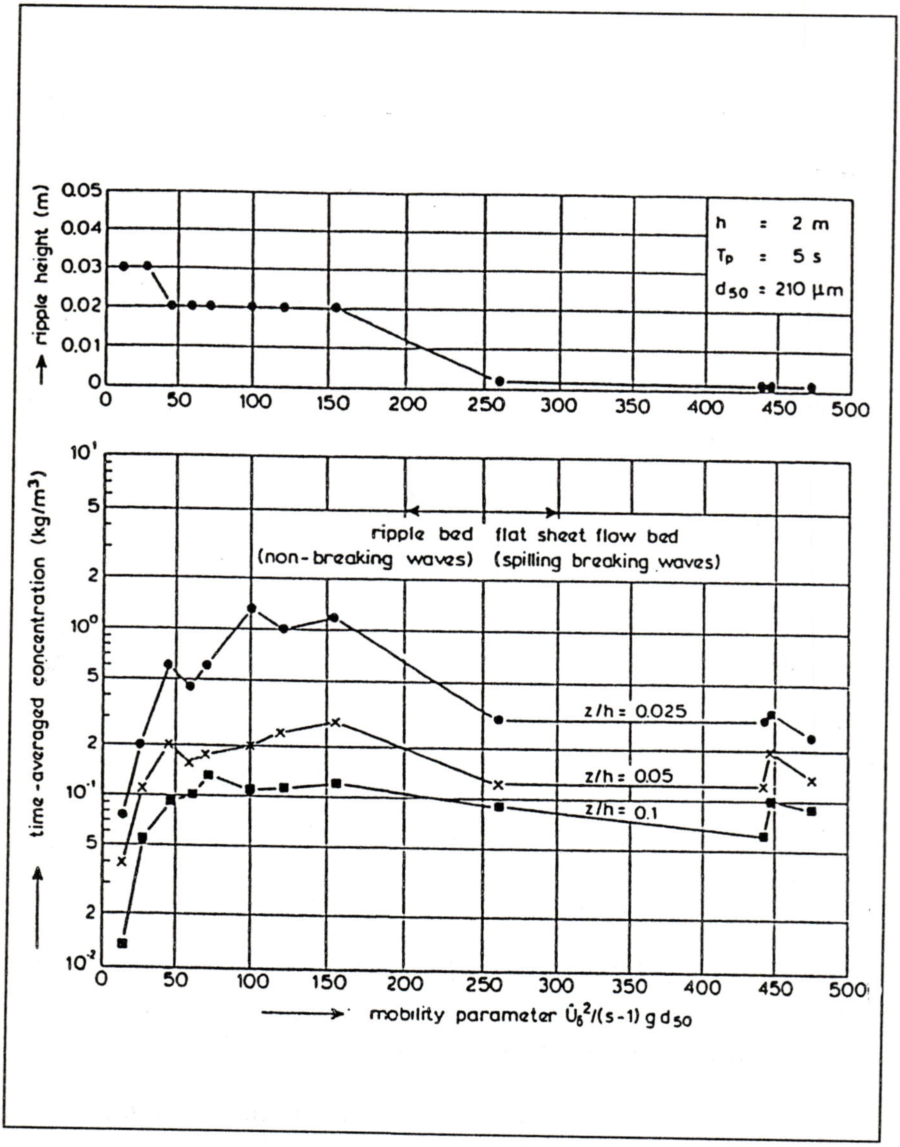

Fig. 14 : Time-averaged sediment concentrations in non-breaking and breaking waves.

The experimental results indicate a simultaneous decrease of the near-bed concentrations and the ripple height when the wave height increases from H_s = 0.69 m (2.21 ft.) to H_s = 1.1 m (3.52 ft.). This can be observed more clearly in Fig. 14 showing the concentrations at three elevations (z = 0.025 h, 0.05 h and 0.1h) as a function of the mobility parameter $\Psi = \hat{U}_\delta^2 / ((s-1) g d_{50})$ in which $\hat{U}_\delta$ is the amplitude of near-bed orbital velocity. The ripple height is also plotted as a function of Ψ. Fig. 14 clearly shows that the concentrations are largest for Ψ = 150 in the presence of ripples with a height of 0.02 m. For $\Psi > 150$ there is a gradual transition from the ripple regime (height ≈ 0.02 m) with relative large concentrations to the smooth bed regime (smooth fiat ripples of 0.001 m) with relatively small concentrations. It is most likely that the ripple-generated eddies, which are most effective in the entrainment of particles from the bed, are gradually disappearing for Ψ = 150 resulting in a gradual decrease of the concentrations, as shown in Fig. 14.

Bosman (1982) and Nap-Van Kampen (1988) observed the following phenomena :

- increasing concentrations for increasing wave heights,
- slightly flatter concentration profiles for increasing wave heights,
- constant concentration gradients at each wave height (for H_s < 0.18 m or 0.58 ft.),
- two layer concentration profiles for H_s = 0.2 m (0.64 ft.) and 0.23 m (0.74 ft.) with constant concentration gradient in near-bed region z/h < 0.15 and increasing concentration gradients for z/h > 0.15,
- decreasing ripple height for increasing wave height (upto H_s = 0.18 m or 0.58 ft.).

These results show slightly flatter concentration profiles for a decreasing ripple heights, which is related to a somewhat less intensive mixing in the near-bed layer. The concentrations in the region z/h > 0.15 are relatively large for wave heights of H_s = 0.2 (8 in.) and 0.23 m (9 in.) in the experiments of Bosman. This may be caused by the effect of occasional breaking of the waves.

2.2.2 Concentrations by Breaking Waves over a Flat Bed

Bosman (1982, 1986) observed two phenomena :

- the near-bed concentrations are approximately constant (≈ 10 kg/m³ or 0.6 lb./ft.³) for increasing wave heights,
- the concentrations at higher levels show a large increase for wave heights increasing from H_s = 0.12 m (0.38 ft.) (non-breakingwaves) to H_s = 0.19 m (0.61 ft.) (plunging breaking waves).

Kana (1979) observed the following phenomena :

- concentration profiles have a similar shape in spilling and plunging breaking waves,
- concentrations are a factor 5 larger in case of plunging breaking waves.

Kroon and Van Rijn (1988) observed the following phenomena :

- the near-bed concentrations are of the same order of magnitude (factor 2) as those of Kana (1979), Steetzel (1987) and Van Rijn (1987) for spilling breaking waves,
- the concentrations at mid depth and near the surface are smaller than those of Kana (1979) but comparable (factor 2) to those measured by Van Rijn (1987) in the large wave flume at similar wave heights.

Based on the available measurements, the following concentration ranges can be given for the surfzone with bed material in the range of 200 to 300 μm and water depths of 1 to 2 m :

sediment concentrations	spilling breakers	plunging breakers
near-bed concentrations	0.5 - 1 .5 kg/m³ (0.031 - 0.092 lb./ft.³)	1 - 5 kg/m³ (0.062 - 0.31 lb./ft.³)
near surface concentrations	0.05 - 0.2 kg/m³ (0.003 - 0.124 lb./ft.³)	0.5 - 1.5 kg/m³ (0.031 - 0.092 lb./ft.³)

The data indicate that the breaker type is a dominating factor in waverelated sediment suspensions. Spilling breakers are less effective than plunging breakers, which is probably caused by the relatively small scale of the eddies generated by spilling breakers. Furthermore, these small eddies are confined to the near-water surface region and do not extend below the trough level (Miller, 1976). Plunging breakers show strong jets penetrating to the seabed resulting in the generation of large sediment concentrations.

2.2.3 Concentrations in sheet flow layer

Observations have shown that bedforms are washed out when the mobility parameter Ψ is larger than about 200 to 250. In that case a thin ($\approx$ 0.03 m or 0.096 ft.) layer of moving sediment particles with high concentrations close to the bed is generated. This is called the sheet flow layer. Time-averaged concentrations in the sheet flow layer have only been measured in wave tunnel experiments (Horikawa *et al.*, 1982 ; Staub *et al.*, 1984 and Ribberink, 1989). Field data are not available. Horikawa *et al.* (1982) used an electroresistance probe to measure the concentrations over a sand bed of 200 μm. Staub *et al.* (1984) and Ribberink (1989) used a pump sampling instrument. A regular sinusoidal oscillatory motion was generated by Horikawa et al and by Staub *et al.*, while irregular oscillatory flow (Jonswap spectrum) was generated by Ribberink. Based on their results, the following characteristic concentration ranges at 0.01 m (0.03 ft.) and at 0.02 m (0.06 ft.) above the initial bed level (measured before the test) are given :

Height above the bed (m)	Sediment concentrations*				
	Horikawa *et al.*, 1982	Staub *et al.*, 1984			Ribberink, 1989
	200 µm, $\hat{u}$ = 1.08 m/s, T = 4s	190 µm, $\hat{u}$ = 1.3 m/s, T = 9.1 s	190 µm, $\hat{u}$ = 1.68 m/s, T = 6.8 s	380 µm, $\hat{u}$ = 1.9 m/s, T = 6.8 s	200 µm, $\hat{u}_{rms}$ = 1.13 m/s, T_p = 8 s
0.01	3 kg/m^3*	10-25 kg/m^3	50 kg/m^3	10-30 kg/m^3	15-20 kg/m^3
0.02	1 kg/m^3	3-10 kg/m^3	20 kg/m^3	3-15 kg/m^3	4-8 kg/m^3

The results of Staub et al and Ribberink are in good agreement for peak velocities in the range of 1.1 to 1.3 m/s (3.52 to 4.16 ft./s) and sediment material of about 200 μm. The concentrations measured by Horikawa et al are a factor 5 smaller at a peak velocity of 1.08 m/s. (3.46 ft./s).

2.2.4 Summary

Summarizing all available concentration data for non-breaking and breaking waves in water depths of 0.5 to 2 m (1.6 to 6.4 ft.) and bed material sizes from 100 to 300 μm, the following characteristics can be observed :

- two-layer concentration profile with rather large concentration gradients in the near-bed layer ($\approx$ 3 ripple heights) and smaller concentration gradients in the upper layer, in case of non-breaking waves over a rippled bed,

---

* m/s → ft/s : 1.08 → 3.46 ; 1.3 → 4.16 ; 1.68 → 5.12 ; 1.9 → 4.16 ; 1.13 → 3.62
kg/m^3 → lb./ft^3 : multiply by 0.062

- two-layer concentration profile with rather large concentration gradients in the near-bed layer (≈ 3 ripple heights) and smaller concentration gradients in the upper layer, in case of non-breaking waves over a rippled bed,
- strong influence of the ripples giving relatively large concentrations in the near-bed layer due to eddy-generated motions,
- increasing concentrations in the upper layers by spilling breaking waves,
- strongly increasing concentrations in the near-bed layer and in the upper layer by plunging breaking waves (factor 5 larger than in spilling breaking waves).

2.3 Sediment Transport Rates

Sediment transport rates under wave action have been investigated by several authors. Most of the experiments were carried out with flat (oscillating) beds. Transport rates were generally small.

Based on these results, the transport rate in the sheet flow regime seems to be dependent on approximately the third power of the velocity : $\overline{q_w} = (\hat{U})^3$.

Sawamoto and Yamashita (1987) did similar experiments in a wave tunnel in the sheet flow regime. Sand material with particle sizes of 200 μm, 700 μm and 1800 μm was used. Coal and plastic material was also used. Based on their results, these investigators have found that the transport rate in the sheet flow regime is related to $(\hat{\ }_{t})^{1.5}$ or $(\hat{U}_\delta)^3$.

Summarizing, it can be concluded that the net transport rate under wave action strongly depends on the transport mode and on the type of bedform :

1. *bed load* or *suspended load* over *flat bed* : net transport rate is always in the direction of the largest peak velocity
2. *dominantly bed load* over *rippled bed* : net transport rate is in the direction of the largest peak velocity
3. *dominantly suspended load* over *rippled bed* : net transport rate is against the direction of the largest peak velocity.

3 Computation of Time-Averaged Concentration Profiles

3.1 Introduction

In the literature various models are proposed to compute the sediment concentration profiles. Most models are based on the time-averaged convection-diffusion equation (Nielsen, 1979, Bosman-Steetzel, 1986, Dally, 1980, Skafel-Krishnappan, 1984), as follows :

$$w_{s,m}\, c + \varepsilon_{s,w} \frac{dc}{dz} = 0 \qquad (42)$$

in which :

c = sediment concentration
$w_{s,m}$ = particle fall velocity in fluid-sediment mixture
$\varepsilon_{s,w}$ = sediment diffusion or mixing coefficient related to the wave motion
z = vertical coordinate

Some models are based on the time-dependent convection-diffusion equation, as follows :

$$\frac{\partial c}{\partial t} - w_{s,m} \frac{\partial c}{\partial z}\left(\varepsilon_{s,w} \frac{\partial c}{\partial z} \right) = 0 \qquad (43)$$

Hagatum and Eidsvik (1986) presented a numerical model based on a two-equation turbulence model to represent the eddy viscosity coefficients. It is questionable whether such a sophisticated improvement is necessary considering the experimental problems related to verification of the concentrations in the sheet flow layer. Analytical solutions of the convection-diffusion equation are given by Nielsen (1979) assuming that the eddy viscosity coefficient is constant in space and time.

Given the complexity of all phenomena (eddy viscosity, hindered settling, turbulence damping) within this thin ($\approx$ 0.01 m) high-concentration sheet flow layer, the predicting ability of these time-dependent models is rather questionable. Verification is hardly possible due to measuring problems in the near-bed region with extremely large vertical concentration gradients in a thin layer ($\approx$ 0.01 m).

Since reliable models to predict *time-averaged* concentrations in waves alone are still lacking, Van Rijn (1984) proposes a new model for *time-averaged* concentrations, which is based on the *time-averaged* convection-diffusion equation and which describes the vertical distribution of the sediment concentrations above the top of the bedform (ripples) or above the edge of the wave boundary layer in case of a plane bed with sheet flow.

The following subjects are described :

- convection-diffusion equation
- particle size and fall velocity
- sediment mixing coefficient
- reference concentration
- computation of concentration profile
- comparison of measured and computed concentration profiles

3.2 Time-Averaged Convection-Diffusion Equation

Usually, the traditional convection-diffusion equation is applied to compute the equilibrium concentration profile in steady flow. This equation reads as :

$$w_{s,m}c+\varepsilon_z \frac{dc}{dz} = 0 \tag{44}$$

in which :

$w_{s,m}$ = particle fall velocity of suspended sediment in a fluid-sediment mixture
ε_s = sediment mixing coefficient
c = time-averaged concentration at height z above the bed

Here, it is assumed that eq. (44) is also valid for wave-related mixing. To verify this, eq. (44) is expressed as :

$$\frac{d}{dz}(\ln\ c) = -\frac{W_{s,m}}{\varepsilon_s} \tag{45}$$

The diffusion concept is valid when the slope of measured concentration profiles represented in a plot of log c versus z are inversely proportional to the fall velocity in case of constant wave conditions. To investigate this, some experimental results presented by Roelvink (1985) are analyzed. Roelvink measured concentration profiles of different size fractions above a rigid bottom consisting of artificial ripples. A limited amount of sediment particles was fed to the (oscillatory) flow. Fig. 15 shows measured concentration profiles for various size fractions in case of a constant wave height. Since the wave height is constant, the local mixing coefficient will be approximately constant assuming that this parameter is not or only weakly correlated to the sediment particle size. As can be observed, there is a tendency for the slopes of the concentration profiles to be inversely proportional to the particle fall velocity, which supports the applicability of the diffusion concept for wave induced mixing. A similar conclusion is given by Van Der Graaff (1988). Nielsen (1984) questions the applicability of the diffusion equation for wave-induced mixing. He presents some data showing that the eddy trapping mechanism is the dominant suspension mechanism rather than the diffusion process. All sand particle sizes, that are small enough to be trapped in an eddy will travel with the eddy until it dissolves and their time-averaged concentrations will become distributed in very much the same way, irrespective of settling velocity.

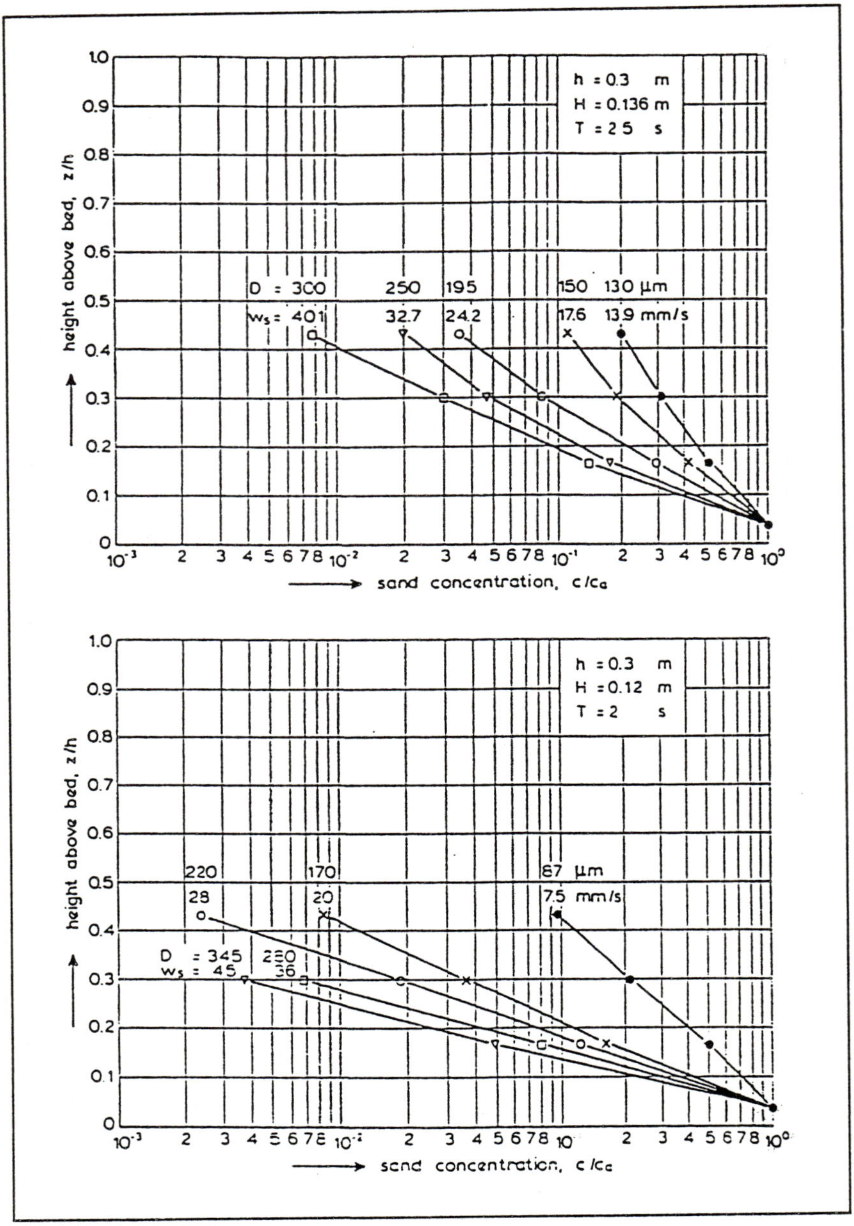

Fig. 15 : Vertical distribution of concentration profiles.

3.3 Particle Size and Fall Velocity of Suspended Sediment

3.3.1 Particle size of suspended sediment

Observations in flume and field conditions have shown that the suspended sediment particles are not uniformly distributed over the depth in case of a nonuniform bed material composition. The coarser particles are suspended in the near-bed region, while the finer particles are suspended in the upper layers. Some information can be obtained from the measurements of Van Rijn (1987) in a large-scale wave flume. Sediment concentration profiles were determined by using a pump sampler to obtain water-sediment samples. The suspended sediment samples of each profile were put together to get a sufficiently large sample for analysis. Fall velocities of these samples were measured in a settling tube with a length of 2 m graphically. The fall velocities were converted to particle size. The ratio of the median suspended sediment size (d_s) and the median bed material size (d_{50}) is shown in Fig. 16 as a function of a dimensionless bed shear stress parameter T. Fig. 16 shows similar data obtained in the surf zone near the coast of The Netherlands (station Groote Keeten, cf. Kroon and Van Rijn, 1988). Since the bed material is quite uniform ($\sigma_s = 0.5 (d_{50}/d_{16} + d_{84}/d_{50}) < 2$) in both cases, the median suspended sediment size d_s is fairly close to the median bed material size $d_{50,bed}$. The data can be reasonably represented by the following equation :

$$\frac{d_s}{d_{50,bed}} = 1 + 0.011(\sigma_s - 1)(T - 25) \qquad (46)$$

in which :

d_s = representative particle size of suspended sediment
$d_{50,bed}$ = median particle size of bed material
σ_s = $0.5(d_{50}/d_{16} + d_{84}/d_{50})$ = geometrical standard deviation of bed material*
T = bed shear stress parameter.

Equation (45) has been proposed by Van Rijn (1984) to represent the suspended sediment size in steady uniform flow. Based on the present results (Fig. 16), this equation is also applicable for oscillatory flow over a movable fine sediment bed.

* The symbol σ_s is used in this chapter even though it is not universally accepted by geostatisticians.

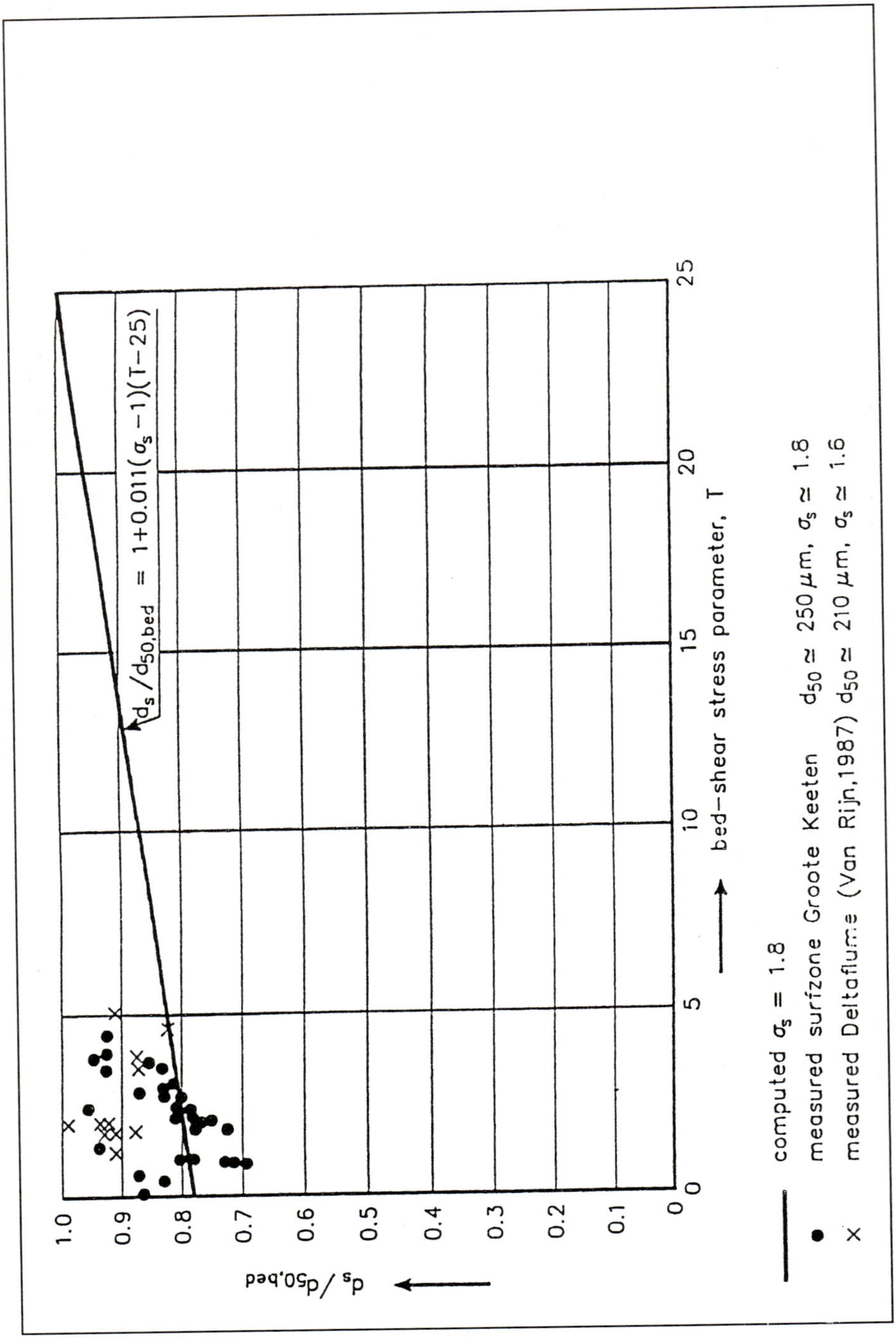

Fig. 16 : Particle size of suspended sediment

3.3.2 Particle Fall Velocity

Experiments of high sediment concentrations have shown a considerable reduction of the particle fall velocity due to the presence of the surrounding particles. This effect is known as the hindered settling effect. For flow conditions with particles in the range of 50 to 500 μm the reduced particle fall velocity can be described by a Richardson-Zaki (1954) type of equation :

$$w_{s,m} = (1-c)^5 \, w_s \qquad (47)$$

$w_{s,m}$ = effective fall velocity in a fluid-sediment mixture
w_s = fall velocity in a clear fluid

Since wave-generated sediment concentrations may be rather large in the nearbed region, the influence of the hindered settling effect has been investigated by carrying out a few sensitivity computations. The sediment concentration profile has been computed (numerically) for a reference concentration of 2,500, 5,000 and 10,000 mg/l (0.0088, 0.177 and 0.353 oz/l). The results, in Fig. 17, show an increase of the sediment concentrations due to the hindered settling effect. Considering all uncertainties involved, the hindered settling effect seems to be of minor importance for concentrations smaller than 10,000 mg/l (0.353 oz/l). Measurements show that concentrations of this order of magnitude do not occur outside the wave boundary layer.

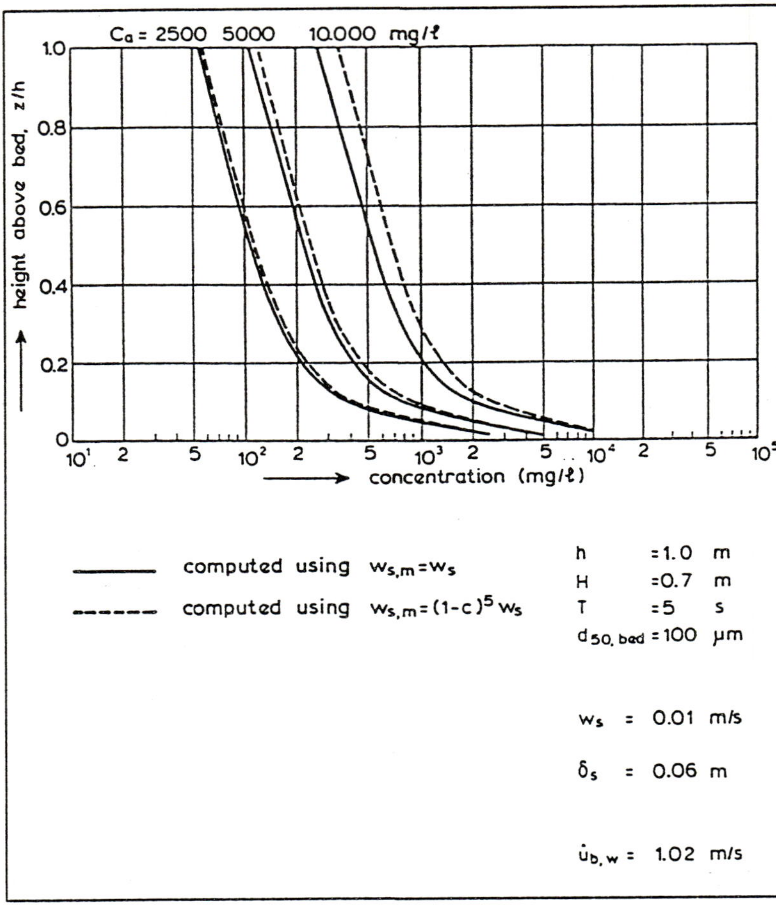

Fig. 17 : Influence of hindered settling effect on concentration profile

4 References

Bagnold, R.A., 1954, Experiments on a Gravity-free Dispersion of Large Solid Spheres in a Newtonian Fluid under Shear, *Proc. Royal Soc.*, Vol. 225A, England.

Bakker, W.T., 1974, Sand Concentration in an Oscillatory Flow, *Coastal Engineering Conference,* Copenhagen Denmark, pp. 1129-1148.

Bakker, W.T., & Van Kesteren, W.G.M.,1986, The Dynamics of Oscillating Sheet Flow, *Coastal Engineering Conference*, Taiwan.

Bosman, J., 1982a, Concentration Measurements under Oscillatory Motion. Delft Hydraulics, *Report M1695-II*, Delft, The Netherlands.

Bosman, J., 1982b. The Influence of Bottom Slope, Water Depth, Breaking Waves, Orbital Velocity and Current Velocity on the Concentration Distribution under Waves and Currents. Delft Hydraulics, *Report M1875*, Delft, The Netherlands [in Dutch].

Bosman , J., & Steetzel, H.J., 1986, Time and Bed-averaged Concentration under Waves, *Coastal Engineering Conference*, Taiwan.

Fredsoe, J., Andersen, O.H., & Silberg, S., 1985, Distribution of Suspended Sediment in Large Waves, *Journal of Waterways, Ports, Coasts and Ocean Engineering*, Vol. III, No. 6.

Hagatum, K., & Eidsvik, K.J., 1986, Oscillating Turbulent Boundary Layers with Suspended Sediment, *Journal Geophysical Research*, Vol. 91, No. C11.

Horikawa, K., Watanbe, A, & Katori, S., 1982, Sediment Transport Under Sheet Flow Condition, *Coastal Engineering Conference*, Cape Town, South Africa, pp. 1335 - 1352.

Kana, T.W., 1979, Suspended Sediment in Breaking Waves, Tech., *Report No. 18-CRD,* Coastal Res. Div., Dep. of Geology, Univ. of South Carolina, Colombia, USA.

Kroon, A., & Van Rijn, L.C., 1989, *Space-Averaging Aspects of Sediment Concentrations in the Surf Zone.* Dep. of Phys. Geography, State Univ. of Utrecht, Utrecht, The Netherlands.

Miller, R.L., 1976, Role of Vortices in Surf Zone Prediction : Sedimentation and Wave Forces. Symp., *Beach and Nearshore Sedimentation*, Dallas, USA.

Nakato, T., Locher, F.A, Glover, J.R. , & Kennedy, J.F., 1977, Wave Entrainment of Sediment from Ripples. *Journal of Waterways, Ports, Coasts and Ocean Div.,* ASCE, No. WW1.

Nap, E, & Van Kampen, A; 1988, *Sediment Transport in Irregular Non-Breaking Waves.* Coastal Engineering Department, Delft Univ. Of Technology, Delft, The Netherlands.

Nielsen, P., 1979, Some Basic Concepts of Wave Sediment Transport. *Series Papers 20,* Inst. of Hydrodyn. and Hydr. Eng., Techn. Univ. of Denmark, Lyngby, Denmark.

Nielsen, P., 1984, On the Motion of Suspended Sand particles, *Journal of Geophysical Research*, Vol. 89, No. C1.

Nieuwjaar, M, & Van Der Kaaij, Th., 1987, *Sediment Concentrations and Transport in Irregular Non-Breaking Waves.* Coastal Eng. Dep., Delft Univ. of Technology, Delft, The Netherlands.

Ribberink, J.S., 1989, Sediment Concentration and Bed Regime Measurements in Large Wave Tunnel, *Report H840.22,* Delft Hydraulics, Delft, the Netherlands.

Richardson, Y.F., & Zaki, W.N., 1954, Sedimentation and Fluidization, Part I, *Trans. Inst. Chem. Eng.*, Vol. 3 : 35 - 53.

Roelvink, D., 1985, Sand Concentrations under Progressive Waves, Delft University of Technology, Coastal Eng. Dep., Delft, The Netherlands [in Dutch].

Sawamoto, M., & Yamashita, T., 1987, Sediment Transport in Sheet Flow Regime. *Proc. Conf. Coastal Sediments*, New Orleans, USA.

Skafel, M.G., & Krishnappan, B.G, 1984, Suspended Sediment Distribution in Wave Field. *Journal of Waterways, Ports, Coasts and Ocean Engineering*, Vol. 110, No. 2.

Staub, C., Jonsson, I.G., & Svendsen, I.A., 1984, Variation of Sediment Suspension in Oscillatory Flow, *Coastal Engineering*, Houston, USA.

Steetzel, H., 1987, Model Investigations Dune Revetments, *Report H298*, part I, Delft Hydraulics, Delft, The Netherlands [in Dutch].

Van Der Graaff, J., 1988, *Sediment Concentration due to Wave Action.* Thesis, Delft Univ. of Technology, Delft, The Netherlands.

Van Rijn, L.C., 1987, Data Base Sand Concentration Profiles for Currents and/or Waves. *Report M1695-04-1,* Delft Hydraulics, Delft, The Netherlands.

Chapter III : Detection of Suspended Sediment

1 Principle

In order to detect and measure concentrations of suspended matter (mostly sediment and chlorophyll), the data from aerial or satellite scanners must be multispectral in nature (Doerffer, 1978; Morel & Prieur, 1977; Liedtke, 1987). Especially the red, green and infrared play an essential role in accurately determining these concentrations. The red portion of the spectrum provides the most consistent information over the broadest range of sediment and atmospheric conditions. The green spectral region is also useful for broad suspended sediment concentration ranges where its behavior can be compared to that of the red band by ratio techniques. A channel in the infrared portion of the spectrum is necessary to extend the analytical range of the sensor system and to ensure that image over-exposure (i.e. image saturation) does not occur at high suspended sediment concentrations. In addition, the infrared band is used to help delineate the shallow water and the land-water interface, for these wave lengths are attenuated within the first few centimeters of the water.

Multispectral scanner data can either be acquired by satellite borne scanners or by a scanner operated in an aircraft. Satellite scanners have the advantage that geometrical corrections to the data can more easily be performed; and in general satellite data are available at relatively low cost. For the study of sedimentary processes in the nearshore zone however, most benefits are gained from an airborne multispectral scanner. First, an airborne scanner can be deployed at the time that best suits the aim of the study (e.g. with respect to the tide). It is also essential, for calibration of the multispectral recordings, that ground truth (i.e. near-surface suspended sediment concentrations) be available of the precise time of the recordings. Such data can more easily be obtained when the time of recording can be planned, as with the airborne scanner. When satellite data are used, one has to rely on empirical formulas in order to get any quantified ground truth.

The corrections that have to be carried out to airborne digital multispectral scanner recordings include :

- a panoramic correction, which recalculates the image plane, that is the result of a central projection, to a straight plane.
- a thermal calibration. This procedure recalculates the recorded values of the thermal band so as to simulate an instantaneous recording (in reality, a certain time elapses between the beginning and the end of the recording during which the earth surface temperature gradually changes).

- a geometric correction, which eliminates the deformation of the image due to the irregularities in the flight path. The correction procedure is based on the use of ground control points.
- a radiometric correction can be applied in order to compensate for atmospheric influences (haze, reflection, ...).

For the quantification of the scanner image, a best fit relationship between the suspended matter concentrations and a combination of several of the scanner values is established. Several relationships (linear, logarithmic, etc.) should be considered. The one with the highest regression coefficient is chosen. This relationship is then used for all the other non-calibrated scanner data to obtain a representation of the sediment concentrations over the entire field of view. In accordance with Ritchie & Cooper (1988), a simple linear relationship can be used. Here, in each point, the suspended sediment concentration C is calculated by

$$C = a + \sum_i b_i B_i \tag{48}$$

where a, b_i = regression coefficients, determined in the picture points where ground truth is available;
i = the number of spectral bands used in the regression analysis;
B_i = pixel value of spectral band i.

A logarithmic relationship can also been applied (Tassan & Sturm, 1986) :

$$\log C = a + b \log X \tag{49}$$

where a, b = parameters inferred from best fits to *in situ* measurements of suspended sediment concentration;
X = a function of the pixel values B_i in the i spectral bands.

Using the best relationship, one comprehensive picture is compiled based on all the bands of the digital multispectral scanner which enter into the equation. Fig. 18 is a black-and-white representation of the compiled color image showing the results in the Zeebrugge region of a flight executed on 4 September 1987, one hour before low water, i.e. at the time of the greatest flow velocities. The sediment concentrations have been represented into 8 classes ranging from less than 0.05 g/l to over 0.40 g/l (0.00177 to > 0.0142).

Fig. 18 : Black-and-white representation of a sediment concentration map of the Zeebrugge outer harbor Belgium. Time of data acquisition : 4 September 1987, one hour before low water. Height of picture represents about 6 km (3.72 mi). A vortex-like area to the north west of the harbor contains relatively high suspended sediment concentrations.

In order to calculate sediment transport rates, assumptions must be made concerning the link which exists between the near-surface sediment transport and the transport integrated over the water column.

In shallow water, one could assume a constant suspension concentration if field data are lacking (e.g. Froidefond *et al.*, 1991). Numerous vertical sediment concentration profiles, measured in the nearshore zone, indicate that in almost any stage of the tidal cycle, the vertical sediment distribution can be described by one of the theoretical sediment distribution functions, of which the following form is most currently used (see section 3.4, eq. 16) :

$$c = c_a \left(\frac{d - z}{z} \frac{a}{d - a} \right)^Z \qquad (50) \text{ (same as 17)}$$

where c = sediment concentration at a distance z above the bed
c_a = sediment concentration at a distance a above the bed
d = total water depth
z = height above the bed
a = reference level, where the concentration is known
Z = suspension parameter

2 The c_a and Z parameters.

Two parameters in equation (50 or 17) are not obvious to determine : c_a and Z. The quantified airborne multispectral scanner recordings however provide an immediate answer to the first parameter, c_a. It was shown in the field measurements that at any given stage of the tide, the amount of sediment suspended in the near-surface part (0-4 m or 12.8 ft.) of the water column does not vary significantly with height. Therefore, the sediment concentration values detected by the airborne multispectral scanner can be assumed to represent the sediment concentration at a fixed level, say 1 m below the water surface, and thus constitute an areal cover of c_a values.

The value of Z mostly determines the vertical gradient of the sediment concentration profile. The most significant parameters influencing the value of Z are w_s, the settling velocity of the suspended sediments, and u_* the bed shear velocity (see section 3.4, chapt. 1, this part). w_s is mainly dependent on the size of the suspended matter. Of this parameter again, the field measurements show no significant changes over a tidal cycle; it is mostly the content of suspended matter that varies. The size of the suspended material is determined on the calibration samples.

The variation of u_* is the main cause of uncertainty in the approach. u_* can be derived from a velocity profile, using the well known Prandtl - von Kármán equation :

$$u = \frac{u_*}{\kappa} \ln\left(\frac{z}{z_0}\right) \tag{51}$$

in which

- u = the velocity at a height z above the bed
- κ = Von Kármán's constant, 0.40
- z_0 = height above the bed where u = 0. When (2) is used for estimating u_* from a set of (u, z) values, z_0 is rather a calculation parameter.

The Prandtl-Von Kármán equation (20), which has a sound theoretical basis (cfr. 3.5), was found to fit the field data of the nearshore test sites very well. On the nearshore locations, where measurements were carried out, it appears that there is a systematic variation of u_* with the tide. The degree of scatter on the trend of u_*, however, is relatively high, and as yet, no clear relationship with factors of supposed importance, such as the significant wave height, could be established. Another factor still to be examined, is the lag time that is expected to occur between a sediment suspending event, such as an increase of u_*, and the actual response, the increase in suspended sediment content. The data gathered hitherto are promising enough to continue this line of thinking.

2 References

Doerffer, R., 1978, Zum Problem der Fernkundung von Substanzen im Wasser mit dem Multispektralabtaster. Bul., 1978 (4), p. 133-138.

Froidefond, J.M., Castaing, P., Mirmand, M. & Ruch, P., 1991, Analysis of the Turbid Plume of the Gironde (France) Based on SPOT Radiometric Data. *Rem. Sens. Environ.* 36 149-163.

Liedtke, J.L., 1987, *Discrimination of Suspended Sediment Concentrations using Multispectral Remote Sensing Techniques.* Thesis submitted in partial fulfillment of the requirements for the degree of Master of Science, Simon Fraser University, USA, 108 pp.

Morel, A.Y. & Prieur, L., 1977, Analysis of Variations in Ocean Color. *Limn. Oceanogr* , 22, p. 709-722.

Ritchie, J.C. & Cooper, C.M., 1988, Comparison of Measured Suspended Sediment Concentrations with Suspended Sediment Concentrations Estimated from Landsat MSS data, *Int. J. Remote Sensing,* 9, 379-387.

Tassan, S. & Sturm, B., 1986, An Algorithm for the Retrieval of Sediment Content in Turbid Coastal Waters from CZCS Data, *Int. J. Remote Sensing*, 7, 643-655.

Chapter IV : Sediment Transport

1 Introduction

It is common practice to make a distinction between sediment transport *parallel* to the shoreline and sediment transport *transverse* to the shoreline. In general the longshore sediment transport, and in particular the gradients in this transport, are responsible for the *long-term* changes in the coastline, whereas the *cross-shore* transport is responsible for the *short-term* variations. In this connection it should be observed that *cross-shore* transport may also cause long-term erosion or accretion of the coast, but this is often difficult to detect because of the different nature of transport phenomena.

Longshore transport occurs in a relatively narrow zone along the coast, and the direction and magnitude is mainly determined by the height, period and direction of the waves apart from the effect of possible tidal currents. The presence of this longshore transport can be easily detected from the development of the coastline and sea bed geometry near river mouths, headlands, groins, harbor moles, etc. The short-term effects of cross-shore transport can also be easily detected, from the changes in magnitude and position of breaker bars, erosion of the dunes during storm surges, etc. As such the effect of seasonal variations in the wave conditions on the coastline and seabed geometry are considered short-term effects. Usually the net longshore transport per meter width is much larger than the net cross-shore transport per meter width. However, the gross cross-shore transport will be far higher. In principle under each wave crest sediment is transported in the shoreward direction and under each wave through in seaward direction. During this motion the longshore current transports the sediment in longshore direction. Therefore the gross cross-shore transport may be of the same order of magnitude as the gross longshore transport. Because of the nature of the cross-shore transport the possible net long-term transport cannot be measured directly. If a long-term change of the coastline cannot be explained by gradients in the longshore transport it may be concluded that a net cross-shore transport exists, but an indisputable proof of that is often hard to give.

2 Longshore Transport

2.1 CERC-Formula

The early ideas of Watts (1954) and Inman and Bagnold (1963) have led to the well-known CERC-formula. This type of approach is energy-based and relates the sediment transport to the energy released with the breaking process of the waves. According to Bagnold's concept sediment is transported by grain-to-grain interaction (bed load) and by the stream via turbulent diffusion (suspended load). In principle Bagnold tried to determine the fraction of the total stream energy spent to transport bed load and suspended load in the form of efficiency factors. Applied to oscillatory flow Bagnold reasoned that the local rate of energy dissipation is responsible for stirring-up the sediment and that an arbitrary flow, for example the longshore current transports this sediment in the direction of the current.

A number of longshore transport models have been based upon the concept described above, of which the CERC-formula is the most widely applied. In fact the model assumes a linear relationship between the longshore wave energy flux due to breaking waves and the consequent longshore transport. The formula can be written as:

$$S = A\ H_O^2 c_O K_{rb}^2 \sin\phi_b \cos\phi_b \qquad (52)$$

where :

S	=	longshore transport due to breaking waves
A	=	constant
H_O	=	deepwater wave height
c_O	=	deepwater wave velocity
K_{rb}	=	refraction coefficient at the breaker line
ϕ_b	=	breaker angle

Different values are given in the literature for the constant A, partly due to the choice of H_O. If for H_O the significant wave height in deepwater is applied, the usual value of $A \approx 0.025$. Possible reasons for the uncertainty concerning the value of A may be the inaccuracies in the data, with respect to both waves and longshore sand transport, on which the model is based. Moreover the CERC-formula does not account for differences in grain size.

It is observed that the CERC-formula is valid only for relatively long and straight beaches, where the alongshore differences in the breaking wave heights are only small. Moreover the formula does not account for currents which are not generated by breaking waves, such as tidal currents. If such currents are important other transport formula should be used.

2.2 Bijker formula

To obtain a model for the longshore transport, which takes into account the effect of tidal or other types of currents, a transport formula can be combined with a model for the longshore current. Bijker (1967) proposed a formula for bed load due to waves and currents, which is based upon the formula of Kalinske-Frijlink for bed load due to currents only. From that a formulation for the increase of the bottom shear stress by waves has been developed. Afterwards Bijker (1968) has added to the bed load a distribution of the suspended load, which is based upon the Einstein-Rouse concentration vertical. This model has until now frequently been applied at the Delft (The Netherlands) Hydraulics Laboratory. Bijker proposed as bed load formula :

$$Sb = b\ D_{50} \frac{v}{C} \sqrt{g}\ \exp\left[- \frac{0.27 \Delta D_{50} C^2}{v^2 \left\{ 1 + \frac{1}{2} \left(\xi \frac{u_b}{v} \right)^2 \right\}} \right] \tag{53}$$

where :

S_b = bed load transport
b = constant = 5
D_{50} = median grain diameter
v = current velocity
C = coefficient of Chézy = 18 log (12d/r)
d = water depth
r = bed roughness
g = acceleration of gravity
Δ = specific density = $(\rho_s - \rho_w)/\rho_w$
ρ_s = density of sand
ρ_w = density of water
μ = $(C/C_{90})^{1.5}$, with $C_{90} = 18 \log (12d/D_{90})$
ξ = $C(f_w/2g)^{0.5}$, with $f_w = \exp\left\{-6.0 + 5.2 (a_o / r)^{0.19}\right\}$
a_o = amplitude of orbital excursion at the bed
u_b = amplitude of orbital velocity at the bed

The suspended transport can be written as :

$$S_s = 1.83\ S_b \left\{ I_1 \ln\left(\frac{33d}{r} + I_2\right)\right\} = 1.83\ Q\ S_b$$

where :

S_s = suspended transport
S_b = bed load transport
I_1, I_2 = Einstein integrals (see section 2.8., ch. 1, this part).

The value between the brackets (= Q) is a function of the relative bed roughness r/h and of the value of $Z_* = w/xv$, with w = fall velocity of sand in stagnant water, x = 0.4, and

$$v_* = v\sqrt{\frac{g}{C}}\sqrt{\left\{1 + \frac{1}{2}\left(\xi \frac{u_b}{v}\right)^2\right\}} \qquad (54)$$

The relation between $r/h, z_*$ and 1.83 Q is given in Fig. 19. When the value of 1.83 Q is known the total sediment transport according to Bijker is found by :

$$S = S_b + S_s = (1 + 1.83Q)\ S_b$$

The value of the constant b in eq. (53) is disputable. The value b = 5 is mostly applied, but sometimes other values are proposed, for instance b = 5 in the breaker zone and b = 1 outside the breaker zone. Occasionally the available measurements allow a calibration of the Bijker formula, which means that the value of b can be adapted to the local situation.

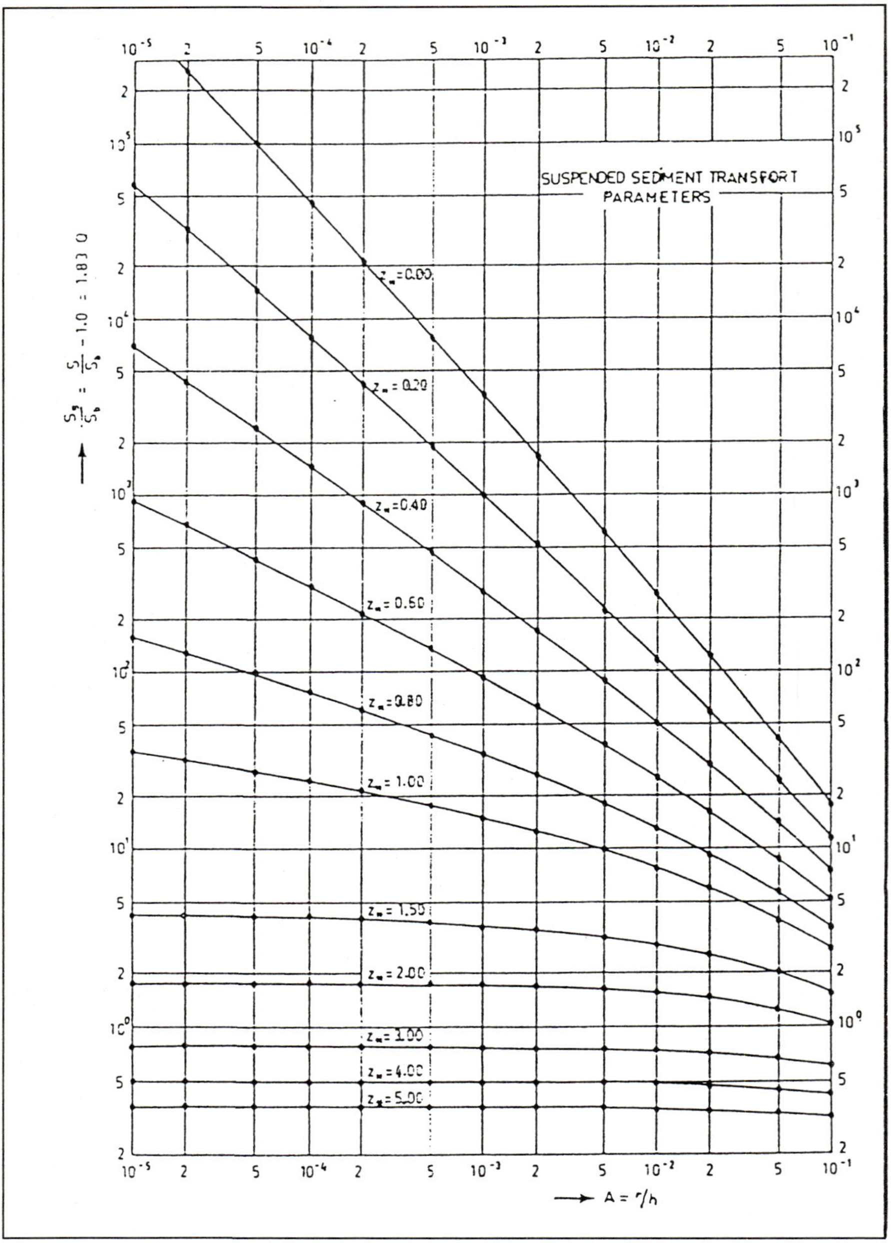

Fig. 19 : Relation between suspended sediment transport parameters

Van Rijn (1985) developed recently a formula for the computation of sediment transport by waves and currents. The model is based upon a large number of measurements of suspended sediment concentrations, both in the laboratory as well as in the field. The consequent formulas for the computation of the resulting suspended sediment transport have to be solved numerically, because of the equations applied for the vertical diffusion of sediment.

Besides the suspended sediment transport a formulation is given for the computation of the bed load transport. The formulas are given in section 2.1, this chapter and chapter 1, this part.

Van Der Graaf and Van Overeem (1979) performed a comparative study between the results of the Bijker formula, the adapted Engelund-Hansen formula and the adapted Ackers-White formula*. Although the CERC-formula is not necessarily correct in every respect ; this formula has been chosen as yardstick since it predicts at least a mean value on the basis of quite a number of prototype measurements. Although this method of comparison has some weakness, Van Der Graaf and Van Overeem concluded that the Bijker formula is a better one than either the adapted Ackers-White or the adapted Engelund-Hansen formula. Furthermore the Bijker formula is, compared to the other formulas, rather insensitive to effects of particle diameter, bottom slope and bed roughness. In using this formula, a possible error in the estimation of the actual conditions results in a slight error in the computed transport.

3 Cross-Shore Transport

3.1 Two-Line Theory of Bakker

Bakker (1968) has developed a two-line theory for the coastal morphology. The coastal profile is schematized in two lines, denoted by the term "beach" and "inshore", which changes as a result of gradients in the longshore**Error! Reference source not found.** transport along both lines.

The interchange of sediment between both lines takes place in the form of onshore-offshore transport. The magnitude of this onshore-offshore transport is supposed to be proportional to the difference between the actual profile and the equilibrium profile (see Fig. 20)

* The adapted Engelund-Hansen formula and the adapted Ackers-White are described in section 2.1., ch. 1, this part.

$$S_y = s_y (\Delta y - W) \tag{55}$$

wherein :

S_y = onshore-offshore transport
s_y = proportionality constant

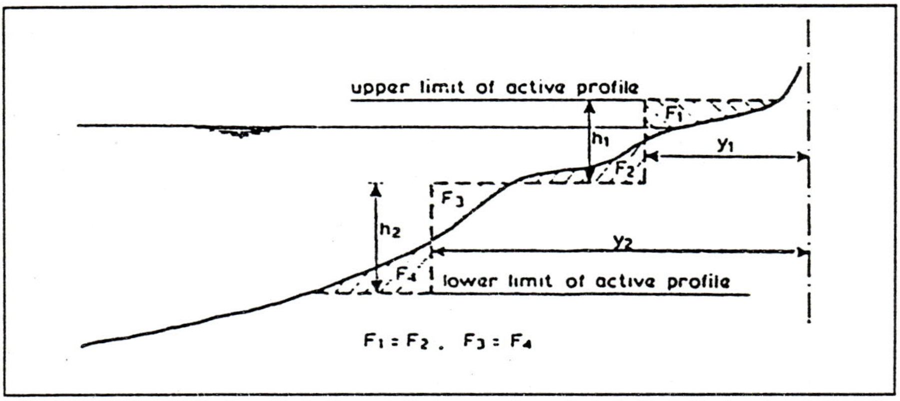

Fig. 20 : Two-Line schematization by Bakker

Δy = distance between beach-line and inshore-line
W = equilibrium distance

In an equilibrium situation $S_y = 0$ and $W = \Delta y$.
The concept of Bakker regarding the onshore-offshore transport forms an essential element in the theory of Swart, discussed below.

3.2 Predictive Model of Swart

Swart (1974) presents a model for the computation of offshore sediment transport on sandy beaches under wave action, and of the corresponding time-dependent and eventual equilibrium beach profiles. Essentially Swart follows Bakker's concept that the offshore transport is proportional to the difference between the actual profile and the equilibrium profile.

Because of the differences in the transport processes Swart has divided the coastal profile in three parts (see Fig. 21) :

a. the backshore, which is the area above the wave run-up limit,
b. the developing profile (D-profile) where a combination of bed load and suspended load transport takes place,
c. the transition area, seawards of the D-profile up to the depth where sediment motion by wave action is initiated.

The schematization of the profile as defined above, is highly affected by the observed coastal profiles in small scale tests. Also the relations derived by Swart are mainly based on the results of small scale tests, while they have been certified with the aid of a few large scale tests in the former wave flume of the U.S. CERC.

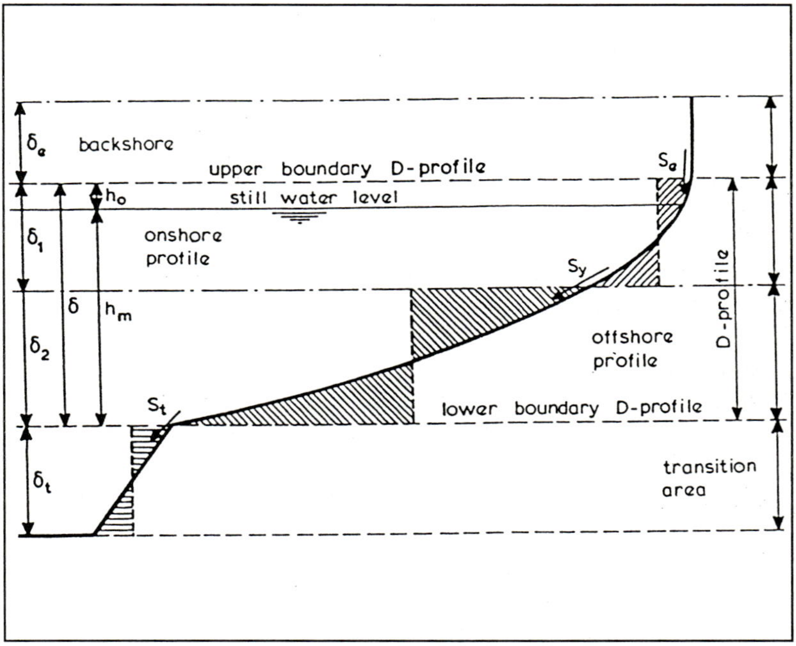

Fig. 21. : Profile schematization by Swart

On the basis of theoretical considerations and a detailed analysis of available data Swart has proposed a number of relations for the computation of the offshore transport and of the equilibrium profile. The upper and lower limit of the D-profile, within which the actual profile development will take place, are defined as follows :

$$\frac{h_o}{D_{50}} = 7644 - 7706 \exp\left(-0.000143 \frac{H_o^{0.488} T^{0.93}}{D_{50}^{0.786}}\right) \tag{56}$$

$$\frac{h_m}{L_o} = 0.0063 \exp\left(4.347 \frac{H_o^{0.473}}{T^{0.894} D_{50}^{0.093}}\right)$$

The total D-profile thickness is :

$$\delta = h_O + h_m \tag{57}$$

The maximum value s_{ym} of the coastal constant s_y can be calculated from :

$$\ln \frac{s_{ym} T}{D_{50}} = 10.7 - 28.9 \left[H_o^{1.68} \left(\frac{H_o}{L_o}\right)^{-0.9} D_{50}^{-1.29} \left(\frac{H_o 2.66}{h_m}\right)^{-0.079} \right] \tag{58}$$

The location of the maximum coastal constant s_{ym} in the D-profile is defined with the aid of the maximum value s_{ym}, its location Δ_{2m} and a dimensionless distribution function s_y/s_{ym} of the form :

(i) for $\delta_2 > \delta_{2m}$ (i.e. landwards of the location of s_{ym})

$$s_y/s_{ym} = \frac{0.93}{1 - 1.01 X_{2.11}} + 0.07 \tag{59}$$

(ii) for $\delta_2 < \delta_{2m}$ (i.e. seawards of the location of s_{ym})

$$s_y/s_{ym} = \frac{0.99}{1 + 1.14 X_{2.11}} + 0.01 \tag{60}$$

in both cases :

$$X = \Delta_m \left(\frac{H_o}{L_o}\right)^{-1} \left(\frac{H_o}{h_m}\right)^2 \tag{61}$$

where $\Delta_m = \left| \frac{\delta_{2m} - \delta_2}{\delta} \right|$

For the computation of the equilibrium D-profile first is calculated the equilibrium distance between the onshore and offshore profiles, with the division between them at the water line (W_r).

$$\frac{\delta}{2W_r} \frac{H_o}{L_o} = 1.51\ 10^3 \left[h_o^{0.132} D_{50}^{-0.447} \left(\frac{H_o}{L_o}\right)^{-0.717} \right]^{-2.38} + 0.11\ 10^{-3} \tag{62}$$

The distribution of the equilibrium length W across the D-profile is fully defined by the value of S_r and a dimensionless distribution function W/W_r:

$$\frac{W}{W_r} = 0.7\Delta_r + 1 + 3.97\ 10^7\ b\ D_{50}^2 \Delta_r 1.36\ 10^4 D_{50} \qquad (63)$$

$$\text{Where } \Delta_r = \frac{\delta_1 - h_o}{\delta} = \frac{h_m - \delta_2}{\delta}$$

and b = 0 for $\Delta_r \leq 0$

b = 1 for $\Delta > 0$

Swart also gives relations for the computation of offshore transport in the three-dimensional case and for the computation of time-dependent offshore transport and D-profile development. The computation procedures being quite complex these relations will not be presented here.

However, Swart (1976) proposed a modification to simplify the computations without affecting the results significantly. Nevertheless, the model has not been used frequently. A reason for that may be that the computational procedure is rather untransparent. A more important reason for the small application of the model, however, is the fact that the results often do not conform to observations in nature. As has been mentioned already, the model is mainly based on the results of small scale tests, and moreover these tests were performed with regular waves. However, secondary waves, which, together with the basic wave always originate from a sinusoidally-moving wave board, can severely affect the onshore-offshore transport, and thus the beach profile formation. Many descriptions regarding profile characteristics found in the literature are probably based on experimental results. Swart realized this aspect and therefore he made a selection, but nevertheless model effects will certainly have disturbed the experimental results he used. Additionally scale effects also play a prominent role in coastal morphologic tests, which makes the extrapolation of the results to prototype conditions hazardous. In this respect the model of Swart only relates to offshore transport, due to the fact that in the small scale experiments only offshore transport occurred. Even when the slope of the profile was quite gentle, no or hardly any onshore transport could be observed, which does not seem to conform with what occurs in nature.

Chapter V : Coastal Morphology

1 Introduction

In coastal areas morphological evolutions are usually reflected in changes in the shoreline position, and these changes are often decisive in taking measures against coastal erosion or not. Morphological developments can be divided into short-term and long-term evolutions. A clear distinction between these two types of evolution as far as time scale is concerned cannot be made. However, in the light of beach erosion and possible protection measures one should usually think in terms of periods of decades in the case of long-term evolution. The time scale of short-term evolution may be quite variable. Sometimes the morphological changes during a single storm may be significant, but also yearly variations in the wave climate may cause developments which can be denoted as short-term phenomena.

In decision making whether to take beach protection measures, the long-term developments are of particular importance. Yet, this decision may be actuated by short-term developments. e.g., a single storm may cause a temporary but unacceptable retreat of the coastline or erosion of the dunes, so that beach protection measures are urgently required.

The cross-shore transport, and thus the variation in the coastal profile, is mainly responsible for the short-term fluctuations in the coastline, while the long-term evolution of the coastline is mainly governed by the gradients in the longshore transport. Therefore attention will first be paid to the development of coastal profiles. Subsequently the long-term morphological evolutions will be treated, and the available mathematical models for computing such morphological evolutions subsequently briefly described.

2 Coastal Profiles

The form of the coastal profile and their form the variations are mainly governed by the cross-shore transports. These transports and the resulting coastal profile, are closely related to wave motion, and because of the variability of the wave conditions in nature, the form of the coastal profile is also quite variable. This variability is most pronounced in the nearshore zone where the cross-shore transports are largest.

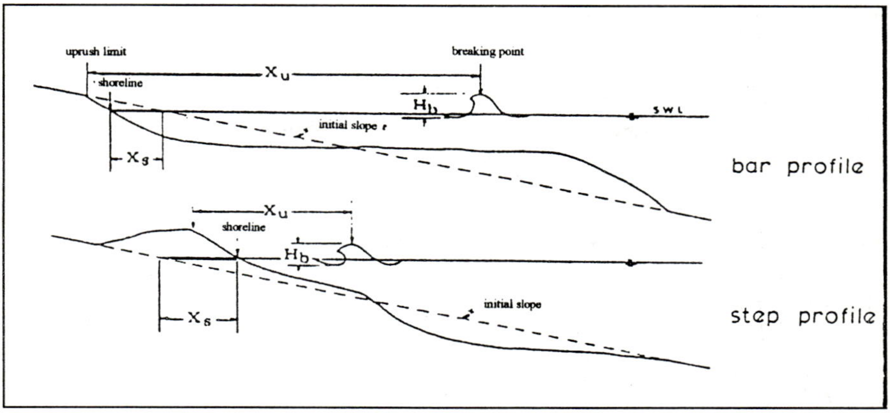

Fig. 22 : Definition of bar and step profiles

2.1 Type of Coastal Profile

Much research has been done during the past decades on the development of coastal profiles under wave attack. This research was mostly directed towards equilibrium profiles, by which certain features of the profile were related to the wave conditions and sediment characteristics. In this respect much attention was paid to the type of profile that developed, and a distinction was made between bar profiles and step profiles (see Fig. 22).

It is noted that *bar* profiles are identified with *eroded* profiles and *step* profiles with accreted profiles. In the literature bar profiles are also called winter or storm profiles, while for step profiles the name summer, ordinary and berm profile are in use. This means that bar profiles can be found during or shortly after storms with offshore transport, whereas step profiles are formed during periods with moderate wave conditions and onshore transport.

Waters (1939) carried out a series of experiments on beach profiles and concluded that the change from step to bar profile was a function of the deepwater wave steepness H_O/L_O. He found the critical value to lie between 0.02 and 0.03, the bar type existing for the higher values. On the basis of model experiments by Meyer (1936) and those by Waters (1939), Johnson (1949) defined the following criteria for the type of coastal profile:

$H_O/L_O < 0.025$: ordinary or step profile
$H_O/L_O > 0.030$: storm or bar profile.

For values of H_O/L_O between 0.025 and 0.030 there is a transition zone where either type of profile might exist.

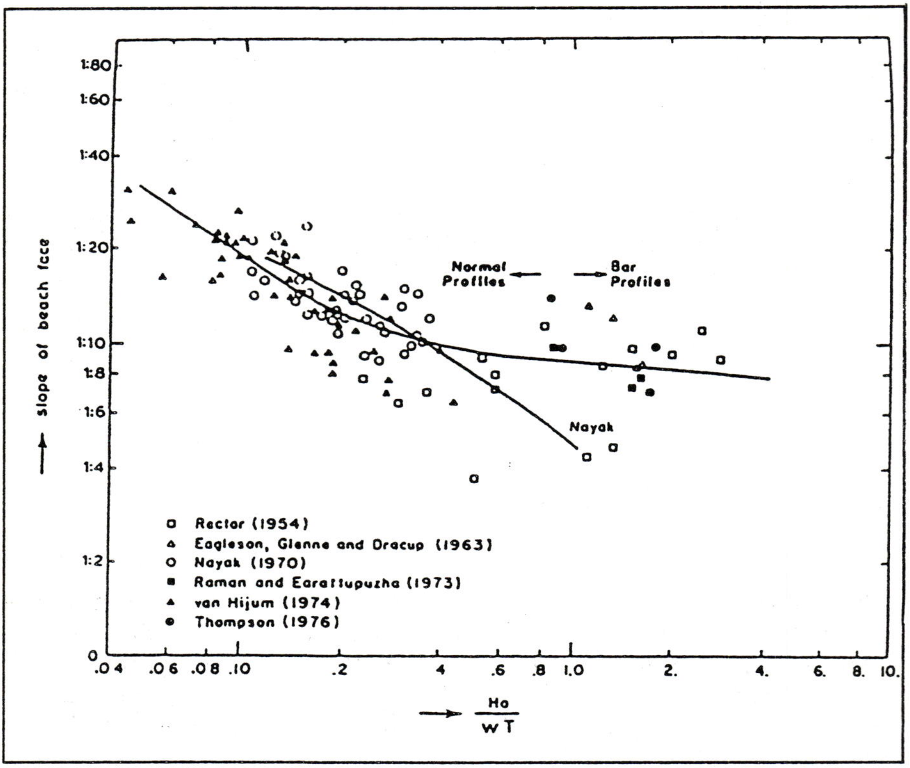

Fig. 23 : Beach slope versus fall velocity parameter

The above tests were carried out with regular waves and constant water level. Watts and Dearduff (1954) observed that water level variation influences the profile development.

Later, Iwagaki and Noda (1962) observed that grain size influences the critical value of the wave steepness for a bar or step profile. On the basis of model experiments they concluded that the criterion for bar formation varies roughly between $H_O/L_O = 0.025$ for $H_O/D_{50} < 300$ and $H_O/L_O = 0.01$ for $H_O/D_{50} > 1000$.

Dean (1973) found that the wave steepness separating a bar profile from a step profile varied with the fall velocity of the sediment w divided by the deepwater celerity $c_O = L_O/T$, thus $H_O/L_O = C\ w/c_O$. This can be rewritten as $H_O/(wT) = C$, where w is the fall velocity of the median grain size D_{50}. Dean concluded that C=0.85. According to the U.S. Army Corps of Engineers, *Shore Protection Manual* (edition 1984), it can be assumed that erosion usually occurs for C>1 and accretion for C<1. In this connection, it is observed that the *Shore Protection Manual* dissociates itself from the opinion that the presence of a berm characterizes an accreted profile and a bar characterizes an eroded profile. On an accreted beach a berm may be absent, while nearshore bars do not directly indicate an eroded beach. Battjes (1974) has shown that bars are connected in complicated ways with breaker processes.

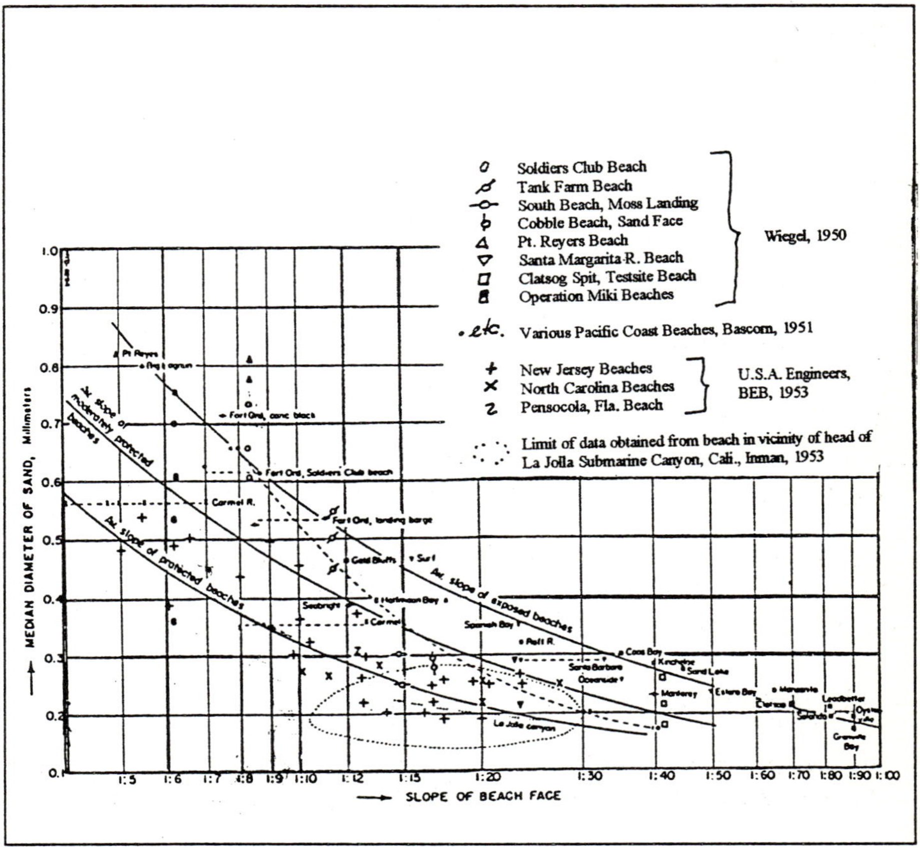

Fig. 24 : Beach slope versus grain size and wave load

Further, Dalrymple and Thompson (1976) concluded that the above parameter $H_O/(wT)$ determines to some degree the slope of the wet beach above still water level, as is shown in Fig. 24.

Earlier, Wiegel (1964) presented relations between the slope of the beach face as a function of the grain size (see Fig. 24). For that purpose he made a distinction between exposed, moderately and protected beaches.

Sunamura and Horikawa (1974) related the occurrence of erosion or of beach profiles to wave steepness, grain size and slope of the profile, for which they found the following relation :

$$H_O/L_O = C\ (\tan \beta)^{-0.27}\ (D_{50}/L_O)^{0.67} \qquad (64)$$

where :

C = constant
tanß = profile slope.

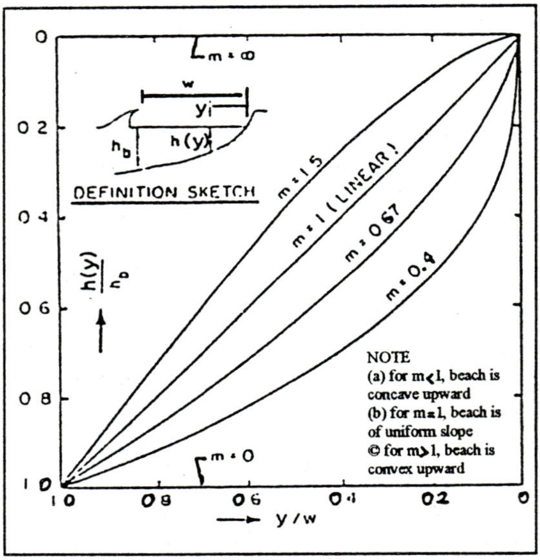

Fig. 25 : Dimensionless profile shape in surf zone for different values of m

For model experiments this represents the initial slope, while for field observations the average slope between shoreline and 20 m (64 ft.) depth is used.

2.2.2 Shape of Coastal Profile

In addition to the studies aimed at finding a criterion for defining the type of coastal profile (bar/step, eroding/accreting), various attempts have been made to describe the shape of the equilibrium profile. On the basis of an analysis of beach profiles along the Danish North Sea Coast and Mission Bay, California, Bruun (1954) found that the equilibrium profile can be approximated by a simple power function, *viz.*

$$d = A\, y^{2/3} \quad (65)$$

The above concept has been applied in an extensive analysis of beach response to storm surge (Dean, 1983a). The conclusion was that eq. (16) is consistent with a spilling breaker model and uniform wave energy dissipation per unit volume of water which means that m = 0.67.

where :

d = water depth,
y = distance from the shoreline and,

A = factor

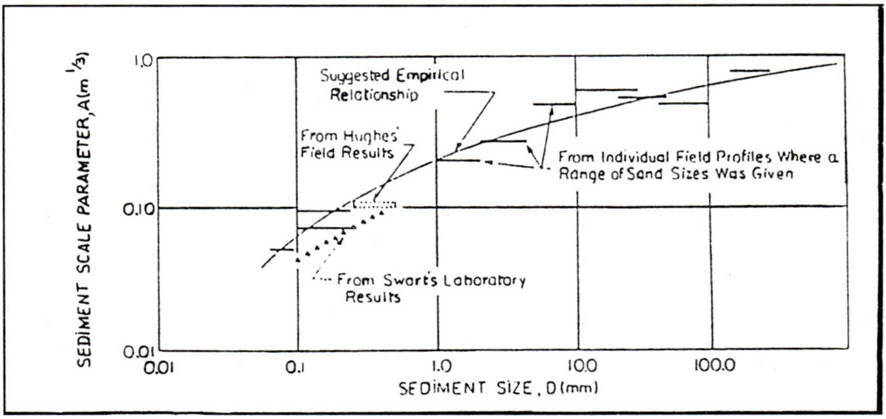

Fig. 26 : Value of A in $d = Ay^m$ versus sediment size

Using physical considerations of the various forces acting in the surf zone Dean (1977) arrived at the following description of the nearshore profile:

$$d = Ay^m \quad (66)$$

where :

A = shape factor, depending on the stability characteristics of the bed material
m = value of the exponent, depending on the force considered (see Fig. 25)

Moore (1982) evaluated published beach profiles comprising a wide range of sand sizes, and found that $A \approx D^{1/3}$, as was reported by Dean (1983b) and shown in Fig. 26. Combination of the above results gives :

$$h = B\, y^{2/3}\, D^{1/3} \qquad (67)$$

where :

B = an unknown factor which is likely to be dependent on several variables, like wave climate, water level variations, coastal currents, etc..

On the basis of extensive experiments on dune erosion during storm surges Vellinga (1983) has derived a mathematical description of the erosion profile. This erosion profile is formed by the sand eroded from the dune and extends up to a depth of 0.75 H_{Os} below maximum storm surge level, see Fig. 27. In a further elaboration of the results, Vellinga (1984) shows that this profile can be satisfactorily described by the power curve $d = 0.08\, y^{0.78}$.

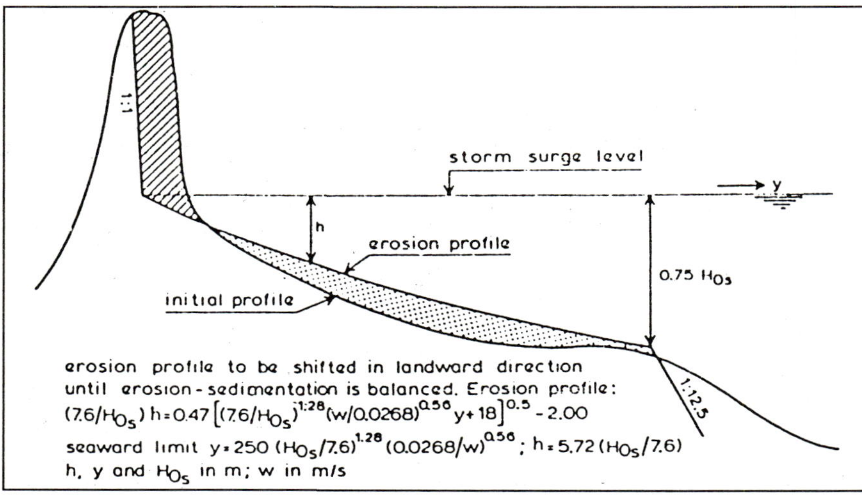

Fig. 27 : Erosion profile after storm surge

After introducing the effect of wave steepness and the effect of grain size in a semi-empirical way, a more general expression of the erosion profile in the nearshore zone is found :

$$d = 0.70\, (H_O/L_O)^{0.17}\, w^{0.44}\, y^{0.78} \qquad (68)$$

where :

d	= depth below water level
y	= distance from the shoreline
w	= fall velocity of the sand related to D_{50}
H_O	= significant wave height in deep water during the storm
L_O	= wave length in deep water = 1.56 T^2

3 Coastal Morphology Models

3.1 One-Line and Two-Line Theory

3.1.1 Introduction

Based on longshore transport and coastline configuration, the theory of Pelnard-Considére (1956) gives the basic equations describing the morphological processes of coastline evolution due to longshore sand transport. These equations lead to the well-known diffusion equation. The fundamental equation may be solved point by point (both in space and time) by means of numerical methods if an accurate description of the wave climate and of its influence on the longshore transport is known.

A computation program has to include two parts:

a. computation of the longshore transport along a straight coastline and of the variations in this transport as a function of the coastline direction;
b. computation of the morphological evolution of the coastline by means of the basic equations.

3.1.2 Basic Equations

For the one-line theory the coastal profile is schematized according to Fig. 28(a). The equation of continuity is derived from Fig. 28(b). The x-axis is chosen along the original coastline and the y-axis in a direction perpendicular to the original coastline in offshore direction. When considering an infinitesimal element with length dx, the equation of continuity yields:

$$\left(S_1 + \frac{\partial S_1}{\partial x}.dx\right).dt + ph\, dx\, dt - S_1\, dt = \frac{\partial y}{\partial t} h\, dx\, dt \qquad (69)$$

or :

$$\frac{1}{h}\frac{\partial S_1}{\partial x} + \frac{\partial y}{\partial t} + p = 0 \tag{70}$$

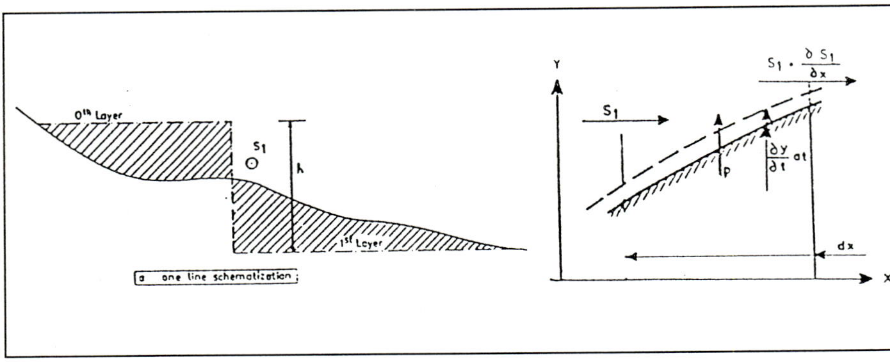

Fig 28 : (a) Profile Schematization (b) Plan view

where :

S_1 = longshore sand transport
h = height of the schematized beach profile
p = transversal sand transport
t = time

The equation of motion may be derived from Fig. 29 :

The following conditions are set :

1. The longshore transport is a function of the coastline direction only and of no other parameters.
2. This function can be differentiated for any of its values.
3. The second order and higher terms may be neglected.

The first condition leads to:

$$S_1 = S_1(\alpha) \tag{71}$$

The theorem of Taylor-Maclaurin, together with the second condition, yields :

$$S_1(\alpha) = S_{O1} + \alpha\left(\frac{dS_1}{d\alpha}\right)_{\alpha=0} + \frac{1}{2}\alpha^2\left(\frac{d^2S_1}{d\alpha_2}\right)_{\alpha=0} + .. \tag{72}$$

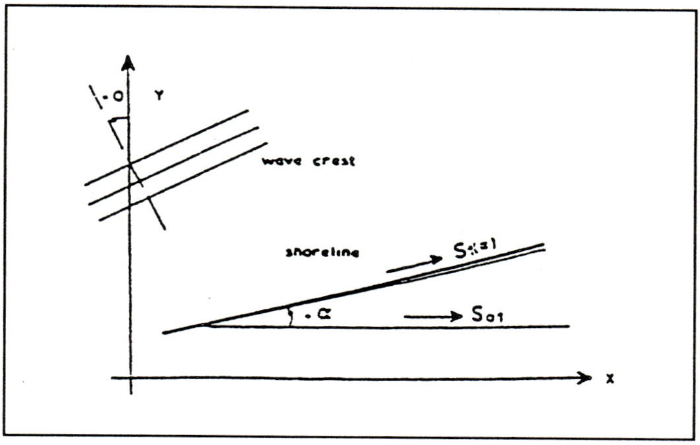

Fig. 29 : Longshore sediment transport

When defining :

$$\alpha \approx \tan\alpha = \frac{\partial y}{\partial x} \quad \text{and} \quad \frac{dS_1}{d\alpha} = s_1 \tag{73}$$

Equation (72) yields :

$$S_1(\alpha) = S_{O1} - s_1 \frac{\partial y}{\partial x} \tag{74}$$

where :

S_1 (α)	= longshore transport as a function of the coastline direction
S_{O1}	= longshore transport along a straight parallel to the x-axis
s_1	= variation of the longshore transport per radian of the coastline rotation
α	= coastline direction with respect to the x-axis in radians

Eq. (74) is the Pelnard-Considére equation wherein the coastal constant S_{O1} and s_1 may be functions of x. When the equation of motion (74) is substituted in the equation of continuity (70) the following relation is obtained (with constant S_{O1} and s_1 , and $p = 0$) :

$$\frac{\partial y}{\partial t} = \frac{s_1}{h} \frac{\partial^2 y}{\partial x^2} \tag{75}$$

This is the so-called equation of diffusion for which analytical solutions can be found.

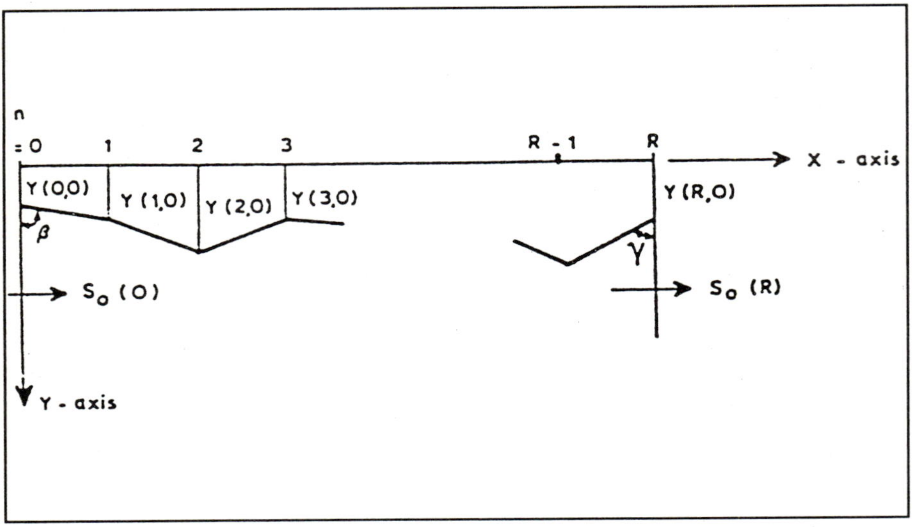

Fig. 30 : Schematization of boundary conditions

The advantage of a numerical approach is that solutions of the complete equation, without the simplification, can be found. The complete equation yields :

$$\frac{\partial y}{\partial t}=\frac{1}{h}\left\{\frac{\partial s_1}{\partial x}\frac{\partial y}{\partial x}+s_1\frac{\partial^2 v}{\partial x^2}-\frac{\partial S_{01}}{\partial_x}\right\}-p \qquad (76)$$

3.1.3 Boundary conditions

The starting condition Y(n,0), is given by the coastline position at time t = 0. Any position may be introduced at each point n of the x-axis. The boundary condition at n = 0, Y(0, t) must be given as a function of time. A time-position function may be given at n = 0, or a longshore transport may be given (constant in time). Both possibilities reduce to a given coastline direction at point n = 0.

The boundary condition at the other end of the computed coastline may be given in the same way. The boundary conditions may thus be summarized as in Figure 30.

As the longshore transports S_{O1} (0) and S_{O1} (R) are time constants, the angles ß and γ of the coastline at the boundaries are time-constants as well. Note that the boundary longshore transports could easily be introduced as time functions, if this function is known.

3.1.4 Computation of the coastal constants

The essential role played by the coastal constants S_{O1} and s_1 could be noticed in the mechanisms described above. Hence, an accurate computation of those values is important and they have to be constant in time but not basically constant in space (along the x-axis).

The longshore transport S_{O1} along a straight coastline can be computed as a function of the wave climate by adding up the transports due to the waves from various directions, taking into account the frequency distribution. The resulting transport then is the coastal constant S_{O1}.

The longshore transport may be computed for each wave direction by means of the CERC formula. It is, however, advisable to check the resulting transport S_{O1} by means of other methods, such as volumes of erosion or accretion based on measurements.

Finally, the longshore transport is a function of the wave height and direction in the breaker zone. These values can be deduced from deep water data by means of refraction and diffraction computations.

The second coastal constant s_1 can be computed by applying a rotation $\Delta\alpha$ of the coastline and calculating the difference in resulting longshore transport ΔS_1 caused by this rotation. The coastal constant s_1 then is equal to :

$$s_1 = -\frac{\Delta S_1}{\Delta\alpha} \tag{77}$$

3.2 Two-Line Theory

The limitations of the one-line theory are numerous. Often it schematizes reality too strongly. This can apply to initial and boundary conditions, as well as the physical wave and beach characteristics. Bakker (1968) was concerned in particular about a coast upon which the longshore sand transport is only partially blocked by groins which were shorter than the width of the breaker zone. Bakker proposed a so-called two-line theory for the solution of this problem. Instead of schematizing a coastline with a single curve, two curves are used (Figure 31).

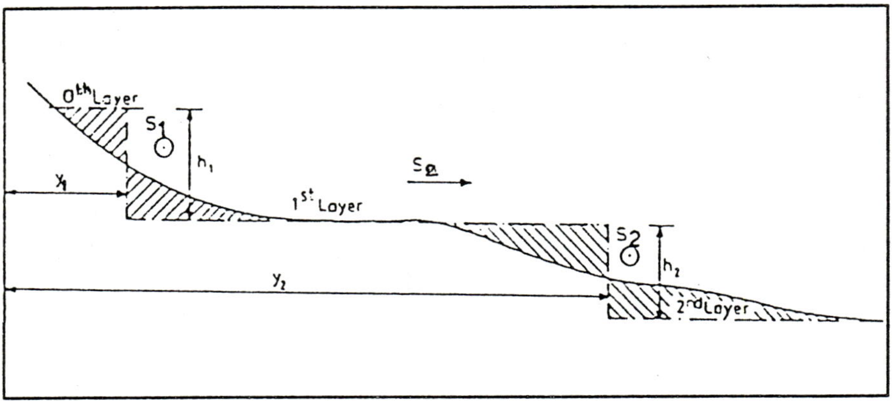

Fig 31 : Profile schematization

The choice of a reference axis for the y distances is arbitrary. The areas of each of the pairs of shaded areas are equal. The longshore transport S_i (i = 1, 2) is directed parallel to the coast out of the plane of the paper, and describes the littoral sand movement in its zone. The horizontal planes are usually selected at elevations, which correspond more or less to flat portions of the total profile. If special structures, such as groins, are involved in the schematization, the horizontal planes are often chosen so that the limits of the boundary conditions correspond with the limits of a transport zone.

Just as with the one-line theory, it will be necessary to develop a continuity relationship and equations of motion for each of the schematization zones. The equation of continuity is more complex than that in the one-line theory since there is supply and removal of sand from two directions. Figure 32 shows an element in plan; the net transport of material into the element is equal to the volume retained for each of the 2 elements - one for each line of the schematization. Thus, for i = 1 and 2 the following relation is obtained :

$$S_i\,dt - (S_i + dS_i)\,dt + S_{yi-1}\,dx\,dt - S_{yi}\,dx\,dt = dx\,dy_i\,h_i \qquad (78)$$

Sand is only transported transversely between the first and second zone so that :

$$S_{yO} = 0 \text{ and } S_{y2} = 0 \qquad (79)$$

The above eqs. (78) and (79) can be written as :

$$\frac{\partial S_1}{\partial x} + S_{y1} + h_1 \frac{\partial y_1}{\partial t} = 0 \qquad (80)$$

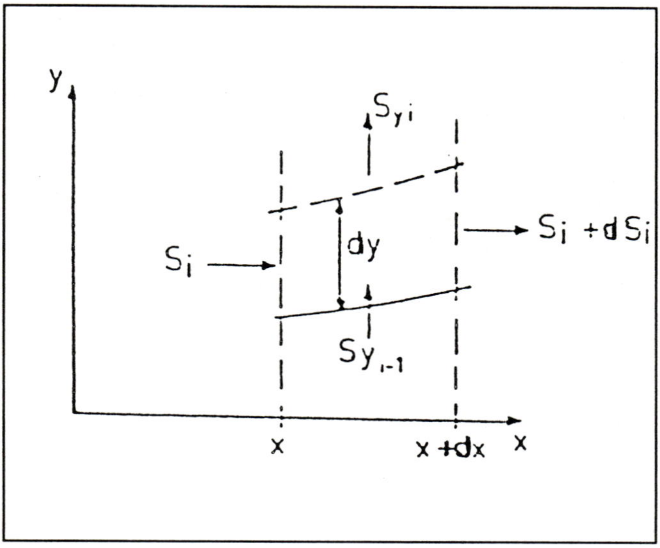

Fig. 32 : Plan View

$$\frac{\partial S_2}{\partial x} - S_{y1} + h_2 \frac{\partial y_2}{\partial t} = 0 \tag{81}$$

the cross-shore transport S_{y1} is defined by Bakker (1968) as:

$$S_{y1} = s_y (y_1 - y_2 + W) \tag{82}$$

where :

s_y = cross-shore transport constant
W = equilibrium distance between beach and inshore.

Similar to the one-line theory the equation of motion for each of the schematized zones is :

$$S_1 = S_{01} - s_1 \frac{\partial y_1}{\partial x} \tag{83}$$

$$S_2 = S_{02} - s_2 \frac{\partial y_2}{\partial x} \tag{84}$$

The above set of differential equations (80) to (84) can be solved numerically or analytically; the latter only for simplified boundary conditions.

3.3 N-Line Model

This type of model has been developed to include the effect of tidal currents in the longshore transport and to obtain a more horizontally two-dimensional picture of coastal processes. This is contrary to the above mentioned one- and two-line models, which consider only the longshore transport due to breaking waves. However, the effect of tidal currents can certainly be important for the development of the nearshore part of the coastal profiles, beyond the breaker zone.

The set-up of the n-line model is as follows. The coastal area under study is represented by a number of coastal profiles. Each coastal profile is schematized to a number (n) of depth contours. Bars and troughs cannot be reproduced, the consecutive lines increase in depth in seaward direction. Wave conditions are determined from deep water conditions *via* refraction and shoaling. The longshore current is driven by wave forces (shear stress due to radiation stresses) and the tidal forces. The distribution of random waves inside the breaker zone is based e.g. upon the early concept of Battjes (1974), according to which the local root-mean-square wave height is obtained by deleting from the wave height distribution the portion of waves which are higher than a certain maximum value.

The sediment transport capacities are computed in sections between the schematized profiles, both in longshore and cross-shore direction. The resulting transport gradients are accounted for in depth changes in the schematized profiles. Consequent discontinuities in the profiles are compensated for by a smoothing procedure in each separate profile.

The Department of Public Works of The Netherlands (RWS) carried out a number of computations with the N-line model to evaluate its possibilities. The main conclusion was, that the principle of schematizing the water- and sediment-transport in a horizontal plane is a sound approach for the development of a morphological coastal model. However, a representation of the coastal profile in a number of lines appears to be too strong a schematization for general cases of curved coastlines, and is only suitable for practically uniform and straight coasts. The ultimate advantage of the N-line model with respect to the KC/KL models is the fact, that besides the waves, tidal or other currents can be taken into account in the computation of the longshore transport. The longshore transport is locally determined, which means that apart from the integrated value, the cross-shore distribution of the longshore transport is determined. In cases with additional structures, such as groins, part of the longshore transport can be blocked.

3.4 Box Models

Such models have also been called "three-dimensional evolution" models, since they attempt to model the coastal zone in three or quasi-three dimensions. The basis of the model is the two-dimensional continuity equation :

$$\frac{\partial Q_x}{\partial x} + \frac{\partial Q_y}{\partial y} + \frac{\partial z}{\partial t} = 0 \qquad (85)$$

where :

Q_x, Q_y represent the total transport of sediment as bed load and suspended load to waves and currents.

z represents the change in elevation of the seabed.

Operation of the model requires that a grid of computation points be placed over the study area, and that sediment transport rates be calculated at each corner of the grid for conditions throughout one or more tidal cycles. The model can, therefore, deal with 'worst-storm' events.

The box model can, in theory, take into account velocity variations over the flow depth due to wave, wind and tidal forces, provided details can be obtained from field or model data. It is usually necessary to assume some form of equilibrium sediment concentration profile over the flow depth and this assumption limits the technique to medium to coarse sands and also imposes limitations on the size of space steps (Δ x, Δ y) that can be used in the model.

3.5 Continuum models

In situations involving large tides and waves together with fine sands, it is necessary to use a continuum approach whereby the depth-averaged or three-dimensional form of the suspended load equation is used to study 'worst-storm' events. The appropriate equations are:

$$\frac{\partial c}{\partial t} + u_i \frac{\partial c}{\partial x_i} - w_f \frac{\partial c}{\partial x_3} = \frac{\partial}{\partial x_i}\left(\upsilon_{si} \frac{\partial c}{\partial x_i}\right). \qquad (i = 1, 2, 3) \qquad (86)$$

where :

u_i are the velocity components in $x_1 = x$, (longitudinal), $x_2 = y$ (lateral), $x_3 = z$ (vertical) directions, respectively ;

w_f = the fall velocity of the sediment ;

v_{si} = the mixing coefficients, which include wave and current effects

Bed level changes are determined from the equation:

$$\frac{\partial q_{b1}}{\partial x_1} + \frac{\partial q_{b2}}{\partial x_2} + (1-n)\frac{\partial z}{\partial t} = D - P \qquad (87)$$

where :

P = the quantity of sediment entrained per unit area, per unit time into the flow ($= w_f c_a$) ;

D= the quantity of sediment depositing on the seabed per unit area, per unit time ($= w_f c_A$) ;

c_a = the concentration of sediment in the mobile 'bed' layer

c_A = the concentration of sediment at level A (some three times the bed ripple height) above the bed;

q_{b1}, q_{b2} = the bed load transport rates in x_1 and x_2 directions.

The depth-averaged form of the sediment equation is :

$$\frac{\partial C}{\partial t} + u_i \frac{\partial C}{\partial x_i} = \frac{1}{h}\frac{\partial (D_i h \partial C)}{\partial x_i} + S \ (i = 1,2) \qquad (88)$$

where :

D_i is an effective dispersion coefficient to allow for changes in flow direction over the water depth ;

S is a course term given b by $S = w_f (c_a - ßC)/h$;

ß is a coefficient relating depth-mean concentrations (C) to near-bed values (C_A).

Bed level changes are again determined from equation (87).

The continuum models are solved on a grid of computational points, as for the box models, but clearly require more data to operate them, particularly for the three-dimensional version. Unfortunately, the cost of operating the models and difficulties in obtaining input data means that these sophisticated models while presently confined to research level, have great potential for the future.

4 Beaches Backed by Dunes

As is known, "storms" erode beaches, which means by combinations of storm tides**Error! Reference source not found.** and stormwaves. To the contrary, they build up with normal or low tides and by swells. The hydrodynamic behavior for the two conditions differs greatly. During storms material concentrations are high, and the material is fed into the wave uprush by eroding dunes or beach berms. Swells on the other hand generate much less concentration of suspension loads. They "shear the material" off the bottom and wash it up on the beach thereby building it up.

Vellinga (1983) established a predictive model for beach and dune erosion during storm surges. In many ways his findings have confirmed certain known facts.

(a) as seen in the field :

(1) A steep slope causes heavier erosion than a more gentle one.
(2) Coarser material on a steep beach washes out faster but it also settles faster.
(3) Sand material when eroded from a dune and from the upper beach during a storm, deposits in the lowest part of the beach and in the nearshore shallow water, thereby establishing a mass balance two-dimensionally and three-dimensionally, assuming uniform material characteristics and a "long" straight shore.

(b) as seen in the laboratory :

(4) Reproduction on a "scale model" using native beach sand only gives qualitative similarity. Using finer sands a better similarity is obtainable.
(5) The present available longshore transport formulas are unreliable, particularly for storm conditions which ignore the differences in erosion and transport caused by differences in tides as well as differences imposed by wave characteristics.

Because of the great predictive-importance of the Vellinga-model for beach and dune erosion, this computational model will be briefly outlined.

In 1972 the Technical Advisory Committee for the Sea Defense (TAW) published a provisional guideline for the computation of dune erosion during storm surges. On the basis of a limited number of beach profile measurements after storm surges, the following erosion profile was defined ;

$$h = 0.415\,(y + 4.5)^{0.5} - 0.88 \qquad (89)$$

in which :

y = distance from the intersection of the erosion profile and storm surge level in m.
h = depth below storm surge level in m.

The erosion profile develops in such a way that the erosion equals the sedimentation. After the storm surge the slope of the dune front is assumed to be 1:1, while the erosion profile has developed to a depth

$$d_b = 1.25\ H_{sb} \tag{90}$$

where :

d_b = breaker depth
H_{sb} = significant wave height at the breaker depth during maximum storm surge level.

At the depth d_b the erosion profile ends vertically. This already demonstrates that the aim was not tried to obtain a physically correct description of the dune erosion, but only to activate a simple computational method.

In order to arrive at a more reliable computational method a large number of experiments with different scales and different types of sand were carried out. Further irregular waves were applied to avoid the effect of secondary waves. A detailed analysis of the measurements resulted into a set of scale relationships for converting the model results to prototype values, *viz.*

$$n_1 / n_d = \left(n_d / n_w^2\right)^{0.28} \tag{91}$$

$$n_t = n_d^{0.5} \tag{92}$$

in which :

n_l = length scale
n_d = depth scale
n_w = scale of fall velocity of sand
n_t = morphologic time scale

The above mentioned scale relations were verified by large scale tests in the Delta-flume (The Netherlands). For standard conditions with H_{OS} = 7.6 m, Tp = 12 s and w = 0.0268 m/s the erosion profile is defined as follows :

$$h = 0.4714\ (y + 18)^{0.5} - 2.00 \tag{93}$$

in which :

y = distance from the intersection of the erosion profile and storm surge level in m.

h = depth below storm surge level in m.

For conditions with other values for H_{OS} and w, but approximately the same wave steepness the erosion profile is

$$\left(\frac{7.6}{H_{OS}}\right) h = 0.4714 \left[\left(\frac{7.6}{H_{OS}}\right) 1.28 \left(\frac{w}{0.0268}\right)^{0.56} y + 18 \right] - 2.00 \qquad (94)$$

The erosion profile develops up to a depth of :

$$y = 5.717 \left[\frac{H_{OS}}{7.6}\right] \approx 0.75 H_{OS} \qquad (95)$$

Below that depth the slope of the erosion profile is 1:12.5 and above storm level it is 1:1. The profile is shown in Fig. 27.

5 Mathematical Modeling of Beaches Backed by Seawalls

As a first step in the description of the model a rectangular coordinate system (x, y) is introduced and the most important are listed. The orientation of the co-ordinate system is chosen so that the x-axis lies roughly parallel to the beach. The beach is assumed to be represented by a single line, y (x, t), where t is time. It is also assumed to have everywhere a constant slope, tan ß, between the swash limit on the crest of the berm, and some underwater contour beyond which profile changes can be assumed negligible. Under this last assumption, the level on the beach profile which defines the line y (x, t) can be chosen in a variety of ways. Ozasa and Brampton (1980) choose Mean Sea Level (MSL) as both the datum to which depths are reduced and the contour to which y (x, t) is measured. The depth below which profile changes are negligible is called D and is usually assumed constant along the beach. In cases where, for example, one part of a beach is consistently more sheltered than the remainder, this condition may be relaxed and D allowed to vary. The height of the swash limit, or beach berm, above MSL is denoted by BH.

The model can now be conveniently handled into two stages. In the first, the beach above MSL is assumed sufficiently wide that any seawall behind it has no significant effect on the alongshore sediment transport. This situation is shown diagrammatically in Fig. 33. and can be represented by the equation :

$$y \geq y_w + \frac{BH}{\tan \beta} \tag{96}$$

where y_w is the distance of the sea wall from the baseline.

Considering sediment motion across two lines δx apart and perpendicular to the base line (y = 0), for a small time interval δt, we obtain for the change in beach volume in that strip :

$$\delta V = \delta t\left(Q - \left(Q + \frac{\partial Q}{\partial x}\partial x\right)\right) \tag{97}$$

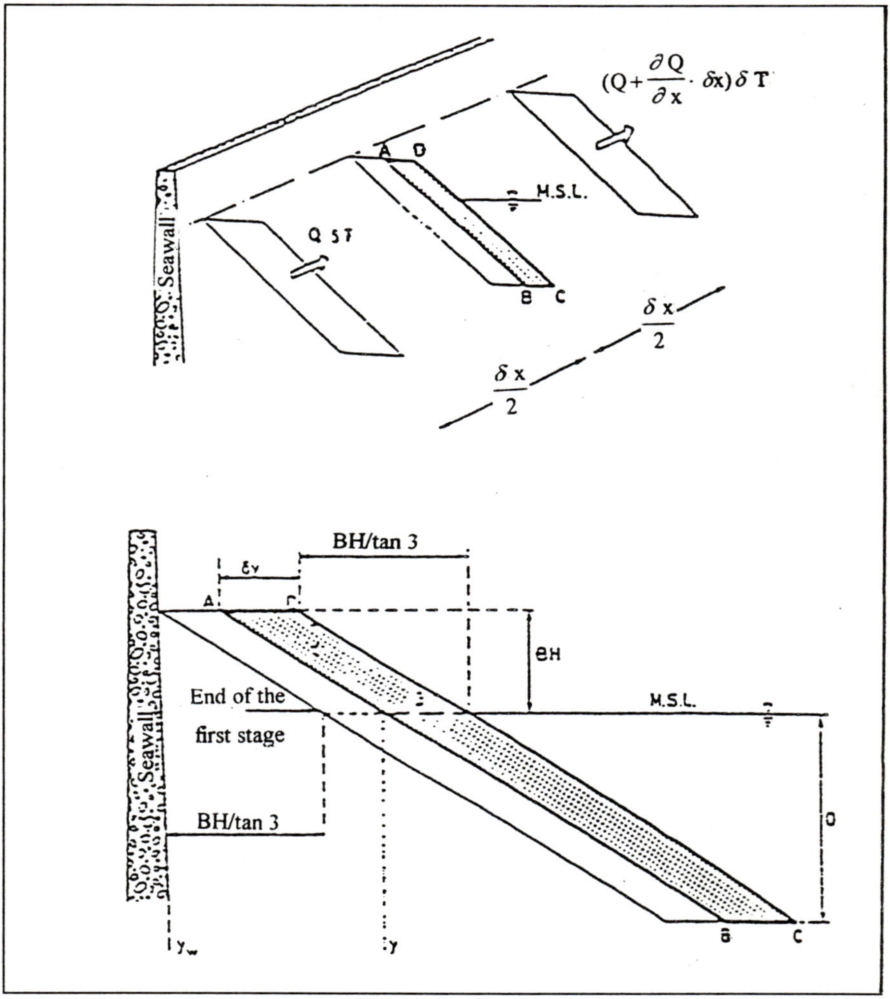

Fig. 34 : Continuity equation (First stage)

where Q is the volume rate of alongshore sediment transport. If we further assume that as the beach either erodes or accretes this profile retains the same shape, tan ß, then we obtain :

$$\delta V = (D + BH)\, \delta x \delta y \qquad (98)$$

where δy is the mean translation of the profile in that strip. Combining these two equations and letting both δx and δt tend to zero, we obtain the continuity equation :

$$\frac{\partial Q}{\partial x} + (D + BH)\frac{\partial y}{\partial t} = 0 \qquad (99)$$

The term D + BH is, of course, the total depth over which the beach profile changes.

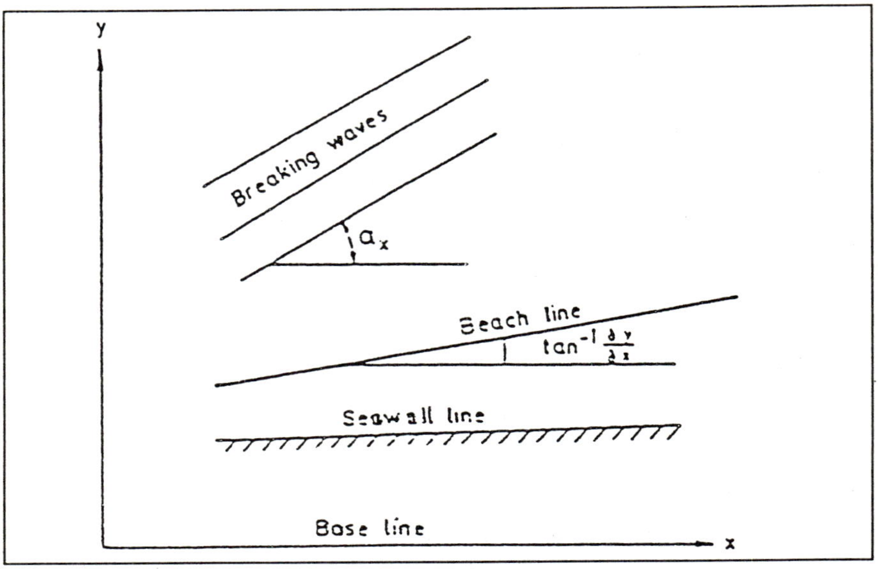

Fig. 33 : Breaker angle

Ozasa and Brampton (1980) used a new formula for the alongshore transport of sediment. In normal situations, alongshore transport on beaches is caused by the oblique breaking of waves. In some case, however, alongshore wave height gradients must also be considered. Such situations often occur when waves diffract around a headland or a breakwater. The equation for the volume rate of alongshore sediment transport used by Ozasa and Brampton takes both these effects into account and is given by :

$$Q = \frac{0.385\left(EC_g\right)_b}{\gamma_s}\left(\sin\alpha_b - 3.24\frac{\partial S_b}{\partial x}\cot\beta\cos\alpha_b\right) \qquad (100)$$

where :

E = the energy density of the waves = $\frac{1}{8}\rho g H^2$

H = the wave height,

g = the acceleration due to gravity,

C_g = the group velocity of the waves,

α = the angle their crests make with the shoreline,

ρ = the density of the water,

γ_s = the specific of beach material in place and where subscripts b refers to breaking wave conditions.

Turning to Fig. 34, we relate the angle α_b to the angle between the breaking crests and the baseline α_x by :

$$\alpha_b = \alpha_x - \tan^{-1}\frac{\partial y}{\partial x} \tag{101}$$

The numerical solution of equations (95), (100) and (101) by a finite difference method is then very straightforward. Before that, however, we need to consider the situation where the beach in front of a sea wall is so narrow that the longshore sediment transport is affected by the wall's presence. This situation can be represented by :

$$y_w \le y < y_w + \frac{BH}{\tan\beta} \tag{102}$$

Although the basic continuity relation still holds, the subsequent eq. (98) and (99) have to be adjusted.

In the absence of exact theory, the necessary changes were based on observations in the following manner. Experience has shown that a beach in front of a sea wall will erode to a certain level (Line BC of Fig. 35) and then reach a state of static equilibrium (providing that the sea wall has not collapsed in the meantime). The level at which this occurs depends considerably on local conditions, but normally the highest part of the beach is about MSL which is why Ozasa and Brampton choose this particular level as their depth datum. In this situation, the alongshore sediment transport becomes negligible.

So the second stage of the model has to represent not only an alongshore transport rate which diminishes to zero, but also reduces the berm height in a similar way. Therefore eq. (99) has to be written as :

$$\frac{\partial Q}{\partial x} + (D + B)\frac{\partial y}{\partial t} = 0 \tag{103}$$

where $B = \tan \beta\ (y - y_w)$, the height of the beach above mean sea level.

The problem that now arises is how to diminish Q as the beach retreats. Two possible methods come to mind. Denoting Q in the first stage by Q_1 and in the second stage by Q_2, we could either set :

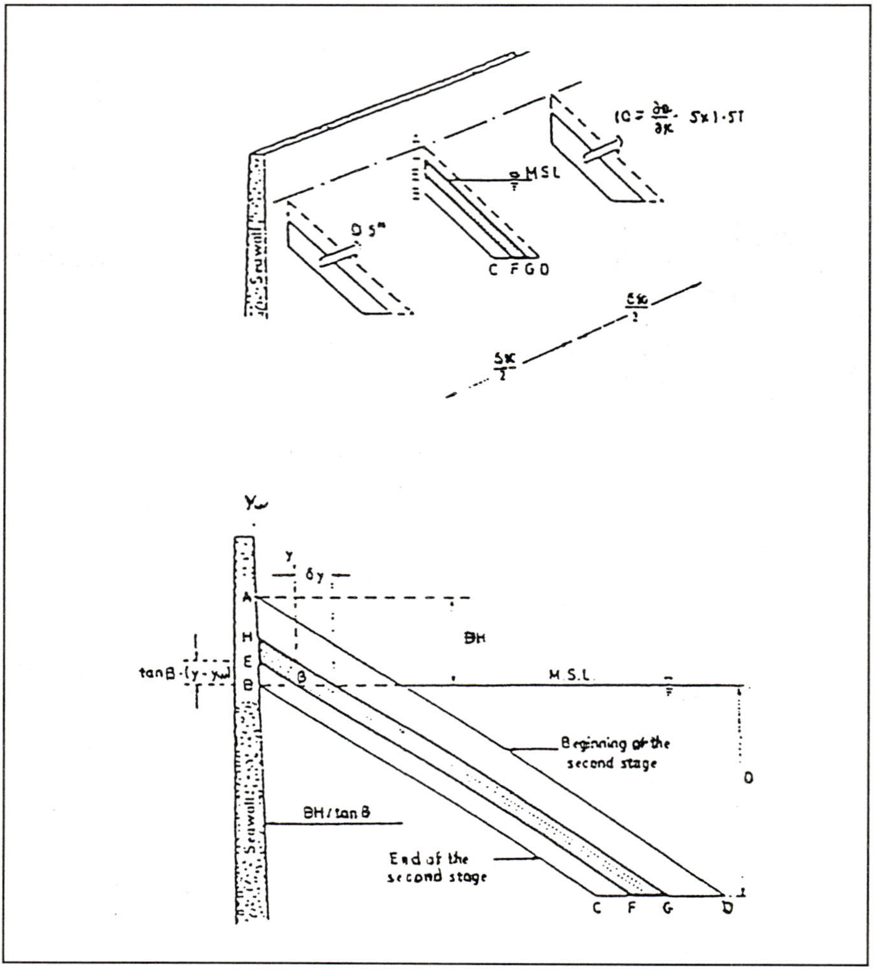

Fig. 35 : Continuity equation (second stage)

$$Q_2 = \frac{B}{BH} Q_1 \tag{104}$$

which diminishes Q by the ratio of the remaining height of beach above MSL to the original berm height or referring to fig. 35, set :

$$Q_2 = \frac{\text{AreaEBCF}}{\text{AreaABCD}} Q_1 = \frac{B}{BH} \frac{2D + B}{2D + BH} Q_1 \tag{105}$$

Both possibilities satisfy $Q_2 = Q_1$, at the beginning of this second stage and $Q_2 = 0$ at the end of it.

The numerical method for this stage is based on equation (100), (102) and either (103) or (105). The differential s equation can be solved using a finite-difference numerical method.

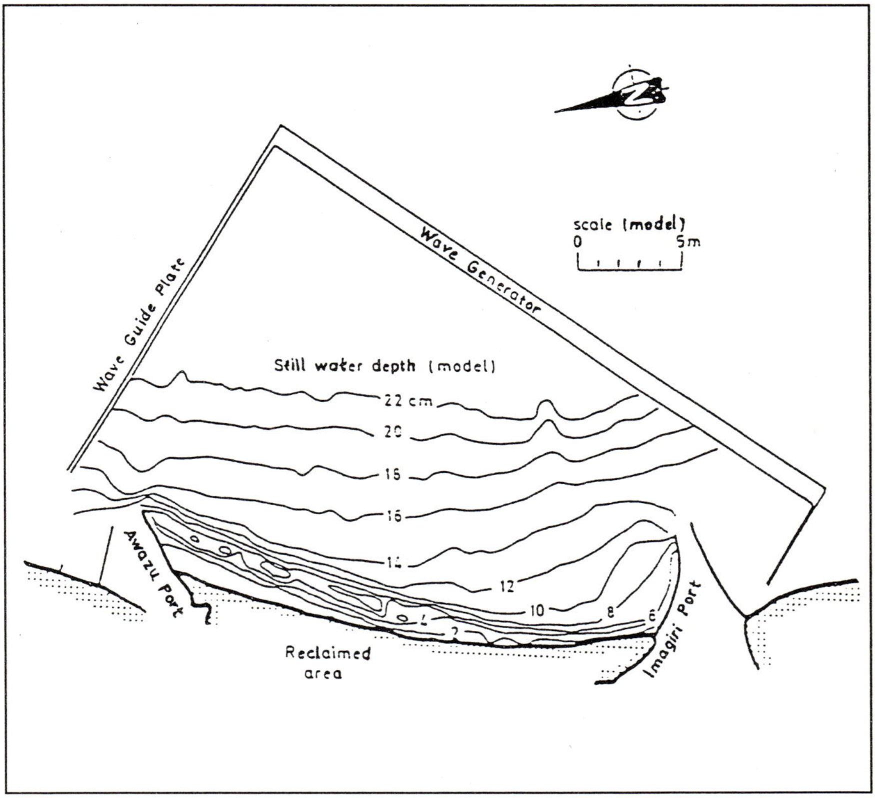

Fig. 36 : Contour map of the physical model experiment

The model of Ozasa and Brampton has been used to try to reproduce shoreline changes which were measured during a physical model study of a bay in Japan (see Fig. 36). The models of the physical model are compared with the predictions of the mathematical model in Fig. 37. Both forms of the alongshore sediment transport formula were tried (i.e. eqs. (103) and (104)) and the results from each are shown, although the difference between the two was always less than 1 cm (0.4 in.).

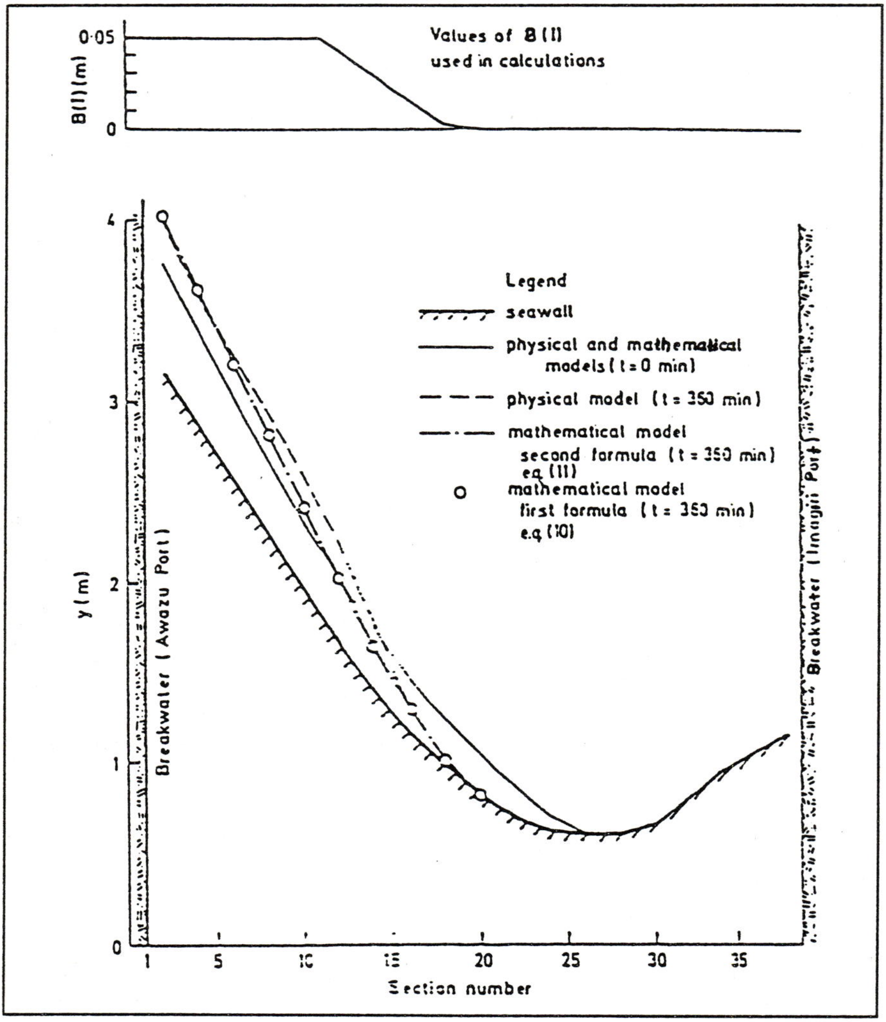

Fig. 37 : Plan of shoreline evolution between Imagiri and Awazu Ports

6 References

Bakker, W.T., 1968, The Dynamics of a Coast with a Groyne System, Proceedings 11th Conf. on Coastal Engineering.

Battjes, J.A., 1974, Surf Similarity, *Proc.* 14th Conf. on Coastal Engineering.

Bijker, E.W., 1967, Some Considerations about Scales for Coastal Models with Movable Bed, *Delft Hydraulics Laboratory, Publication No. 50.*

Bijker, E.W., 1968, Littoral Drift as Function of Waves and Current, *Delft Hydraulics Laboratory, Publication No. 58.*

Bruun, P., 1954, Coast Erosion and the Development of Beach Profiles, U.S. Army Beach Erosion Board, *TM-44.*

Dalrymple, R.A., & Thompson, W.W., 1976, Study of Equilibrium Beach Profiles, *Proc.* 15th Coastal Engineering Conference.

Dean, R.G., 1973, Heuristic Models of Sand Transport in the Surf Zone, *Proc.* First Australian Conference on Coastal Engineering, Sydney, May 1973.

Dean, R.G.,1977, Equilibrium Beach Profiles : U.S. Atlantic and Gulf Coasts, University of Delaware, Newark, Ocean Engineering Technical Report No. 12.

Dean, R.G., 1983a, Shoreline Erosion Due to Extreme Storms and Sea-level Rise, University of Florida, Coastal and Oceanographic Engineering Department, Tallahassee, FL.

Dean, R.G., 1983b, Principles of Beach Nourishment in : *CERC Handbook of Coastal Processes and Erosion* (ed. Komar, P.D.)

Deigaard, R., Fredsoe, J., & Hedegaard, I.B., 1986, Mathematical Model for Littoral Drift, *J. Waterways, Ports, Coasts and Ocean Eng.*, ASCE, 112, 3:351 - 369.

Delft Hydraulics Laboratory, 1978, Berekeningen van de Kustlijn ten Oosten van Zeebrugge (Computations of the Coastline East of Zeebrugge), *Report R1301.*

Hulsbergen, C.H., Bakker, W.T., & van Bochove, G., 1976, Experimental Verification of Groyne Theory. *Proc.* 15th Coastal Eng. Conf. ASCE, pp. 1439 - 1458.

Inman, D.L., & Bagnold, R.A., 1963, Littoral Processes, In *: The Sea: Ideas and Observations,* Vol. 3, Interscience, New York.

Iwagaki, Y., & Noda, H., 1962, Laboratory Study of Scale Effect in Two-Dimensional Beach Processes. *Proc.* 8th Coastal Engineering Conference.

Johnson, J.W., 1949, Scale Effects in Hydraulic Models Involving Wave Motion, *Trans. American Geophysical Union.* Vol. 30, No. 4.

Kraus, N, 1987, Effects of Sea-walls on Beaches : a Literature Review. *Proc.* Cong. Coastal Sediments, New Orleans.

Moore, B., 1982, Beach Profile Evolution in Response to Changes in Water Level and Water Height, Master of Science *Thesis*, University of Delaware, Newark, DL.

Muirwood, A, & Fleming, C., 1981, *Coastal Hydraulics*, MacMillan Book Co., Ltd., New York.

O'Connor, B.A., Morcos Fanos, A, & Cathers, B. 1981, Simulation of Coastal Sediment Movements by Computer Model. In: *Proc. 2nd Int. Conf on Eng. Software* (ed: Adey R.A.) pp: 554 - 568.

O'Connor, B.A, 1985, Coastal Sediment Modelling. in: *Offshore and Coastal Modeling* (eds. Dyke, P.G., Moscardini, A. & Robinson, E.H.), Springer Verlag, New-York: pp: 109 - 136.

O'Connor, B.A., & Nicholson, J., 1989, Modelling Changes in Coastal Morphology. In: *Proc.* ASCE Int. Symposium on Sediment Transport Modelling (ed. Wang, S.Y.) pp: 160 - 165.

Ozasa, H., & Brampton, A.H., 1980, Mathematical Modelling of Beaches Backed by Sea-walls, *Coastal Engineering*, 4:47 - 63.

Pelnard-Considére, R., 1956, Essai de Théorie de l'Evolution des Formes de Rivage en Plages de Sable et de Galets. Société Hydrotechnique de France. *Actes* IVes Journées de l'Hydraulique, Paris.

Perlin, M., 1978, A Numerical Model to Predict Beach Plan Forms in the Vicinity of Littoral Barriers, M.C.E. *Thesis*, University of Delaware, Department of Civil Engineering, Newark DL.

Sunamura, T., & Horikawa, K., 1974, Two-dimensional Beach Transformation due to Waves. *Proc.* 14th Coastal Engineering Conference.

Swart, 1974, Offshore Sediment Transport and Equilibrium Beach Profiles, Thesis, Delft University of Technology (Netherlands).

Swart, 1976, Predictive Equations Regarding Coastal Transports, *Proceedings 15th Conf. on Coastal Engineering.*

U.S. Army Corps of Engineers, 1984, *Shore Protection Manual.* Coastal Engineering Research Center, Dept. of the Army, Waterways Experiment Station. PO Box 631, Vicksburg, Mississipi, 39180.

Van der Graaf, J., & Van Overeem, J., 1979, Evaluation of Sediment Transport Formulae in Coastal Engineering Practice, *Coastal Engineering*, 3:1 - 32.

Van Rijn, L.C., 1985, Initiation of Motion, Bedforms, Bed Roughness, Sediment Concentrations and Transport by Currents and Waves, *Delft Hydraulics Laboratory, Report 5487-IV*

Van Rijn, L.C., 1986, Sedimentation of Dredged Channels by Waves and Currents, ASCE, *J. Waterways, Port, Coast. and Offshore Eng*, 112, 5:541 - 559.

Vellinga, P., 1982, Beach and Dune Erosion during Storm Surges, *Coastal Engineering,* Vol. 6, no. 4.

Vellinga, P., 1983, Predictive Computational Model for Beach and Dune Erosion during Storm Surges, *Delft Hydraulics Laboratory, Publication no. 294.*

Vellinga, P., 1984, A Tentative Description of a Universal Erosion Profile for Sandy Beaches and Rock Beaches, *Coastal Engineering*, Vol 8, No. 2.

Watanabe, A., 1985, Three-dimensional Predictive Model of Beach Evolution around a Structure, *Proc. Water Wave Research Conf,* Hanover, West-Germany: pp: 123 - 141.

Watts, G. M., 1954, *Laboratory Study of Effect of Varying Wave Periods on Beach Profiles*, U.S. Army Beach Erosion Board, TM-53.

Watts, G.M., & Dearduff, R. (1954*), Laboratory Study of Effect of Tidal Action on Wave-formed Beach Profiles,* U.S. Army Beach Erosion Board, TM-52.

Wiegel, R.L., 1964, *Oceanographic Engineering*, Prentice-Hall, Englewood Cliffs, N.J.

PART IV

RETREATING SHORELINES

Chapter I : Coastal Erosion

PREFACE

Coastal erosion, of beach and cliff alike, is not only the result of a complex interaction of natural processes, it is furthered, markedly, by man's actions. The so-called anthropic factor takes several aspects. It results from a general migration towards coasts and ensuring demands for space and habitat. The explosion of tourism and water-related recreation has increased the stress placed on the coastal zone. A search for easily accessible materials, sand and gravel, has contributed to exploitation of marine aggregates. Marine coastal mining has come of age. Railroads and highways have been built close to shorelines. Dunes have been bulldozed to accommodate developments. It has been reported that not a single stretch of the Italian coastline has been left unmodified by man.

Railroad trestles across gullies and cañons, or crossing channels, can be built to span channels without altering much drainage patterns (example A.T. and S. RR along the California coast). However, when highway construction requires installation of culverts to control surface runoff, consequences are considerable as erosion of gullies is intensified. In California new canyons were cut as a result. Bluffs can be degraded by cuts made in rocks or drain pipes laid along former drainage paths. Bluff-top grading for construction, removal of protective vegetation and case-hardening surface materials affect erosion progress.

Beach mining for sand or gravel or cobbles results evidently in the removal of a volume of material but, additionally, and as a consequence sand mining contributes to beach erosion. This involves off-shore as well as onshore sand removal and has been illustrated by a long list of cases ranging from Lebanon and Israel to Morocco and the United States. On the Oregon Coast (Siletz Spit) 72,000 m^3 (2.543 million cu.ft) of sand were removed from Schoolhouse Creek beach : reduction of sand volume reduces its buffer effectiveness for the land. Prior to the mining operations the beach was stable. The mining permit has not been renewed since 1972. The demand for construction and road paving material has encouraged off-shore dredging. Beach exploitation is common. The environmental consequences have been recently reviewed by Charlier (1990). However, destruction of coral reefs to supply ornamental material is equally damaging, both ecologically and dynamically, as it enhances beach erosion by removing "natural breakwaters".

The chapters that follow stress the natural factors of coastal erosion, an inexorable phenomenon, and leave aside man's impact, except where directly pertinent. It will consider causes and processes of erosion, beach response and approaches to alleviate the problem.

Groins were probably built in the United States by private parties to protect their property, while the Federal Government showed only concern for its own property. After World War II, both the individual States and the Federal Government became involved in controlling erosion. According to Per Bruun, "Structurally the art of coastal protection suffered shortcomings [in the United States] compared to the low countries in Europe. [...] the difference between the European low countries lies "in the scale" and in "the degree of involvement". The European is [...] complex ... and the American is relatively simple".

Coastal protection became increasingly a concern for governments during the mid-18th and 19th centuries. Massive engineering works best characterize the approach, but in the more recent decades, artificial nourishment has gained in favor, the more so since cost assessments and improved techniques make it an attractive alternative.

1 Introduction

Coastal erosion, of beach and cliff alike, is not only the result of a complex interaction of natural processes, it is furthered, markedly, by man's actions. The so-called anthropic factor takes several aspects. It results from a general migration towards coasts and ensuring demands for space and habitat. The explosion of tourism and water-related recreation has increased the stress placed on the coastal zone. A search for easily accessible materials, sand and gravel, has contributed to exploitation of marine aggregates. Marine coastal mining has come of age. Railroads and highways have been built close to shorelines. Dunes have been bulldozed to accommodate developments. It has been reported that not a single stretch of the Italian coastline has been left unmodified by man.

Railroad trestles across gullies and cañons, or crossing channels, can be built to span channels without altering much drainage patterns (example A.T. and S. RR along the California coast). However, when highway construction requires installation of culverts to control surface runoff, consequences are considerable as erosion of gullies is intensified. In California new canyons were cut as a result. Bluffs can be degraded by cuts made in rocks or drain pipes laid along former drainage paths. Bluff-top grading for construction, removal of protective vegetation and case-hardening surface materials affect erosion progress.

Beach mining for sand or gravel or cobbles results evidently in the removal of a volume of material but, additionally, and as a consequence sand mining contributes to beach erosion. This involves off-shore as well as onshore sand removal and has been illustrated by a long list of cases ranging from Lebanon and Israel to Morocco and the United States. On the Oregon Coast (Siletz Spit) 72,000 m^3 (2.543 million cu.ft) of sand were removed from Schoolhouse Creek beach : reduction of sand volume reduces its buffer effectiveness for the land. Prior to the mining operations the beach was stable. The mining permit has not been renewed since 1972. The demand for construction and road paving material has encouraged off-shore dredging. Beach exploitation is common. The environmental consequences have been recently reviewed by Charlier (1990). However, destruction of coral reefs to supply ornamental material is equally damaging, both ecologically and dynamically, as it enhances beach erosion by removing "natural breakwaters".

The chapters that follow stress the natural factors of coastal erosion, an inexorable phenomenon, and leave aside man's impact, except where directly pertinent. It will consider causes and processes of erosion, beach response and approaches to alleviate the problem.

1.1 The Beach

Beach materials originate far inland, carried by rivers and streams to their resting place, from coastal formations erosion, and sometimes by onshore movement of deeper water sediment. Beach sediments vary in sizes from fine sands to cobbles, they can be arenaceous, volcanic or biogenic (shell-, coral reef fragments), rarely silt or mud. Sand particles sizes and distribution, sand composition, berm type, slopes of foreshore and inshore zone, off-shore bars are as many beach characteristics. Geomorphological traits include lagoons, inlets and barrier beaches. Dunes built by winds blowing over foreshore and berm carry landward beach sand, and constitute a natural levee sometimes anchored by grass, bushes, even trees. They provide a reservoir of beach sand and thereby a protective barrier against wave attack

A beach must constantly adjust to wave energy and water levels. It must sacrifice some beach, occasionally dune, to provide material for the building of an off-shore bar. The beach berm, built by waves to the highest elevation that can be reached by normal storm waves, is another protective barrier, whose value is reduced when its sand is carried off-shore.

Overtopping of a dune results in landward transport of beach and dune sand, thence lost to the dynamic beach system.

1.2 Causes of Beach Erosion

The term coastal erosion applies to the shoreline proper and to a strip of seafloor immediately bordering on the coast. The most "visible" manifestation is coastline recession, a phenomenon which is our primary concern here. Cliff erosion in hard rocks is less obvious than for unconsolidated beaches. Coastal erosion - and accretion - is subject to significant changes.

Natural causes of shore erosion include sea-level rise, variation in sediment supply, storm waves, waves and surge overwash, deflation, longshore sediment removal and transport, and sorting of beach sediment. Such erosion is exacerbated by anthropic actions.

Other factors responsible for erosion are winds, tides, subaerial processes, petrologic nature of the coast, climatic, chemical, mechanical conditions and biological agents.

Anthropic causes of erosion include removal of subsurface resourses leading to land subsidence, erection of structures interrupting transport of material, that reduce sediment supply, by improving inlets, channels, dredging and implanting "hard" coastal protection structures, concentrating wave energy on the beach, increasing water level variations, modifying nature's coastal protection arrangements and "mining" the beach for its materials or for minerals they contain.

Classical examples of coastal erosion by landsliding are provided by the Vesuvius slope and the Rissa event in Norway. Such coastal landslides can be due to man's interference. Removal of groundwater and lateral support by wave erosion cause repeated damage to beach line property in Seal Cove (California).

The coast of Bulgaria, north of Varna is another example of erosion by landslides. Material that slides into the sea acts as temporary natural breakwaters.

Erosion is not due everywhere to a rise in sea-level. The phenomenon appears to play a more important role, for instance, on the East and Gulf coasts of the United States than on its West coast. Major causes of erosion are waves, storm surges, nearshore currents, and tides. Spring tides bring about higher water levels. Storm surges have a similar effect ; their impact on erosional conditions are particularly important on hurricane-prone coasts. Rip currents promote erosion as well. The usual response of a beach is sand loss, migration off-shore and berm-width decrease. Critical conditions for erosion involve wave steepness, sediment fall velocity and beach-bed slope.

The supply of beach sediment is affected by an area's coastal geology which determines the initial conditions for littoral processes. Pre-existing natural beach topographies influence sediment transport within the surf zone. A close relationship exists between scales, frequencies and intensities of surf-zone motion and present morphology. In a beach's sediment budget the losses, or debits, include longshore transport downdrift, wind transport out of area, off-shore transport, deposition in submarine gullies, troughs and canyons, solution and abrasion, and mining.

2 Waves

Waves are of several types. Each has a specific impact on the coastal lands upon which such waves break. Tsunami are rare and, therefore even if very destructive, they are not a significant factor in coastal erosion.

2.1 Wind-Waves

Wind-generated waves can be a primary cause of erosion themselves, or, due to the energy they have stored, they can create various nearshore currents and develop specific sand transport patterns. The amount of energy stored depends on the duration of the storm that produced the waves and the extent of the fetch area. In the wave energy (E) equation (1.), ρ is the water density, g the acceleration due to gravity and H the wave height :

$$E = \frac{1}{8}\rho g H^2 \qquad (1)$$

If then C_n is the group velocity of the waves, in which n is the rate of energy transfer, a function of water depth (n = 1/2 in deep water, n = 1 in shallow water), the flux of energy (P) is

$$P = Ec_n \qquad (2.1.)$$

or

$$P = (\frac{1}{8} n g H^2)\, C_n \qquad (2.2.)$$

The term wave power has been used in studies dealing with harnessing it to produce power : P is the power per unit crest length delivered to the coast when waves break upon reaching the nearshore.

Designating by h the water depth, the velocity C is given by :

$$C = \sqrt{gh} \qquad (3.1.)$$

$$C = \sqrt{g(h+H)} \qquad (3.2)$$

Waves are bent as they near shore, due to bottom topography, and with a the angle between wave crests and shoreline or isobath, (2) becomes

$$P = (E\,C_n)\cos\alpha = \text{constant} \qquad (4)$$

The position of a breakers zone on a given beach can be determined with a fair approximation by

$$\gamma = \frac{H_b}{h_b} \cong 0.75 - 1.2 \qquad (5)$$

where γ is the ratio of breaker height (H_b) to depth (h_b), and with H_b predicted by following relationship between known deep-water wave height (H_0) and period (T) :

$$H_b = 0.39 g^{\frac{1}{5}} (TH_0^2)^{\frac{2}{5}} \qquad (6)$$

This value can be adjusted to include wave refraction effects. The nature of the waves breaking is thus a function of their initial cambre (or steepness) given by H/L and the beach slope, which is of paramount importance for the value of γ.

The width of a breaker zone, an area in which waves of varying heights break at specific water depths, depends on beach slope and wave size range.

Beach-sediment's grain size, and waves energy level are determinant in over-all beach slope : the coarser, the steeper the slope, as breaking occurs at the foreshore's (beach face's) base. Beach profile morphology is influenced by wave energy level, as high-waves shift sand off-shore, thus eroding the beach berm and redepositing the material in the breaker zone thereby building off-shore bars.

Seasonal profiles (cf. further section 3) can develop as wave heights vary and are referred to as swell profile and storm profile. These changes are very important : the storm profile lets the wave swash attack more directly the coast. It behooves us to recall that broken waves, on small gradient beaches, become bores which cross the surf zone and become swash as they rush up the beach face (figs. 1 and 2).

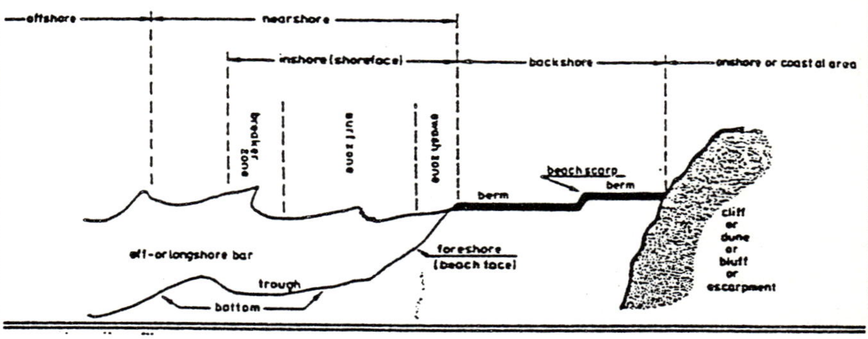

Fig. 1a : Beach profile terminology

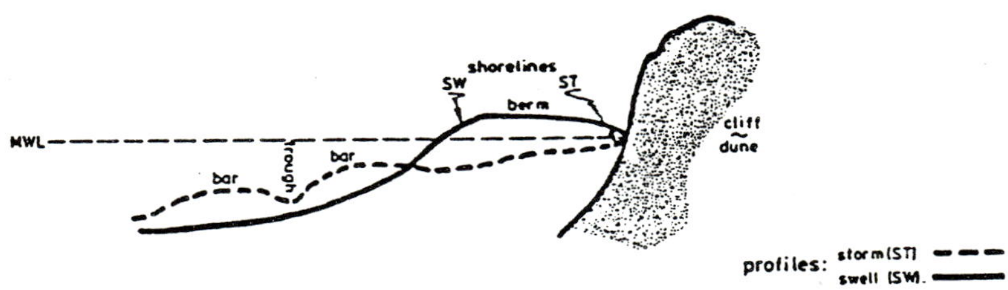

Fig. 1b : Storm and swell profiles

2.2 Storm Waves

Storm waves arriving at relatively steep beaches, very variable in direction, in numerous wave trains, will result in a crest reaching the beach almost every second. As a consequence large volumes of water are constantly poured onto the beach face, which becomes saturated ; water table and beach face coincide, backwash nearly equals uprush. Material is carried seaward, by the downflow, from the beach face surface, furthermore undermined by the large hydraulic jump at its foot (fig. 2).

Storms with a recurrence interval of a year can erode as much as 4200 to 21.000 m³/km (10,000 to 50,000 cu.yd/mi), a large amount, but modest compared to longshore transport that can surpass, for the same area 765,000 m³/yr (1,000,000 cu.yd/y).

A storm can thus remove in record time a large segment of a beach because at the mean water-line the increased ground-water flow goes seaward, its up- flow tendency provides a quicksand effect. An off-shore bar can even be built as the sediment-loaded return flow, of above normal density, travels outwards near the bed, a process encouraged by the action of onshore winds creating a strong surface shoreward drift.

Material accumulation on the bar persists till it has reached a height sufficient to break most incoming waves. The volume of the material required depends upon the off-shore depth to be filled. Bar formation may increase beach erosion : water keeps circulating from beach to bar to beach with a surface inward flow and a storm-profile apron bed outward flow. Material is kept suspended but the bar is consolidated at the beach's expense. A continued pattern of storms nets a beach erosion.

Though the foreshore and the beach face are considered synonymous by some authors, U.S. Army Corps of Engineers texts point out that during storms the beach face (*sensu stricto*) is moved shoreward by the cutting action of the waves on the profile and cuts a scarp into the berm.

As a result of a heavy storm, or several mild ones, the backshore may be completely eroded, allowing waves to erode dunes, cliffs, or even the land behind the beach.

2.3 Standing Waves

Standing waves, as well as storm waves, promote erosion. The beach slope determines the degree of wave reflection, and the coarser the material transported the steeper the beach face ; the greater the reflection, the stronger the wave capacity to suspend and roll sediment. Reflection at walls (man-made) and cliffs (nature-built) create angled wave-trains, and survival of beaches at their front is very difficult.

Any condition promoting water saturation of the beach, destabilizes it. Among these are disposal of storm waters from catchments, sub-beach rock shelves concentrating ground water, seaward run-off, longshore walks blocking temporary uprush of water during storms. Accumulation of water against a vertical wall generates standing waves which increase the rate of sand removal.

2.4 Swell Waves

Swell waves are beach builders as they transport material landward, cover the beach face in an uprush, water percolates down through the sand to the water table, and diminished speed reduces carrying power so that suspended material is deposited (fig. 2).

Beach accretion will be pursued as long as material is available off-shore to feed incoming breakers. Material deposited is slightly compacted.

It is not far fetched to link this phenomenon and the feeder berm technique recently developed. Once swell waves can no longer provide material, the beach profile becomes virtually static from berm to off-shore ; they compact the beach face by pounding it.

An oblique persistent swell will move sediment downcoast.

Swell waves return sand to the beach, often by bar decapitation. While the beach is healed, rip currents, a swift flow fed by longshore currents on either side, return to the sea water accumulated in the trough.

The intensity of erosion is a function of swell profile steepness. The more time swell waves have to sweep off-shore material to the beach, the greater the beach recession the next storm season.

"On land" structures should only be built at such distance of the shoreline that the beach acts as a "bank" that can "feed" the bar.

As R. Silvester stressed *at libitum*, sufficient width of sedimentary material must be available for bar formation to defend a coastline against erosion.

Waves reaching an "obstacle" will react in a variety of ways the two extremes being fully dissipative (in case of flat shallow beaches that have large sub-aqueous sand storage) and highly reflective (in case of steep beaches with small storage).

2.4.1 Edge Waves

Reflective beaches are associated with beach cusps and subharmonic (having twice the period of the incident waves) edge waves. Shallow beaches are associated with wide surf zones of turbulent dissipation.

A surf-scaling parameter provides morphodynamic distinctions.

If a_b is the breaker range

ω the incident wave radian frequency, with T the period $\omega = \frac{2\pi}{T}$

g acceleration due to gravity

β beach/surf zone gradient (or slope)

ε the surf-scaling parameters

then

$$\varepsilon = a_b \, \omega^2 / g.\tan^2\beta \qquad (7)$$

Total reflection occurs for e < 1, but as long as e < 2 to 2.5, strong reflection will still take place, thence standing wave motion, surge, resonance. Waves dissipate energy at $\varepsilon > 2.5$ and for $\varepsilon > 20$ there are spilling breakers. When $\varepsilon > 30$ to 300, a storm profile has been reached, longshore bars are present and the dissipative stage is attained. It must be understood that dissipative and reflective elements may co-exist at some stages.

"The most pronounced longshore segregation of dissipative and reflective regions occurs in association with the 'transverse bar-and-rip topography" (Wright and Short).

Though beach erosion accretion depends greatly on waves' height and steepness, a major role is also played by the state of the beach immediately before, and while, an event takes place. Hence it is possible that specific wave conditions erode reflective beaches while they accrete dissipative ones nearby. In theory the fully reflective beach state is the optimum condition for accretion, yet it is most susceptible to erosion in fact, because it lacks subaqueous sand reserves and because reflection may lead to growth of subharmonic edge waves which promote run-up, thence beach face scarping, overtopping, cutting of the berm. Large, spaced erosional beach cusps form. As long period, not very steep swells, excite the subharmonic edge waves, this type of erosion requires minimal energy.

Storm waves' maximal energy is at the breakpoint and the minimum at the shoreline ; yet, storms often result in huge erosional losses at the latter. Edge waves are a type of motions that dominate inner surf zone processes.

Unlike incident waves, limited by breaking to the surf zone, edge waves can grow during storms and may dominate the zone. The broad aspect of the morphology after a storm is usually consistent with edge wave generation.

It could lead one to conclude that the dominant storm erosive mechanism be edge waves.

2.5 Tidal Effects

Tides, and superimposed storm surges, rise the water level, thereby flood low elevation coastal land and by pushing landward the surf zone, let waves attack directly set-back areas. Perigean spring tides, particularly at syzygy, raise considerably water levels entraining erosion. So do storm surges. Areas not affected by wave attack become suddenly subjected to it. Coastal dikes have been built as protective structures.

2.6 Nearshore Currents

Nearshore currents are created by wave breaking. Waves may strike a shore perpendicularly or at a more or less significant angle. One end of the spectrum of angles is waves parallel to shore - and dominant nearshore currents are a cell circulation with rip currents going seaward - the other with $\alpha_b > 5°$ to $10°$ - and longshore current is continuous along shoreline.

Rip currents are concentrated jets able to carry water seaward through the breaker zone, very noticeable when long, high waves produce wave set-up on the beach, *viz.* a superelevation of the water surface over normal surge elevation due to onshore mass transport of the water by wave action alone.

Some other localized sea-wall flowing currents are due to concentrated flows down gullies in the foreshore (beach face), and still others are caused by interaction of wave- and edge-wave phenomena.

Generally, rip currents carry sand off-shore, even beyond the breaker zone and create embayments of different extension in the process. When these cut through the entire beach berm, substantial erosional effects ensue.

A longshore current, being continuous, it can carry (littoral drift) materials over quite long distances. Littoral drift value can vary widely from storm conditions to normal situations. Any structure placed in its way will cause accumulation on one side and erosion downdrift.

A frequently used littoral drift formula is given by (9) in which Q_S is the volume in m³/day units, and P_e, improperly named the longshore component of wave power, is in W/m units, evaluated with the square root-mean-square height (H_{rms}) :

$$P_e = (EC_n)_b \sin \alpha_b \cos \alpha_b \tag{8}$$

$$Q_S = 6.8\, P_e \tag{9}$$

It permits to estimate sand quantities topped by hard structures, the effects of sediment dredging on the inner (*viz.* shallow) continental shelf on adjacent beaches, and effects of longshore variations in the littoral drift.

2.7 Sea-Level Changes

Sea-level changes are caused by tides and storm surges, but we are accustomed to use the term in connection with a world-wide rise in the level of the oceans, a serious treat to coastal zones. Indeed, over the last 20,000 years that level has risen by more than 100 m (328 ft).

Sea-level changes are not a XXth century scourge. They have occurred rather frequently during geological history. Major level variations, perhaps during the Cenozoic, certainly thereafter, have been the result of glaciations and warming-up stages, and the contemporary level is comparable to that attained during earlier interglacial periods. A rise of 100 m (328 ft) since the Wisconsinian cf.ms a conservative estimate. Annual variations are related to long-term weather "tides". Longer trends are still discussed as to their causes. It has been suggested that the eustatic level rise may reach 1 cm/year (0.4 in/y). In some places the rate is double : in Louisiana subsidence (both natural and man enhanced) worsens the situation. The erosion at a given site is gauged by the net-sea-level change, the sum of the eustatic rise and local land-level changes. Indeed in some areas subsidence has compounded the effects of the global change (Louisiana, Galveston TX) and in others (Juneau, AK) land rise results in a lowering of the mean sea-level.

Global warming taking its toll, some authors (V. Gornitz *et al.*, 1982, Global sea level trend in the past century : *Science* 215, 1611) project a 40 to 60 cm (± 14 in to ± 22 in) sea-level rise by 2050 ; it could even be 1 m (35 in) or more in areas where subsidence is important. Some have cf.n Charleston NC under water by 2020.

3 Coastal Profiles

Profiles are two-dimensional vertical sections showing elevations-to-distance variation. Coastal profiles, usually perpendicular to the shoreline, include shelf-, nearshore, and beach profiles. The latter two are of major importance in coastal protection schemes.

Profiles are needed from both zones because modalities are clearly different in one and the other.

While hydrographic charts are adequate to make a profile for deeper offshore zones, recent surveys are necessary for waters 20 m (66 ft), and less, deep.

Profiles are needed because of long-term stability of the swell profile and short-term transposition from swell to storm beach profile.

The envelope of a profile defines fluctuations of sand level at a site, data that is needed for beach nourishment as well as groin design.

The nearshore profile is usually measured aboard with an echo-sounder or a lead-line. The beach profile is surveyed by leveling and taping techniques. the latter is less affected by such errors as sounding, spacing, closure and sea bottom temporal fluctuations, than the nearshore profile.

3.1 Beach Profile

Profiles are two-dimensional vertical sections showing elevations-to-distance variation. Coastal profiles, usually perpendicular to the shoreline, include shelf-, nearshore, and beach profiles. The latter two are of major importance in coastal protection schemes.

Profiles are needed from both zones because modalities are clearly different in one and the other. While hydrographic charts are adequate to make a profile for deeper offshore zones, recent surveys are necessary for waters 20 m (66 ft), and less, deep.

Profiles are needed because of long-term stability of the swell profile and short-term transposition from swell to storm beach profile.

The envelope of a profile defines fluctuations of sand level at a site, data that is needed for beach nourishment as well as groin design.

The nearshore profile is usually measured aboard with an echo-sounder or a lead-line. The beach profile is surveyed by leveling and taping techniques ; the latter is less affected by such errors as sounding, spacing, closure and sea bottom temporal fluctuations, than the nearshore profile.

When a beach is subjected to constant wave action for a sufficiently long time, an essentially stable beach profile is established. The form of this profile is a function of the wave parameters (usually wave height and wave length) and the size characteristics of the beach material (usually the average grain size and the relative proportions of difference grain sizes).

Several workers (Swart, 1974 ; Dean, 1984 ; Vellinga, 1984) have sought to establish a mathematical relationship between these parameters.

In general, a proportionality has been found between the depth of water and the distance from the limit of uprush. Although an average profile related to normal conditions is often referred to, the profile of the beach is never fully fixed because of variations in wave action changes considerably and the limit of uprush (storm surge level) rises up the beach which is then eroded. The new profile can be calculated By inserting the appropriate values for the wave parameters into the mathematical relationships.

The beach profile is the intersection of the ground surface with a vertical plane ; it may extend from the top of the dune line to the seaward limit of sand movement. A profile perpendicular to the shoreline reflects the action of littoral processes ; all features are not necessarily shown, though dune, berm and foreshore are. Across a beach, the profile adapts to imposed wave conditions.

The typical beach profile left by a severe storm is a simple, concave-upward curve extending seaward to low-tide level or below.

It should never be lost sight of that beach profiles are a momentary picture, not a long-term condition. Seasonal profiles, storm and swell, can develop earlier.

3.2 The Equilibrium Profile

The equilibrium profile being produced by a specific wave climate and the type of coastal sediment, the construction of a structure on-shore or across a river will affect wave height, sedimentation and if beach nourishment is carried out, sediment nature. A "probable" equilibrium profile will help determine fill volume needed to achieve stability.

The equilibrium beach profile, an idealized condition, practically never attained, is a useful interpretation tool.

It has been mathematically represented by :

$$h(x) = A\,x^{m} \tag{10}$$

with h the water depth and x its distance from offshore, A and m parameters depend respectively on sediment size and shape. Values of A = ($length^{1/3}$) and m = 2/3 are accepted. A, a scale parameter, varies with $Ø_{grain}$ according to a gentle convex parabole : for A (in m $^{1/3}$), Ø (in mm) is 0.1 and 1000, 0,1 and 1.0.

A steeper than normal beach profile section indicates "erosion", a flatter than normal "accretion". Several factors enter in the calculation of a beach profile : mean energy per unit area of sea, mean sediments characteristics, wave length or wave period. A statistical value for energy fluctuations throughout the year (e.g. mean of highest third of peak energy) will have to be used.

To obtain a "normal" or "standardized" profile, measurements should be made at the end of a lengthy swell period. The different nature of predominant waves necessitate separate calculations for oceanic (usually "open") and enclosed sea margins.

Material will be retained as compared to the original material ; inversely, a fractional value of R indicates how many times less often such fill will have to be replaced than the original materials.

This model establishes the frequency of renourishments needed to maintain a stable beach. On the other hand, the fill-factor model attempts usually to determine the quantity of fill material needed for a unit volume of grain size distribution as the natural material. Both aim at providing beach stability, but the latter method considers the original material as optimal for the beach and thus thinks in terms of such size material as fill, thus to maintain the status-quo.

3.3 Theoretical Analysis

A theoretical analysis has been given by Silvester (*op. cit.*)

Sitarz [J.A. Sitarz, 1963, Contribution à l'étude de l'évolution des plages à partir de la consistance des profiles d'équilibre. *Trav. Centre Etud. Rech. Oceanogr., 10-20*] has developed equations, based upon energy principles, for profiles beyond the breaker-line to the limit of active bed disturbance. He has determined a relationship for the width of the surf zone for swell and storm-wave profiles, plus the cross-sectional area of the water in this zone, from which its average depth can be derived. The offshore profile is given by :

$$x = ky^2 \quad \text{where} \quad k = 0.95/[(s-1)^{1/2} D H^{3/2}] \qquad (11)$$

where x is the distance seawards from the breaker line ; y is the depth below SWL (Still Water Level) ; s is the ratio (density of sediment)/(density of seawater) = 2.65 ; D is the median sediment diameter at some appropriate depth (in mm) ; and H is the wave height just offshore from the breaker zone (in m).

The above equation indicates that the milder slopes occur with larger values of k, which in turn result from finer sediment or smaller wave heights. The usual value of s (= 2.65) is applicable to most sand situations. It appears strange that wave period does not enter this relationship, since it determines in part the water-particle oscillation. Also, the mass-transport "velocity" at the bed is strongly dependent upon period or depth to wave length ratio. Taken overall it is period that dictates the refraction pattern and hence the proportion of the deep-water height occurring near the breaker zone.

To complete the swell-built profile it is necessary to continue the offshore bed slope or curve up the beach face to the berm. The beach face is best measured at the site, but some general values are graphed in fig. 2 for three conditions of exposure. The D (mm) in fig. 2 represents median diameters of sediment particles in the swash zone. The berm height should be measured where possible, since calculations contain too many variables for practical purposes.

The width of the surf zone (Z) is given by :

$$Z = BH^{3/2}/[(s-1)^{1/2}D^{1/2}] \qquad (12)$$

where H and D have dimensions similar to those in equation 11, although D represents in this case the median diameter of sediment in the beach face (near the SWL).

For the swell-built profile factor B above has the value 43.5, whilst for the storm-wave surf zone it is 75.0. In this latter case, a full spectrum of waves is implied, so that H must be representative of this. Bruun [P. Bruun, 1963. Longshore currents in one and multibar profiles relating to littoral drift. *Proc. 8th Conf. Coastal Eng., 211-247*] has employed $H_{1/3}$ for such a purpose.

The cross-sectional area of water in the storm beach surf zone is given by Sitarz (*op. cit.*) as :

$$A = (3.4g^{1/2}H^2_{1/3}T_{1/3})/[(s-1)^{1/2}D^{1/2}] \qquad (13)$$

where $T_{1/3}$ is the significant wave period ; g is 9.81 m/sec² ; D is the median sediment diameter across the surf zone ; and $H_{1/3}$ and s are defined as before.

Having established the swell profile existing at any time, either completely by survey, or partly by survey and partly by computation, the width Z and depth d_s of the probable storm profile can be overlain so as to predict the likely beach degradation. In this respect the slope of the storm beach face, which is very steep, is of little consequence compared to the other relevant dimensions.

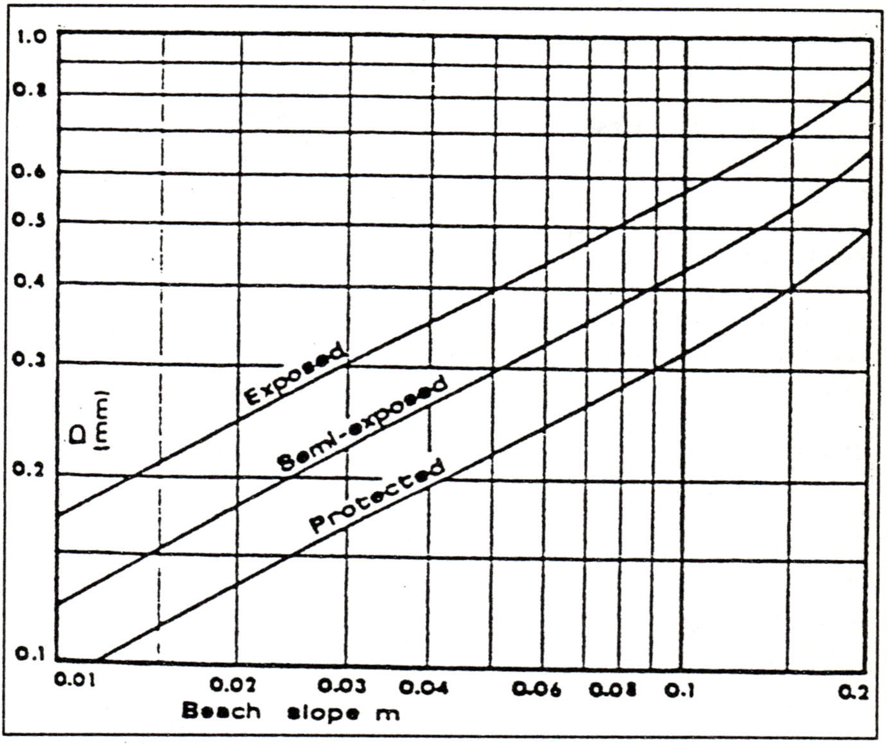

Fig. 2. : Slopes of beach face for different exposures of waves

3.4 Profile Nourishment

In Chapter II, several methods of coastal protection will be discussed ; among those, artificial nourishment is in front. However, such nourishment can be accomplished according to several options. Profile nourishment is one such option analyzed.

4 Coastal Erosion, a Worldwide Phenomenon

Coastal erosion is a world-wide occurrence principally along sea shores but noticed also along many a lake shoreline. It has been reported in the literature for several decades (e.g. Anonymous, 1957 ; Charlier 1955). It encompasses considerable economic consequences, endangers for instance valuable properties.

The rise in water level is a major cause of the phenomenon but anthropic actions have also played an important part.

Stopping coastal erosion and halting shoreline retreat is considered impossible. Reducing the rate of erosion and restoring endangered areas, on the other hand, falls within the reach of coastal engineers.

In February 1989, three streets, three blocks and seven rows of houses were abandoned in Sargent Beach TX, after a losing battle against erosion. In Amélie, a subdivision of Soulac, France several planned streets are now in an area covered by the sea and the present church is doomed to be engulfed within a few years.

The United States government has, for several decades, provided "catastrophe" insurance for constructions destroyed by storms or placed in jeopardy by coastal erosion. It has financed, to a considerable extent, protective devices such as groins, sea-walls, breakwaters, and has undertaken artificial dune building and beach nourishment. However, in a rather recent policy reversal, it now has embarked on a "laissez-faire" approach, that is an attitude of "let nature take its course". Several states have adopted a similar attitude. Retreat form the shoreline is advocated, and sometimes financially encouraged (as in the State of Michigan). Construction set-back lines are advised or even enforced.

Such measures are conceivable where summer cottages line the shore, or even where touristic interests and investments are modest. On barrier islands off the Carolina coast there are hardly any takers for $ 250,000 houses discounted at one third off. A situation no different from that in the late fifties along the Atlantic coast of Long Island (N.Y.).

However, one can hardly imagine letting Miami Beach FL, Atlantic City NJ or Ocean City MD slowly "slide into the sea". The economic consequences would be of such magnitude that protective, and restorative measures, are required. If "hard" structures do not offer the appropriate solution, then "soft medicine" might be prescribed. The alternatives will be discussed further on.

A major effort to assess the damage brought about by erosion along the coasts of a large sector of Europe is known as the CORINE "Coastal erosion project". A part of the CORINE (Coordinated Information on the Environment), an activity of the European Union (EU), it is an inventory of coastal erosion problems and high risk coastal areas in the EU.

The shoreline is described as a set of contiguous coastal segments with three groups of variables : morpho-sedimentological characteristics of the coastal zone ; evolutionary trend of the shoreline (erosion, stability, accretion) ; presence or absence of coastal defense works. The coastline is broken up into homogeneous segments on the basis of morpho-sedimentological units. The smallest coastal segment can have a length of 100 m (328 ft) (e.g. pocket beaches, small harbors, embayments).

The following features and characteristics have been retained :

Morpho-sedimentological features : rocky coast, beaches (unconsolidated sediments) ; tidal marshes, waddens, polders ; artificial coastal segments.

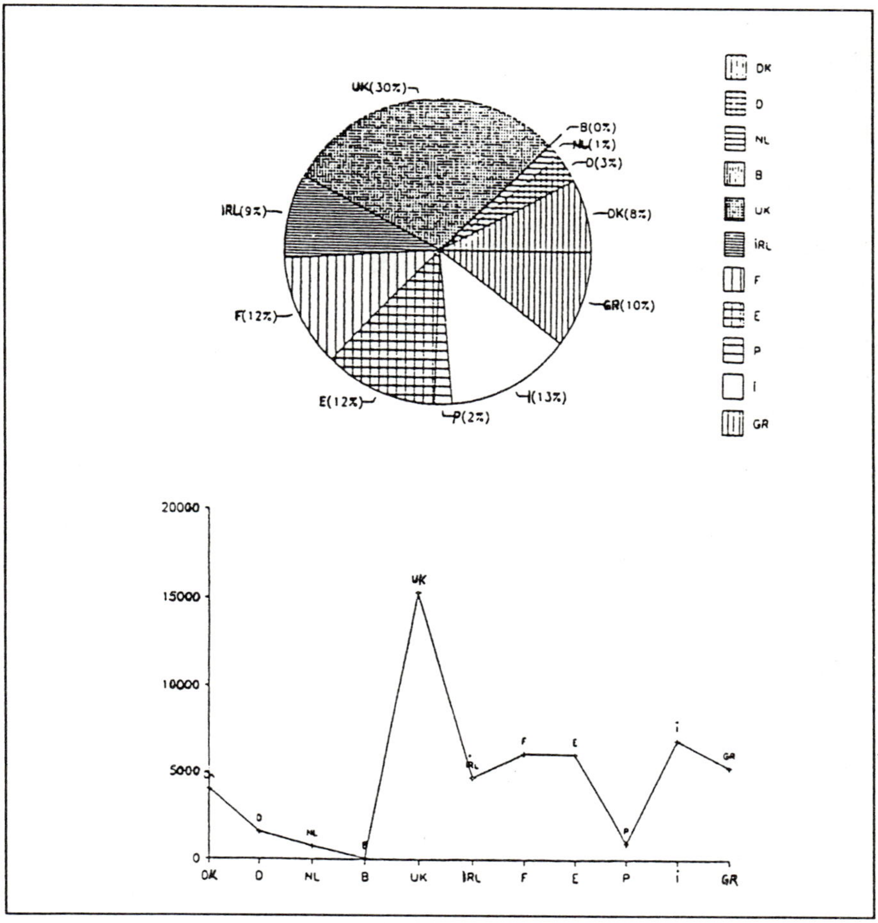

Fig. 3 : Distribution of the length of the coastline within the eleven coastal countries of the European Union.

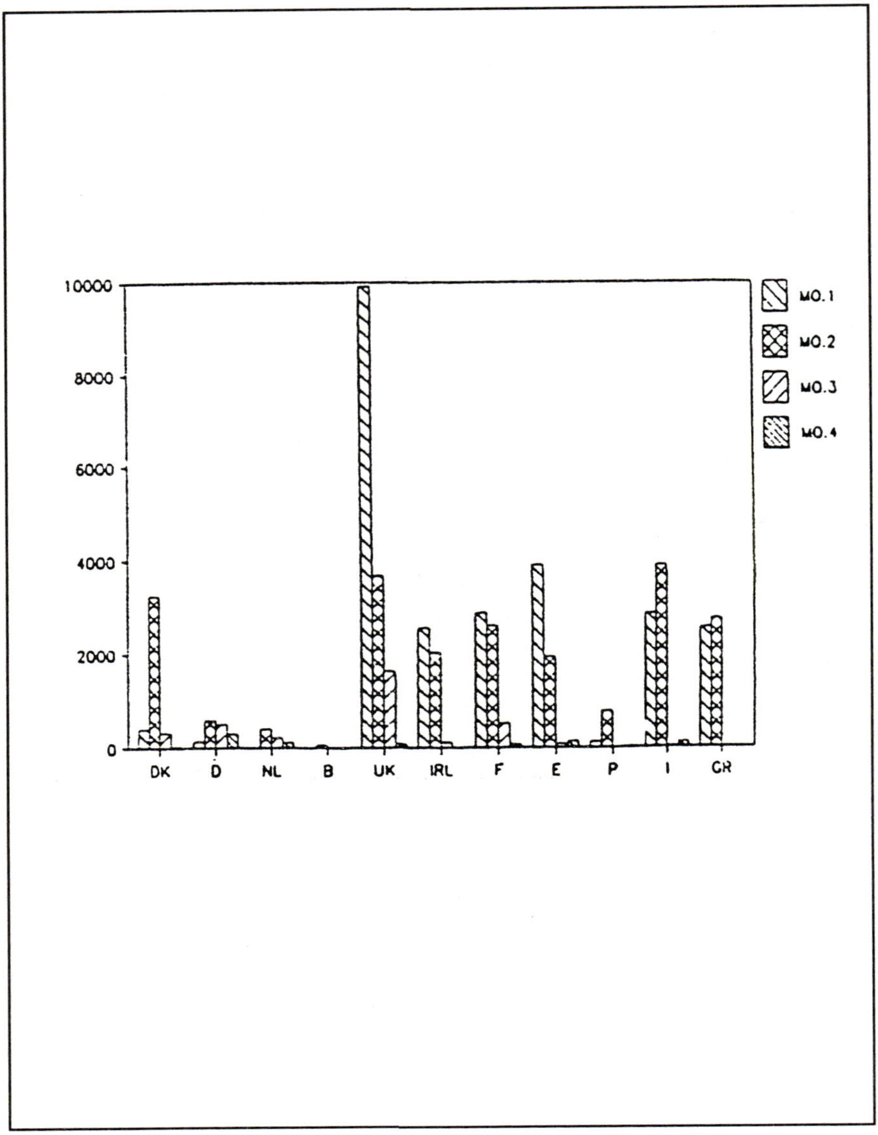

Fig. 4 : Littoral of the European Union distribution of morphological classes.

<u>Shoreline evolutionary trends</u> : trend unknown ; stability ; erosion ; accretion.

These have allowed to determine current erosion in the EU countries.

Preliminary results thus show that a significant part (28%) of the coastlines of the EU suffers from erosion.

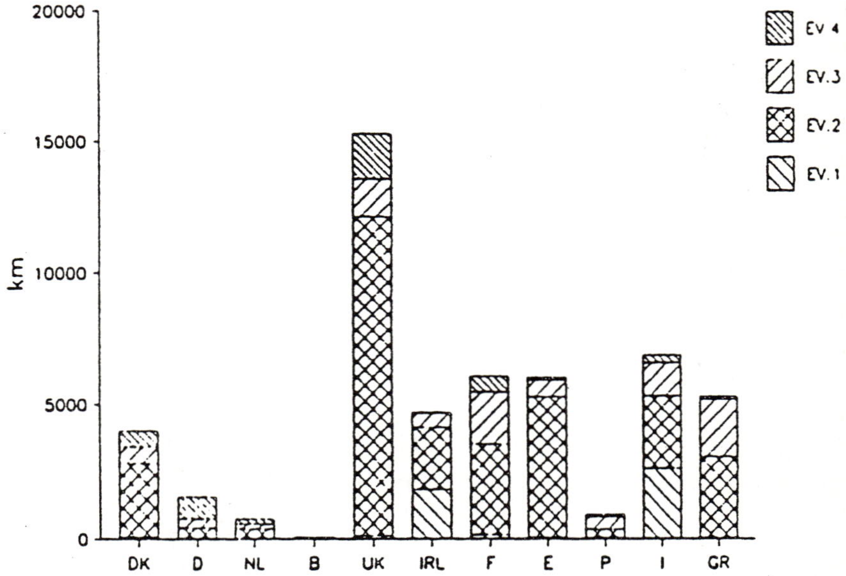

Fig. 5 : Coastal evolution

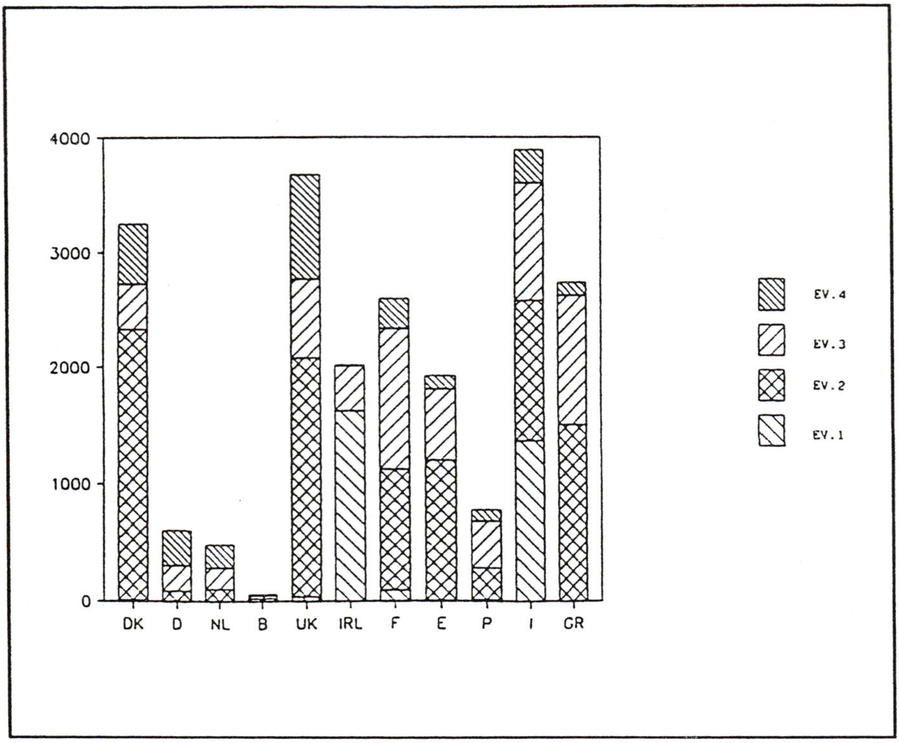

Fig. 6.1. : Beach evolutionary trend in the European Union

	Length (km)	Erosion (%)
Denmark (DK)	3.253	12
F.R. Germany (D)	605	36
The Netherlands (NL)	407	45
Belgium (B)	64	40
United Kingdom (UK)	3.680	19
Ireland (IRL)	2.019	19
France (F)	2.600	46
Spain (E)	1.929	32
Portugal (P)	782	51
Italy (I)	3.894	26
Greece (G)	2.739	41

Table 1 : Beach erosion in the EU.

Chapter II : Coastal Protection

1 Introduction

1.1 Beach Dynamic Processes

Unconsolidated beaches result from a building process linked to accumulation processes and appear as sandy beaches (less than one third of world beaches) or muddy beaches (either colonized by mangroves or in sheltered districts). The principal types are dunal systems, estuarine environments, littoral spits, lagoons and ponds.

Beaches are built of gravel, shingle, shingle upper/sand lower, mixed shingle/sand or sand, silicious, carbonaceous or volcanic.

A shingle beach can be re-orientated and drawn down rapidly by storm waves, but very little of the material is drawn out beyond the breaker line, and smaller and less steep waves tend to restore the profile. The slope of a shingle beach in the inter-tidal zone is typically about 1 : 8 which may be drawn down to about 1 : 12 in extreme storm conditions.

A sand beach, on the other hand, continues to flatten under prolonged storm conditions with consequent continuous removal of material from the beach head. Material woves offshore as well as alongshore, and the beach requires much longer to return to a "normal" profile. It is possible for extreme storms to draw the beach sand into a water depth where milder waves do not influence the sea bed and from whence the material cannot be returned to shore. If this situation is combined with lack of material carried alongshore, artificial recharging of the beach may be required. For a comprehensive assessment of groin performance, it is necessary to consider the envelope of possible beach profile variation as well as the "normal" profile.

Today the majority of beaches, due to natural and/or anthropic action, are thinning or retreating. On the French Atlantic coast some 850 km (527 mi) of littoral lose about a meter (3.28 ft) each year and world-wide sea-level increases on the average from 12 to 15 cm (4.8 to 6 om) a year[1].

If in cases of mild erosion the best action is to take none, human action or ignorance of shoreline processes often necessitates decisive measures, particularly where economic values are at stake. Which erosion control measures have to be taken depends on the type of shore and land-use. Alternatives include construction of hard structures, artificial and induced natural beach replenishment, traffic restriction, vegetation protection, retreat and outright human occupancy abandon.

On the Lower Médoc coast of France the persistence and amplitude of coastal erosion demands decisive action, on a scale larger than for the other affected sectors of the Aquitany littoral. A retreat of 250 m (820 ft) has, for instance been recorded at Les Arros over the last century and a half, and south of Soulac 3.3 m (11 ft), on the average, are lost per year. Occasionally in a single month the dune's foot moves inland by 35 m (117 ft) (north of Amélie) (photo 1).

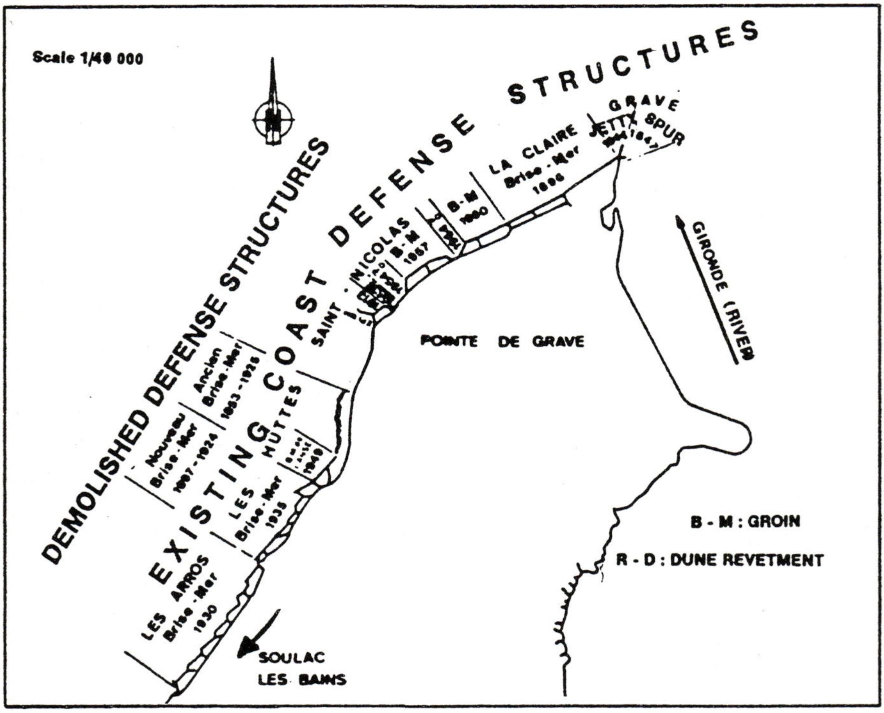

Photo 1 Shoreline protection - Pointe de Grave (after Laboratoire Central d'Hydraulique de France, 1979).

The Belgian coast has been subjected to severe erosion throughout history and human settlements have disappeared during historical times[2]. The Dutch coast has been similarly affected. So has the American shoreline retreated with barrier islands particularly endangered. Even the famed beach of Monte Carlo needs attention (fig. 7).

The Holderness Coast (Yorkshire, England) has retreated 1.6 m per hundred years for the last 2,000 years.

East of Ponce, on Puerto Rico, the coast made up variably of beaches, mangrove swamps, alluvial plains and rip rap positioned by man is eroding on 50% of its length.

Coastal erosion which is quite severe in the area of the Rosetta mouth of the Nile and the Ras-el-Bar, Egypt, has been shown by Apollo-Soyuz photographies.

In Italy, massive settlement of coastal regions, added to touristic and industrial use, enlargement of harbors, construction of coastal defenses, pumping of underground water and reclamation works have accelerated natural subsidence. Inland modifications for a variety of purposes have compounded the ensuing problem of Italian beaches[3]. According to a report of the Commissione di Studio Sistemazione Idraulica e Difesa del Sudo of 1970-1974, only rare stretches of the Italian coasts are still subject to natural conditions.

On the coast of Nigeria, south of Lagos and Bénin, coastal erosion causes beach retreat at a rate of 5 m (16.4 ft) a year. On the Indian Ocean, the East Coast of Africa has beach erosion problems, to a large extent man-made : dynamite fishing, sand mining, salt extraction, irrigation and building of coastal structures have contributed to beach erosion. They have led a.o. to degradation of the beach, destruction of nesting sites for turtles, destruction of houses and hotels and blocking of river channels (W.M. Kudoja, 1995, "Causes of beach erosion on the East Africa Coast", *Bordomer Abstr.*).

If natural causes bring about coastal erosion, and, undoubtedly, sea-level rise is leading to the "drowning of beaches", anthropic action plays a major role in shoreline retreat[4]. Coastal erosion and beach retreat are due to sea-level rise, storms, hurricanes and anthropic action. Man may accelerate the process by having created a greenhouse effect on earth and additionally by local actions. Roads, board-walks and buildings stand in the path of shifting sands.

Studies conducted at the University of North Carolina have revealed an accelerated deterioration of salt marshes along the coast of South Carolina. Here, the marshes already threatened by a relative sea level rise that averages 1.9 mm year over the last fourty years, are placed in additional jeopardy by sand removal for beach nourishment on the nearby developped barrier islands.

The International Geographical Union's Commission on coastal changes examined during the 1972 to 1984 span the world's coasts. It evidenced the important local effects of structural works and coastal land reclamation (e.g. polderization) on sea levels, and warned of the indirect effects of the climatic changes resulting from increased pollution and modifications in the global atmosphere.

Harbor structures, for instance in Seal Cove (California), by causing shore erosion, through their impact on the shoreline adjacent to them, are contributors to slope failure.

Using his observations at South Bethany, DL, Westgate warned those who want an unimpeded view of the sea, flattening dunes, and perching their buildings close to the edge of the beach that "geologic evidence points to increasing movement of the foredunes of the coastal beaches toward land, making additional development so close to the water a risky business[14].

In Tunisia, harbor building at Tabarka disturbed natural sand accretion and caused substantial transformations. A similar consequence resulted from portuary works at Bizerte. Though no loss of beach space has been reported so far along the Moroccan coast, removal of beach material for construction purposes between Rabat and Casablanca may create future problems and new sources of sand and gravel are sought for (personal communication with M. Bensaid, Int. Symp. of Comm. World Geol. Map, Sub-Comm. on Appl. Geol., Trondheim, Norway, May 1986)[4]. The tourism development plans of the Moroccan coasts ought to give due consideration to erosion potential along the Gibraltar Straits, particularly the stretch from Tangiers to Ksar-es-Seghir. Other caution requiring areas include the Fnidek to Tetouan region and the steep coast west of Oued Laou.

Off the coast of Lebanon and Israel, and too close to shore, dredging for construction material led to severe beach territory loss. Efforts were made here and elsewhere to counteract severe coastal erosion (fig. 7 and 8). Similar situations arose on Brunei, and Indonesia is alarmed over intensive coastal erosion.

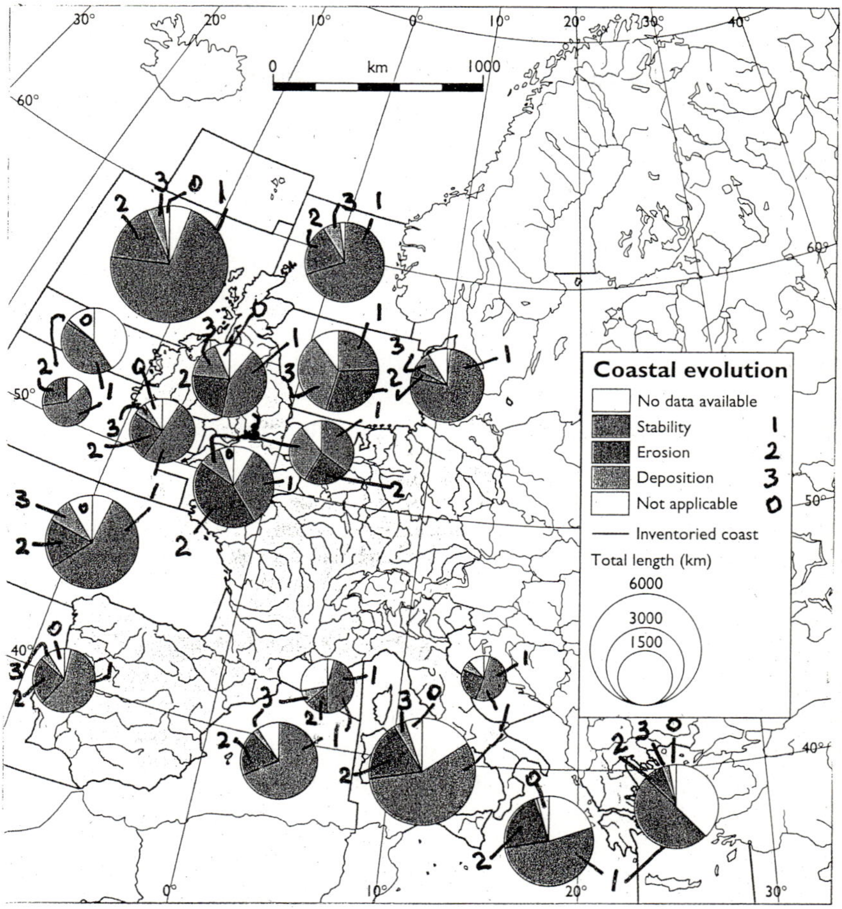

Source: CORINE

Fig. 6.2 : Coastal erosion evolution in Europe based upon an evaluation resulting from various documents of the European Environmental Agency.

If, on Bali, wave attack and natural causes are the culprits of the shrinking of the beaches of Uluwatu and Tanah Lot, and shore-line retreat at Siyut and Lebih can be ascribed to river mouth shifting, losses of beach width on Southern Bali are due to man's activities : tourism having made spectacular advances on this island, lengthening of the runway at Ngurah Rai airport appeared advisable, but 800 additional meters of runway jutting out at sea led to serious damage at Kuta Beach. Traditional coral extraction, and ensuing gradual destruction of the reef, have brought about beach regression at Batumadeg, Sanur, Nusa Dua, and intensified the phenomenon at already damaged Kuta Beach. Both coral and sand are used for ornamental and/or building material in local construction : removal of river material is the principal cause of erosion on Gumbrich Beach. The touristic demand for hotel rooms "on the beach" has led to the siting of such facilities and the constructions interfere with natural maintenance processes. Remedial action is urgently needed.

Extraction of coral poses equally severe, and solution-demanding problems, on the Maldive Islands.

A typical example of coastal erosion problem in developing countries is provided by Bénin. Geophysical factors, but far more the construction of the Akossombo (Ghana) barrage and of the harbors of Lomé (Togo) and Cotonou (Bénin), are responsible for the spectacular retreat of beaches on the Gulf of Bénin. Here erosion has been monitored since 1908 although observations of shoreline retreat were made since 1901 : the coastline migrated inland by at least 20 m (66 ft) between 1892 and 1900. A dramatic acceleration of beach loss has occurred since 1960, when the Ghana barrage was constructed on the one hand, and, on the other important harbor improvements were undertaken at both Lomé and Cotonou. One of the authors observed, in 1972 the extensive damage resulting from sand removal for construction, hotel construction and urbanisation projects. The barrage is a sediment trap and longshore currents now remove littoral sand and erode near-shore ocean floor. At Lomé the harbor jetty impedes sand transport, sediments accumulate on its west flank and there is no beach nourishment east of it.

Only fronting the Tropicana Hotel (Lomé), has erosion abated, apparently due to the outcropping of calcareous cemented sandstone (beach rock) which, once denudated, acts as a natural breakwater.

At Cotonou, near Grand Popo where the beach gained some 7 m (22.5 ft) upon the sea at the turn of the century, erosion is acute because of jetties and the Kpémé wharf. Renewed accretion, however, has been noticed near Sèmè where quarrying of sand has been stopped.

When groins were built at Lagos to counter harbor silting, sand accumulated west of Lighthouse Beach Groin, and the strand grew about 400 m (1320 ft) seawards, while on the east strand the shoreline retreated 1,250 m (4063 ft) in about 60 years. Wave energy has increased in spots, as a result of greater depths that developed near groins. Along this segment of the Nigerian coast, beach retreat has averaged from 5 to 9 m (16 to 30 ft) per year since 1970; however, the problem is not confined to beaches in Lagos State.

A natural breakwater may build, and result in beach stabilization by reducing wave power, where accumulation of removed material takes place at relatively shallow depth. Erosion varies in time and space : along the American Atlantic shore it is winter-intense and during the summer accretion may instead take place[4]. Quantification must take into account volume of sediment moved at tidal, seasonal and long term scale and transport capacity of waves, currents, winds[5].

Gently sloping sandy beaches are far more vulnerable than gravel beaches, and rocky coasts resist better to the waves' onslaught. Seasonal changes are marked and the damage caused by storm waves in the winter may be compensated during the calm summers when no other factors intervene. This is due to long-period waves in the summer. However, too often the coastal dune barrier, is tampered with, breached or even more frequently, in industrialized countries, unhesitatingly flattened by bulldozer !

And the recent popularizing of dune buggies and other vehicles has increased the damage (Mahoney, 1980; Primack, 1982, *op cit.*). Pathways for storms are thus created and the dune no longer carries out its beach protective role.

Storm waves riding on an abnormally high sea level of a storm surge can cause erosion, sedimentation, and short-term flooding. Storm waves erode foredunes and bluffs and carry sediment offshore and alongshore to beachways. High surges from large coastal storms transport sediments from the beach and foredune onto the back of the barrier, resulting in upward and inland gain in these areas as the ocean front is eroded. Damage to and danger for buildings is considerable.

Dunes play the role of stabilizers sometimes "feeding the beach" sometimes receiving beach sand. Overwash may actually start new dune building.

During quiet periods sand is returned onshore to the beaches, but not to the dunes ; storms cause landward progress of dune erosion. In Rhode Island this can be 7 m (22.5 ft) during a single major storm.

Counteractions include the building of sea-walls, and the like but these are very expensive and interfere with the sand supply of adjacent areas; another measure calls for replenishment, by new sand, of the beach; and even more drastic measures require the setting back of all buildings from the foredune crest[6].

Mud coasts, common in the tropics, constitute a somewhat particular case. Destruction has been generally man-induced.

The 860 km (533 mi) east coast of Peninsular Malaysia consists of a straight sandy coast in the north while the southern half is comprised of a series of large and small hook - and spiral-shaped bays. The west coast, 1,100 km (682 mi) long, is made up of low elevation coastal plains formed from a deep marine clay stratum. The 1,800 km (1,116 mi) coastline of Sabah is characterized by rugged formation, and many bays of a variety of sizes and shapes, sandy beaches and mud coast, a situation similar in 1,040 km (645 mi) coastline of Sarawak.

Large tracts of the mangrove fringed coastline of the west coast of peninsular Malaysia have been reclaimed for agricultural development through construction of coastal embankments (coastal bunds or dykes) to prevent tidal flooding. Hinterland drainage is regulated through a series of tidal control gates. Bunds, at least 400 m from the seaward edge of the mangrove area, are protected against waves by mangrove buffer, but the protection is lost when mangroves vanishes due to erosion. [Midum and Lee, 1989, Mud coast protection - Malaya : C*oast. Zone '89 I, 806-820*].

There are two conceivable approaches to effecting coastal protection in mangrove-fringed coasts, to wit stabilizing the eroding scarp by means of revetment aligned approximately parallel to and positioned at adequate distance seaward of the threatened bund. Eventually supplemented with replanting of mangroves in the intervening space, if site condition permits, and direct bund protection of the threatened bund with surface armoring to develop a functional reveted structure.

The first approach is in effect a composite scheme in which the escarpment protection acts to absorb/expend the energy of the incoming waves by triggering off early wave breaking, while the mangroves filter the smaller secondary waves passing unbroken over the escarpment crown of reformed waves regenerated landward of the escarpment. Here, the minimum width of the mangrove thicket required to effectively attenuate the secondary/reformed waves is 50 m (164 ft). The second approach is used when the eroding shoreline has encroached very close to the bund line. The mangrove tree trunks, their protruding roots and the dense underbush throughout the mangrove thicket acts as a wave filter. Waves up to 1 m (3.28 ft) are completely attenuated by mangrove thickets 30 to 50 m (98 to 164 ft) wide. Where mangroves have been severely denuded, direct bund revetment remains the only feasible option short of bund retreat. The bund protection comprises two layers of armor rocks with individual weight heavy enough to remain stable under the action of the breaking waves. Below the cover layer are smaller rocks forming the secondary layer. Further down, a geotextile layer forms the separator between the rock mass and the *in situ* ground.

In tropical climates, mangroves may colonize inlets or part of estuaries. They retain mud particles, although it is not known to what extent, and are thus an accretion agent.

Mahin mud Beach in Nigeria retreated 200 to 600 m (600 to 1800 ft) between 1964 and 1981. Profiling between 1984 and 1987 revealed continuing climate, sediment characteristics, shore and offshore topography, land use practices and lack of sediment starving. [*A..C. Ibe et al,* 1989, Erosion management strategies for the Mahin Mud Beach, Ondo Sate, Nigeria : *Coastal zone '89, I, 821-835*]. Superimposed are regional subsidence and global eustatic rise. Area lack of construction material require a search for erosion control alternative strategies such as re-afforestation program and installation of a variety of non concrete offshore breakwaters. The Mahin Beach, 75 km (46.5 mi) long between Lagos and the Benin River, has one of the fastest erosion rates anywhere in Nigeria, though accretion occurs at the western end. Homes had to be moved inland at least twice in the last 5 years.

Erosion here endangers coastal settlements, accelerates salt water intrusion, modifies the hydraulic regime, and threatens oil wells and facilities. Settlers contributed to the problem by clearing the adjacent mangrove, decimating the mangrove forest along shoreline to erect buildings, making fish traps, boat tying-ups, cooking, construction drying racks, etc. Yet mangroves' network of aerial roots is particularly effective in stabilizing unconsolidated sediment, dampening tidal currents and trapping sediment. Wherever the mangrove has been cleared, the coastline inflects. The custom of digging small canals to the sea creates artificial inlets which enhance erosion, while livestock, particularly pigs, destabilizes the soil and accelerates erosion.

Oil and gas extraction, onshore and offshore, accelerates the natural subsidence of the coastal geosyncline. fluid extraction reduces fluid pressure in the reservoir, leading to an increase of "effective" stress (or grain to grain stress) in the systems, compaction results, hence the basis subsides. The rise in sea level and subsidence compound the problem.

Remedial options include reafforestation, public enlightenment program to preserve the mangrove, damping wave impact with non concrete floating breakwaters that also would trap and retain sediment. Such breakwaters could be made of used tires, disused oil filled hoses linked together to form an underwater "forest" and artificial sea weeds, all of which can be assembled off the site and then transported and "dumped". Artificial seaweed appears to be the most disputable solution.

An erosion study involves taking profiles "cross-island" or up to a point beyond the dune barrier, whichever the case may be, nearshore profiles and sediment samples.

The overall rate of limestone erosion has been estimated at 1 mm/yr (0.04 in/yr). But in tropical and subtropical areas, it is probably much higher, due, partly, to the forms created ; for erosional reefs a fig. of 10 mm/yr (0.39 in/yr) has been advanced even if waves are not particularly strong. Subtidal notches of 5 m (16.4 ft) (horizontal) were measured by Neuman in Harrington Sound (Bermuda). This type of subtidal erosion is enhanced be biological agents that bore (worms, pelecypods and sponges) or browse (echinoids). Boring can cause an erosion rate of 14 mm/yr (0.56 in/yr) and browsing of 6 mm/y (0.24 in/yr).

1.2 Beach Appeal

Beaches gained importance as a focus of attraction for holiday resorts ever since the twenties, and the democratization of travel increased their accessibility to ever larger numbers of people particularly after World War II. Though facilities were already well developed, the economic growth in the sixties increased the space demand on adjoining land and use of the beach to the very edge of the sea. While this phenomenon was particularly evident in the industrialized countries it manifested itself very rapidly in developing countries as well, even in quite remote areas of the world[7]. Affordable long distance travel fostered the "discovery of the pristine beaches of developing countries".

Holiday makers cause serious local damage on intertidal habitat along the British coast, particularly Dorset, where comparison is possible between exploited coasts and adjacent coasts controlled by the Ministry of Defence and are closed to the public. On the latter there is abundance of species rare or absent on adjacent beaches. Damage is also caused by divers. The Isle of Lundy, Skamer, Farne, have been proposed as nature reserves, a measure also recommended for parts of Tor Bay and the Isles of Scilly[8].

However, demand for ocean-side space did not originate only with holiday-makers and tourists, it came with equal force from energy- and manufacturing industries, commercial enterprises, and from the traditional vocations of agriculture, animal husbandry and fisheries[9].

Little attention, however, had, in the planning, been paid to possible negative consequences on beach evolution and marine erosion. Harbor extensions, or outright port creation, whether for industrial purposes, or for marinas, led to extensive dredging of access channels, construction of dams on rivers, particularly in estuaries, and/or large scale retrieval of bottom material for construction works with the concomitant risk of important reductions in bed load material, material thus withdrawn from the natural supply of sand needed to resupply the beach[10].

Beaches, unconsolidated littorals, are in constant evolution. Continued surveillance, e.g. sedimentary dynamic study of the area, impact study prior to any new modification implementation, on the shoreline, is required. Coast defense against erosion, accretion diminution, hightide dry beach area shrinking, has heretofore induced engineers to construct protective structures. In several countries such counter measures meant building of groins, detached breakwaters or similar construction works (Belgium, The Netherlands, Great Britain, Italy, U.S.A., ...)[11].

In the United States due to the absence of eminent domain on several coasts, governmental help, until rather recently, was excluded. This resulted in the building of make-shift breakwaters, which were often inefficient, highly temporary and eyesores[12].

Hoping to stop drastically erosion-induced problems, here local authorities, there private developers, often built sea-walls, dams or other similar "hard" defenses, yet, these in turn while providing some local relief, increased erosion in spots because of reflection, currents, overtopping, scouring and other dynamic phenomena.

During the last decades, beach nourishment, *viz.* artificial accretion, has been called upon as a coastal protection method, a "soft", in contrast to "hard medicine". Such an approach has been followed in Belgium with the beach replenishment project at Knokke-Heist, near the Dutch border. A beach renourishment project involving sluicing of about 8.4 million m^3 (11.2 million cu.yd) of sand was executed along the eastern part of the Belgian coast between Zeebrugge and the Dutch border[13]. Further nourishment plans have been undertaken on the west coast (Nieuwpoort vicinity) and at De Haan (a.k.a., Le Coq).

Completed in 1979, the east coast works covered a stretch of 8 km (5 mi), which meant a supply of rather more than 1,000 m^3 (1,313 cu.yd) of sand per running meter (400 cu.yd/ft). The beach itself was widened by about 100 m (325 ft) ; in some places it required the construction of an entirely new beach.

A few kilometers to the east, extensive works had been undertaken to considerably expand the harbor of Zeebrugge. The neighboring resort of Heist had no "beach" left at high tide ; next to it, the so-called beaches of Duinbergen and Albert-Strand where actually remnants of the dune barrier ; and Knokke-Zoute itself had only a few yards of dry beach left at high tide. Heavy storms were carrying the meager supply of sand seaward.

Once completed, the beach renourishment follow-up required a comprehensive survey program to observe the beaches' further evolution, to ascertain the accuracy of the study prognosis and to enable any new measures to be taken in good time. At present, survey data covering several years are available (De Moor, 1982; Kerckaert, *et al.*, 1985; *op. cit.*).

1.3 Beach Defense

Hurricane Gilbert's (Yucatan peninsula 14-15 September 1988) storm damage was increased by notching of dunes removal of forests, narrowing beaches with seawalls or jetties and by siting development in low elevation and "too close" localities. Damage to development was lessened by wide beaches, undisturbed dunes and forests, construction at high elevation and building setbacks. [E. Robert Thieler Daved Bush, and Orrin H. Pilkey, 1989, Shoreline response to Hurricane Gilbert : lessons for Coastal Management : Coastal Zone '89 I, 765-774.]. It was the most powerful storm in the Western Hemisphere in this century with peak wind speeds of 322 km/h (200 mi/h) at 3 km (1.86 mi) altitude and 282 km/h (175 mi/h) at the surface.

The traditional approach has been to construct hard coastal defense structures such as groins, breakwaters, seawalls, revetments, tetrapods and the like. Often they have been destroyed, all have required maintenance, many have merely displaced the problem from one site to the other. More recently another approach, either by itself, or in conjunction with the conventional measures, has been tried : beach nourishment by artificial means. The "soft" approach is not new since records mention beach nourishment immediately after the end of World War I near Santa Barbara, California (Charlier *et al.*, 1987). However, techniques have been refined and source material varies more in origin ; the U.S. Army Corps of Engineers has used dredgings in beach restoration works.

Artificial beach rebuilding has, in some instances, even become beach creation, as has been the case, for instance in Monte Carlo, Monaco (Charlier *et al.*, 1987).

Perhaps the largest successful beach restoration project was undertaken along the Eastern Coast of Belgium involving some 8 million m³ (283 million cu.ft) of material. The matter has been discussed by Kerckaert *et al.* (1982, 1985), Charlier *et al.* (1987) and others (e.g. De Moor, 1979, 1980). The core of the beach nourishment approach is discussed herewith (Roovers *et al.*, 1981).

In Japan, several approaches have been tried. Though success is claimed hard ?? reports are not readily accessible in the literature.

The country suffered from serious beach erosion for decades. Kawakami, Kumagai, Yamashita and Tanaka provided at Bordomer '95 three typical examples of beach erosion and countermeasures. Effords to counter erosion has centered on three methods : building detached breakwaters, artificial reefs and headlands. Kaike beach, built of Hino River sediment, famous for its iron sands, progressed 180 m (560 ft) in 30 years from 1868. Iron production declined from 1920 on, and beach erosion set in. Jetties, constructed in 1947 soon proved unable to stem it, nor was a subsequent seawall. Installation of a detached breakwater in 1971, led to formation of a tombolo behind it and beach erosion stopped.

The Niigata coast has eroded since the 1890's ; a weather station constructed 100 m (328 ft) landward from the shoreline in 1928, collapsed in the water in 1949. Major beach erosion causes are thought to be the construction of the breakwater of Niigata harbor, the discharge into the Okouzu flood channel and ground subsidence due to extraction of underground water to extract natural gas. A seawall was then constructed, but beach restoration was only achieved by the positioning of a detached breakwater. However, lately an artificial reef has been installed (fig. 6a).

Kashima-nada is one of the longest sandy beaches in Japan. It has been eroded due to the influence of breakwaters at Oarai and Kashima harbors. To counteract, the so-called "Headlands" method was applied. It consists in headlands dividing a long beach into small pocket beaches, e.g. 1 km (0.62 mi) long and aims at confining the littoral sand drift in a pocket beach.

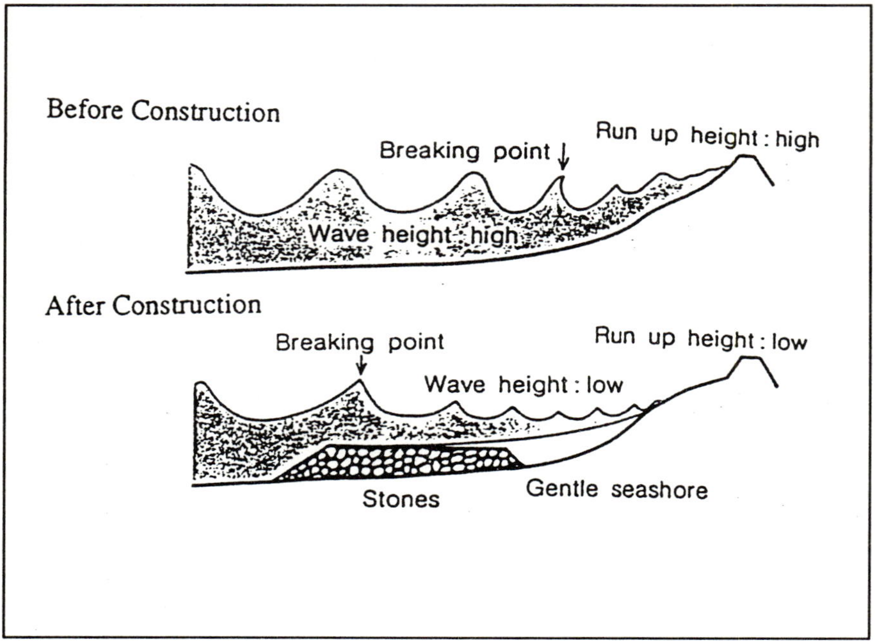

Fig. 6a : Artificial reef effects (T. Kawakami *et al.*)

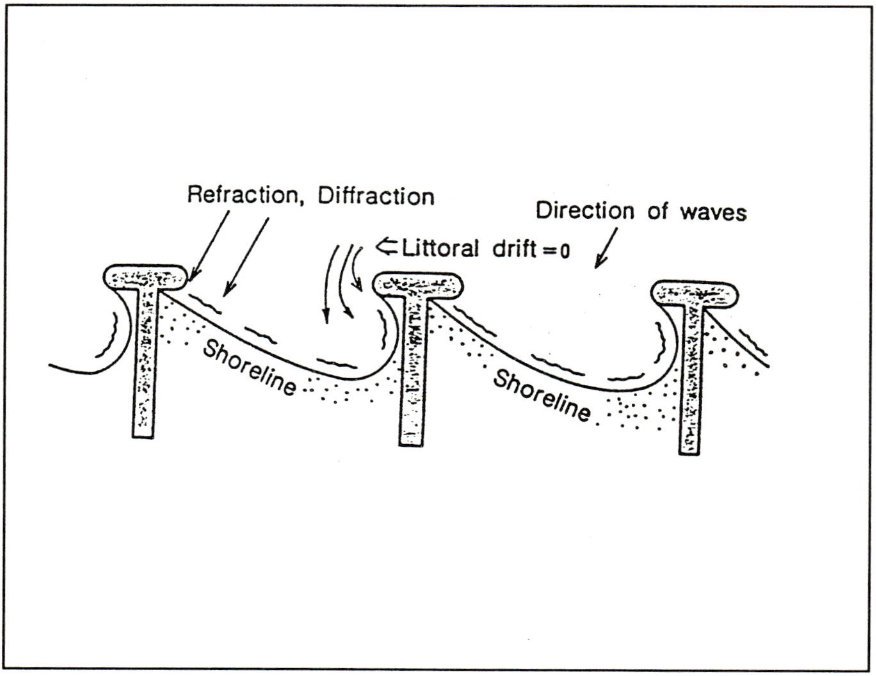

Fig. 6b : Concept of Headlands coastal protection (T. Kawakami *et al.*)

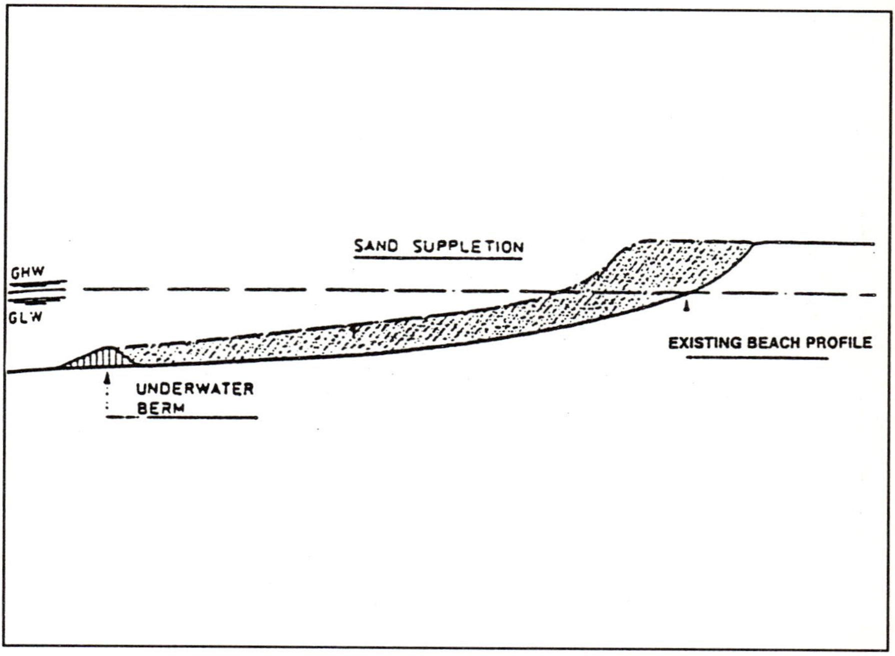

Fig. 7 : Some beach restoration possibilities
Suspended beach (Monte Carlo, Monaco)

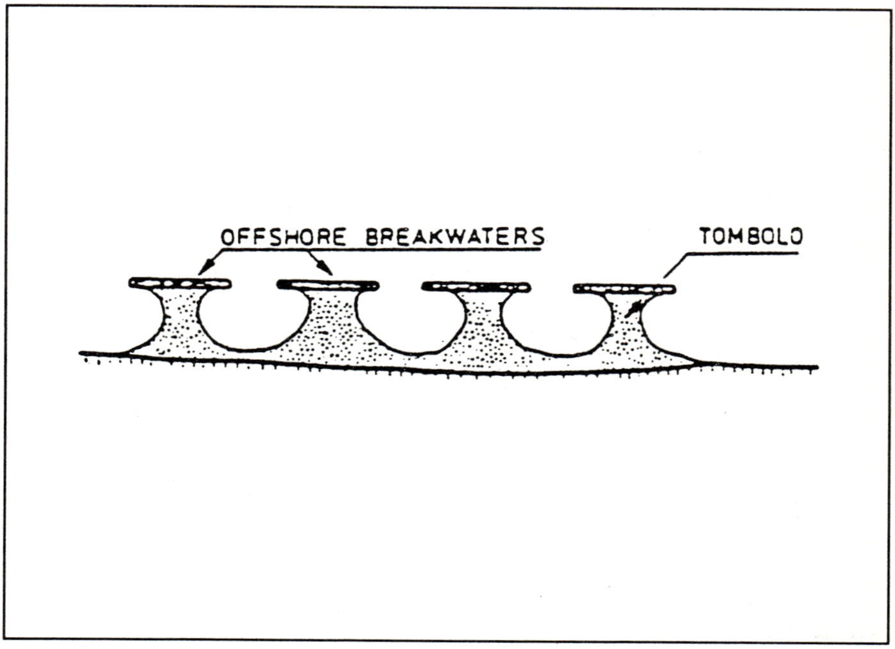

Fig. 8 : Some beach restoration possibilities
Offshore breakwaters (Tel Aviv, Israel)

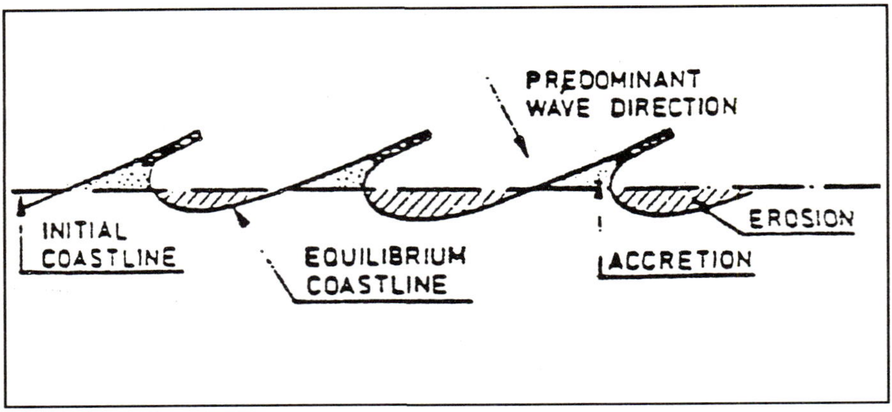

Fig. 9 : Some beach restoration possibilities
Headland system (Singapore)

2 "Hard" Coastal Defense Works

Common hard defenses are groins (set perpendicular to the coast), detached breakwaters (built parallel to the coast), sometimes submerged, and sea-walls. Each of these protective structures has a specific function.

The most commonly used materials for defensive works construction along North Sea shores are quarrystone or rubble. Smaller sized stones form the structure's core which is armored by the larger-sized material.

Scrap tires, often used for floating breakwaters, can also serve as construction material for fixed breakwaters. Treated timber piles are driven into the sea bottom and tires are piled on them (fig. 10). In areas of moderate tidal range and low gradient bottom slopes lean concrete filled burlap bags can be used to build a permeable breakwater. Filter material to prevent settling of the structure must be placed underneath (fig. 10).

New types of concrete units use as underlying filters geotextiles. They can also be used for the "filter cloth" of breakwaters built from bags filled with lean concrete mix (fig. 10).

Groins also can be built of treated timber, or with metal pipes of steel or aluminium or with fuel drums or concrete filled bags (fig. 12, 13).

A flat concrete block, the HARO®, was recently developed for protection of at-sea structures against wave action. It has a large central opening and both of its short sides are widened at the base ; corners are asymmetrically tapered in plane. It achieves thus a high degree of porosity, when used in two layers, and good stability. Its high porosity and lesser weight allow a 30% reduction of the amount of required concrete. The decreased wave run-up permits a reduction in breakwater crest height. The new units have been used in the Zeebrugge harbor (Belgium) extension works.

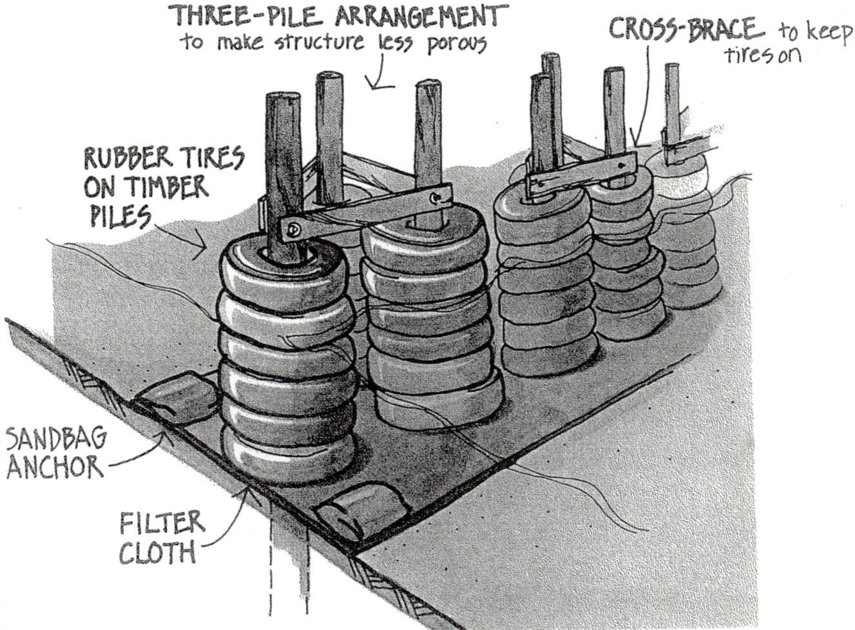

Fig. 10 : Fixed breakwater built of car tires

Fig. 11 : Lean concrete breakwater

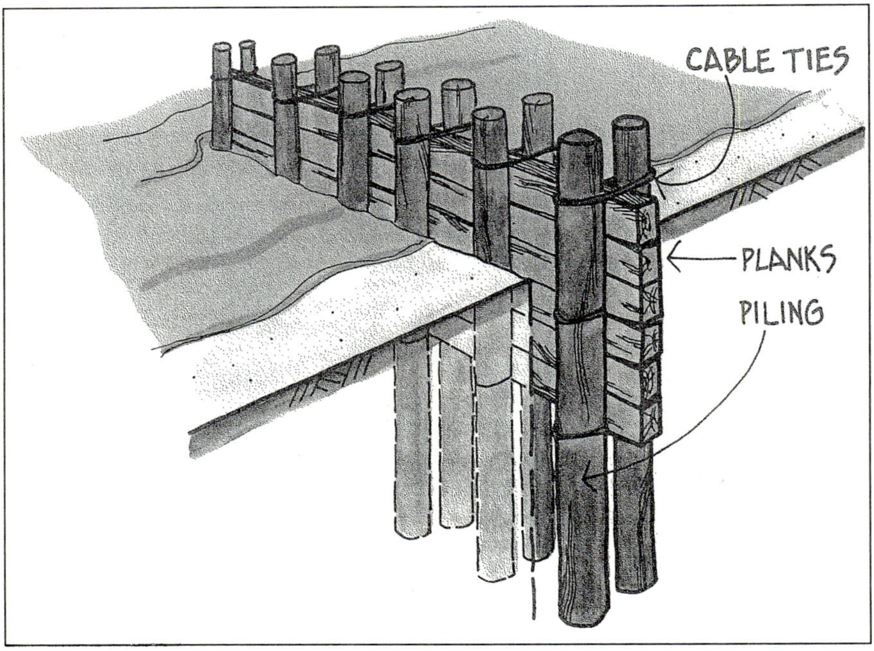

Fig. 12 : Groin made out of timber

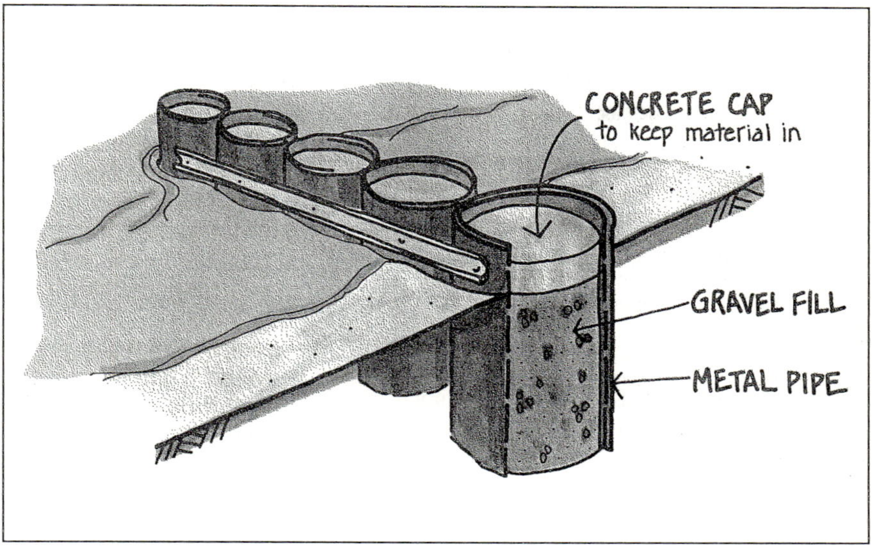

Fig. 13 : Groin made out of metal pipes

2.1 Structures

2.1.1 Groins

Groins (groynes), short, squat structures, jutting out to sea, are built to reduce the longshore sediment transport in a certain beach area and capture passing sand (fig. 6 and 7, 8 and 9). If not well designed (length, height) they can cause at the leeside severe erosion, extending over several kilometers along the coastline.

The latter, in turn, bring about the need for an elongation of the defense system over more and more coastline (cf. Belgium, The Netherlands) (photo 2).

The term groin (or groyne) is usually applied to fingerlike structures placed perpendicularly to the shoreline ; groins are commonly built as a field along an eroding coast (fig. 16).

A field of groins will maintain a stable beach, when filled to capacity with sand but will not provide full protection against storm-driven waves. However, it usually triggers downdrift erosion which can be somewhat mitigated by using beach fill in the groin field. When placed too close together groins cause waves to stir up material and current will carry material offshore.

Wave energy is absorbed by sand movement, seawards or landward, and rushing up gradually sloping sand expanses.

Sand trapped by groins may be severely missed downdrift, and the latter areas continued to be stripped of their sand. A wider beach is built just above the groin, but no further.

To counteract longitudinal sand transport which resulted in silting of Lagos harbor entry near Victoria Island (Nigeria), two groins were constructed between 1907 and 1916 (fig. 17). The actual consequence was sediment accumulation west of Lighthouse Beach's groin but east of the harbor, near Victoria Beach's groin erosion took a considerable toll with the shoreline retreating 1.25 km (0.775 mi)[14]. To redress the situation groins were lengthened and artificial nourishment of 1 million tons of sand per year carried out. Yet, shoreline retreat has continued, depending on the site, at a rate of 5 to 9 m (16 to 30 ft) a year.

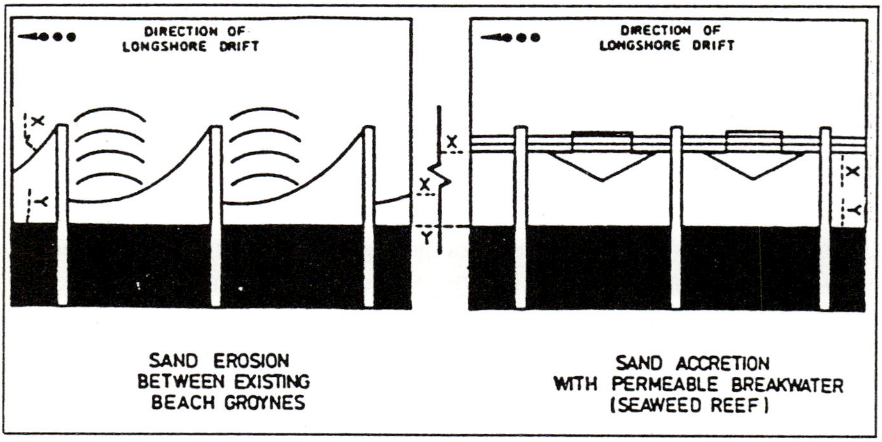

Fig. 14 : Beach stabilization

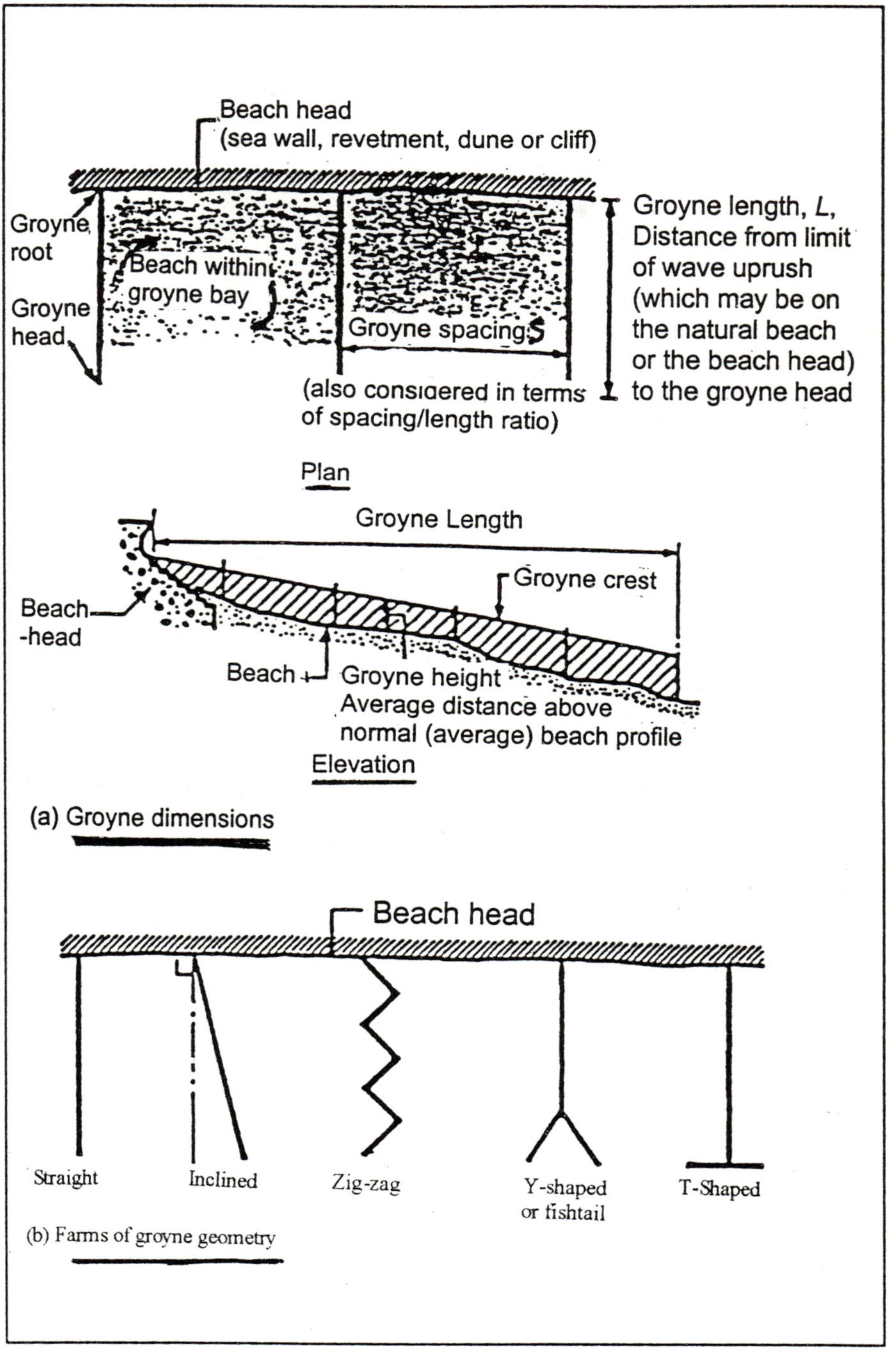

Fig. 15 : Definition sketch for groin terminology (Flemming)

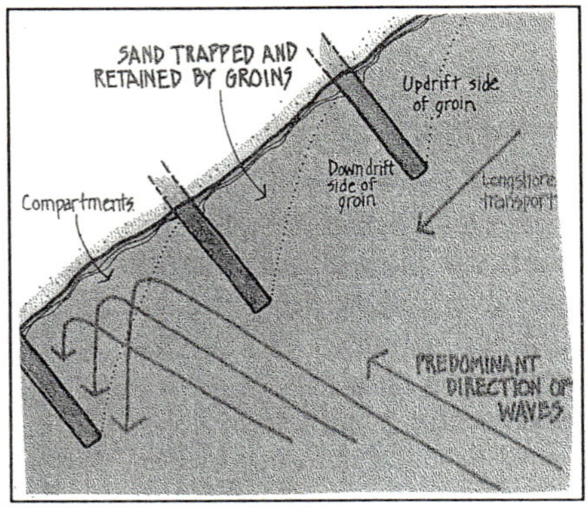

Fig. 16 : Groin fields

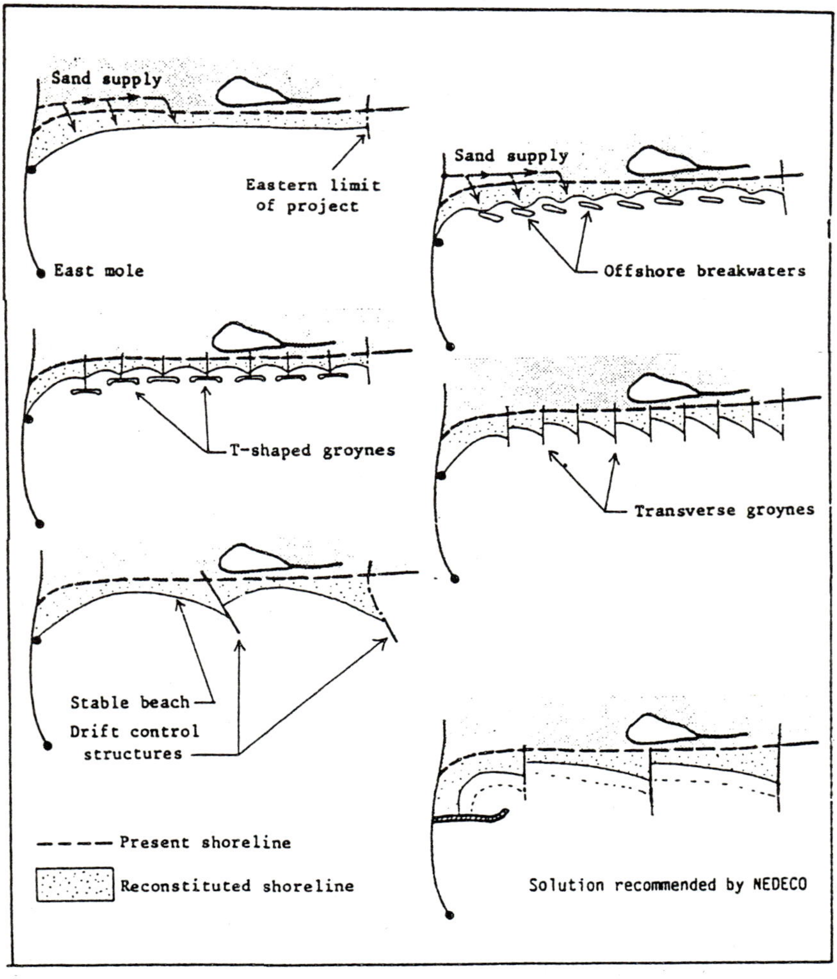

Figure 17: Victoria Island Beach: Proposed shoreline protecion schemes (after M. Akle, 1985).

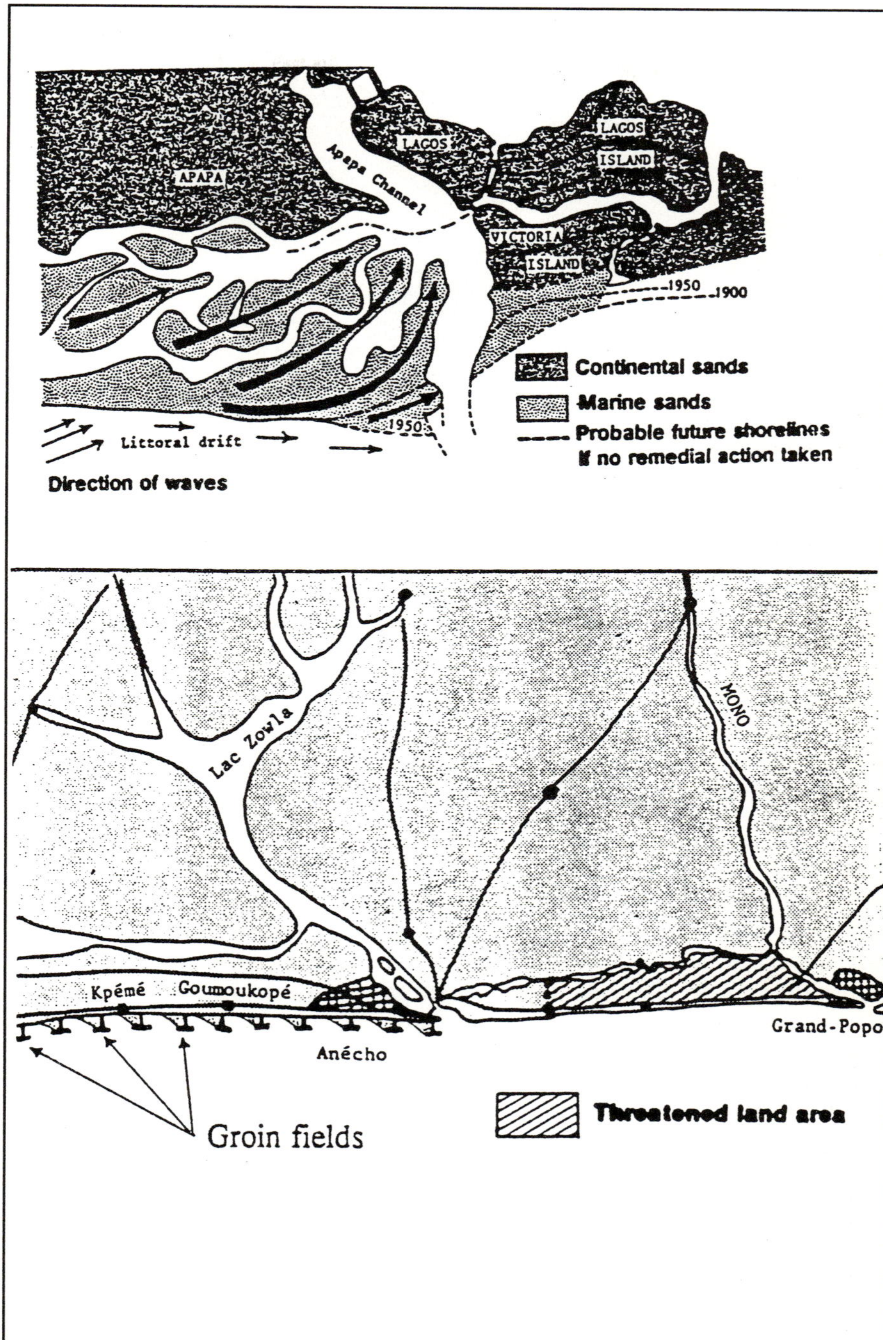

Fig. 18 : Proposed protection scheme for Kpémé and Anécho (Togo) [after M. Akle, 1985].

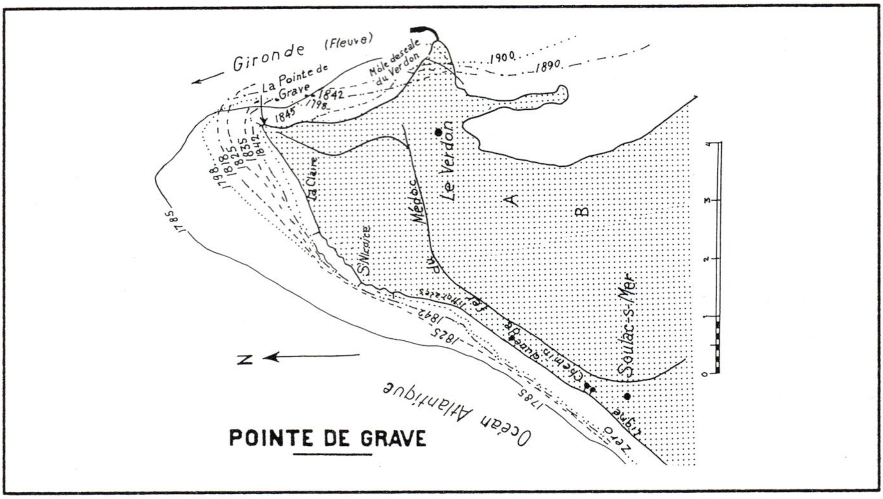

Fig. 19 : Arrest of alongshore drift by groin system

At the Pointe de Grave near the Gironde mouth on the Médoc coast shore protective structures were destroyed, new ones built (fig. 19). Erosion monitored for close to 150 years took away respectively 100 m (325 ft) at Les Huttes, 250 m (813 ft) at Les Arros and 145 m (472 ft) at Soulac south of which city 10 m (32.5 ft) are removed yearly since 1978. Hard defense structures apparently have not stemmed the process except at Les Arros where, since 1930, groins are repeatedly rebuild, damaged and again reinforced (photo 1).

There is considerable hesitation in building more groins and other solutions become favored. But, the investment made over the years in groins must not be viewed as entirely lost. Wave action will transform an existing groin, henceforth left untouched, into a sand hoft. The process can be slow or fast as waves determine it. Wave direction not being necessarily invariable, the sand is carried in different directions and eventually distributed around the hoft. The groin-become- hoft exerts hence a stabilizing effect on the beach as part of the sand passes from long- to short- distance transport.

Construction of groins has still been recently recommended by the Laboratoire Central d'Hydraulique de France for Togo (Kpémé and Anécho). The feasibility study admits, however, that severe erosion will result for near-by areas in Bénin (fig. 18).

Apparently unfavorable effects of groins may not occur. Everts and co-workers reported that at Masirah Island (Oman) a recently built groin does not cause sediments to accumulate, nor favors erosion, rather it creates a slightly more stable beach[15].

Assuming that a longshore current moves from east to west, groins are to be built first at the far western end of the shore to be protected. The field can then be gradually extended eastward. Additionnally, an appropriate quantity of sand must be placed on the eastern side of each groin - a technique called backfilling that insures a continued westward flow of sediment.

In the former Democratic Republic of Germany 40 km (24.8 mi) of sea walls were built and 150 km (93 mi) of coast lined with groins.

In Japan 50-ton concrete tetrapods have been used to protect the east coast of Honshu. Tetrapod barriers have been built at Silt (Fed. Rep. Germany).

A well designed groin system can arrest or slow down the alongshore drift of material and, by building up material in the groin bays, stabilise the foreshore and protect the coastline (cf. fig. 19) and deflect strong tidal currents away from the shoreline.

Groins may also help, in some situations, to hold material on a beach that has no natural supply and has been artificially nourished ; control seasonal shifts of material alongshore and hence swings in shoreline within a bay ; reduce the long term erosive effect of wave activity in an area of coastal defence by accumulating beach material in front of prominent hard beach heads such as sea walls, revetments, bulkheads and cliffs. This requires an adequate supply of material moving alongshore ; improve the extent and quality of an anemity beach and increase the depth of beach material cover to an otherwise erodible sea bed soil. (Flemming, C.A., 1990, Principles and effectiveness of groins. In : Pilarczyk, A. (ed.), *Coastal protection* ; Rotterdam, Balkema).

2.1.2 Jetties

Jetties are more massive than groins. Their principal function is to redirect tidal flow to either protect a beach or a navigation channel. However, construction of jetties results in groin-like effects by redirecting littoral drift. They are occasionally used to cause by-passing. By-passing consists in dredging sand from behind the upstream jetty and bringing it via pipeline to behind the downstream jetty[16].

Port structures obstruct sediment drift and sediment accumulates updrift, causing sometimes siltation of the port inlet. In the United States, Mexico, India and elsewhere sediment by-passing, by means of pumping is frequently practiced.

2.1.3 Breakwaters

Detached breakwaters (parallel to the coast) are offshore barriers intended to reduce wave action on the existing coastline. They absorb wave energy before waves reach the shore. This effect will not only reduce onshore - offshore transport (especially during storm conditions) but will, in certain circumstances reduce littoral drift and create tombolos due to wave diffraction effects (cf. Mediterranean coast in Italy). However, in regions with significant tidal ranges, such as the North Sea, these structures would normally have their crest at mean or low water level. Under severe storm conditions (extreme high water levels) the water layer above the crest will no longer abate incoming waves, which will keep on attacking and eroding the beach area. Furthermore, the wave attack on the structure is heaviest under these conditions.

Such breakwaters have been constructed in great numbers to protect the beaches and resorts from Venice's Lido to Rimini. There were, by 1982, 2305 detached breakwaters in Japan.

The height of a breakwater poses serious problems because if too low inadequate protection will ensue and if too high interference with the shore processes will result in generating new problems (fig. 20).

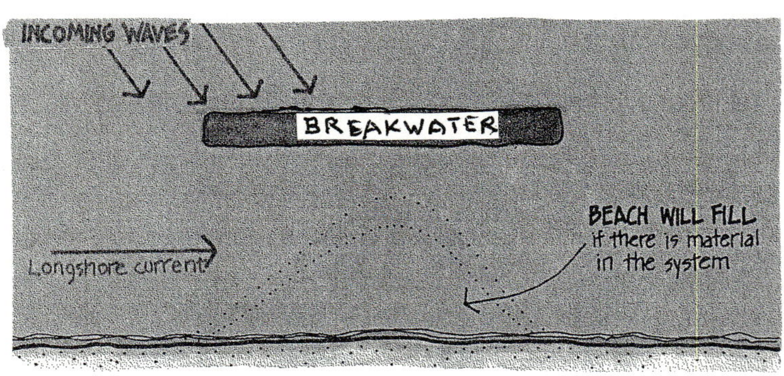

Fig. 20 : Breakwater Overhead view

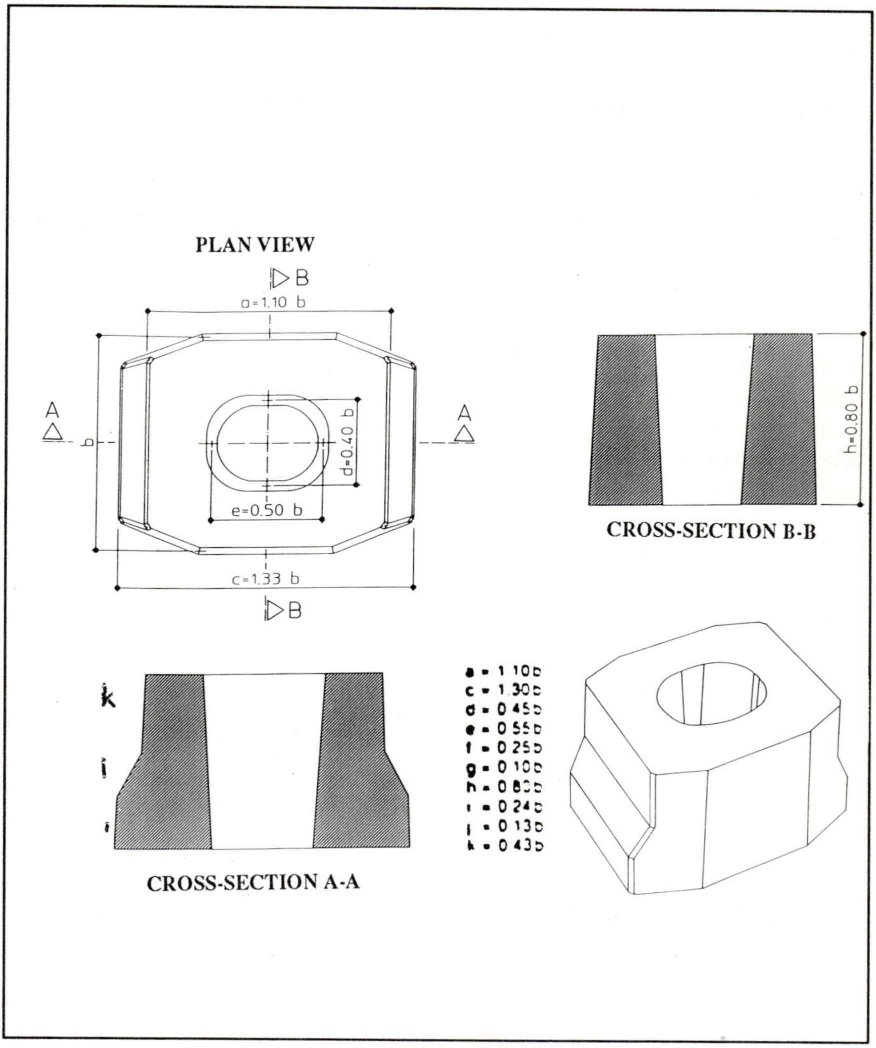

Fig. 21 : The HARO

Damage to and protection of rubble mound breakwaters (talus breakwaters) is a major concern. It has been said, accurately, that no sooner is such structure completed its destruction starts. While they perform generally nevertheless well, some major failures have occurred in various locations throughout the world (J. De Rouck and C. P. De Meyer, 1987, Port structures. Their care and maintenance : *PIANC Bull.* 58, 133-168).

Case examples of breakwaters and sea-walls adapted for specific sites, in Great Britain, have been discussed by R.B. Thorn and A.G. Richards (1981, *Sea defence and coast protection works* : London, Thomas Thelford). Others of relatively recent date, in Spain and in The Netherlands (IJmuiden) are described by N. De Meyer and P. Soete (1970, *Golfbrekers met talud : hun berekening, vormgeving en uitvoering* : Gent, Rijksuniversiteit, Fac. Toegepaste Wetenschappen). More recently P. Bruun and S. Kjelstrup (1981, Practical views on the design and construction of mound breakwaters : *Coast. Engng* 5, 171-192), J. De Rouck (1991, *De stabiliteit van stortsteengolfbrekers. Algemeen glijdingsevenwicht. Een nieuw deklaagelement* : Leuven, Katholieke Universiteit Fac. Toegepaste Wetenschappen) and J. Goda (1985, *Random seas and design of maritime structures* : Tokyo, Univ. of Tokyo Press) provided good reviews and some new ideas on design and stability of rubble mound breakwaters (Fig. 11). De Rouck studied in depth the HARO, mentioned in this text.

Some reservations concerning the use of HARO blocks may validly be made : the cost is not slight. However, the block has a good hydraulic stability and is light weight ; this allows a smaller size, consequently a thinner layer, thence a grater porosity and a cut in quantity of concrete of roughly one third. The reduced weight permits use of a smaller crane thus of a narrower dam. Finally a lower dam crest is only needed (fig. 21).

If the HARO provides greater linking facility and far higher hydraulic stability, even greater resistivity to thermic variations, it has a somewhat lesser wave reflection coefficient and wave run-up than the traditional concrete cube. The best coefficient of run-up is apparently reached by accropods (0.95 vs. 0.85).

In comparative cost units of an armor layer, dolos and accropods run around 75, tetrapods 90, against 70 for the HARO, the grooved cube being taken as the 100 basis.

In the long run utilizing HAROs proves economical.

Breakwaters come in a great variety. Land-backed structures can be either smooth (concrete, stone pitching) or rough (randomly piled rockfill). Rough structures dissipate wave energy strongly because they are permeable. Concrete constructions have been occasionally perforated to reach the same result. Toe depth dictates the speed of approach of the water, thus orbital velocity of particles on impact, which is quite important.

Overtopping, *viz.* water passing over the structure, influences the degree of wave reflection and consequently the standing wave's extent. The flatter the slope the greater the influence of roughness. The amount of scour at the toe of a permeable breakwater is allied withy degree of reflection, while the wave run-up is with the standard wave developed which itself depends on toe water depth and, if small, the slope of bed leading to the structure.

Wave transmission seaward can occur with marine structures that have water on both sides (water-backed). They thus result in reflection, dissipation and partial energy transmission landward. Permeable breakwaters fall in this class, with waves of lesser height transmitted. So do submerged ones which allow overtopping and reduce amplitude of waves.

Composite breakwaters, common in Japan, are rock mounds shaped to be a foundation for monolithic concrete modules (boxes) floated into position, sunk and filled with sediment or rock. A similar technique was used on Kislaya Bay (Russian Federation) to construct the tidal power station.

Pile arrays may be considered "open breakwaters". That are series of closely spaced piles, in a linear or diagonal arrangement, protecting against onslaught of waves or as the support of a jetty. They attenuate the impact of waves.

A major project in Japan has been the development of effective permeable offshore breakwaters at depths between 10 and 20 m. Six types of self sustained permeable offshore breakwaters have been considered. [Takaaki Uda an Y. Murai, 1988. Development of a self sustained permeable offshore breakwater. *Proc. Techno-Ocean '88 Symp.* (Kobe, Japan), 146-153].

Mobile breakwaters rest on the sea-bed but may be easily removed. In the same class fall pneumatic and hydraulic breakwaters. The first opposes currents generated by near-bed released air bubbles which drag water to the surface as they themselves rise. The flow divides and part goes towards incoming waves. The latter two systems are more effective against short-period- and deep-water-waves (J.L. Evans, 1955, Pneumatic and similar breakwaters : *Proc. R. Soc. London*, A-231, 457-466) ; J.A. Williams & R.L. Wiegel, 1962, Attenuation of wind waves by a hydraulic breakwater : *Proc. 8th Conf. Coast. Engng*, 500-520). They are energy costly to operate.

Among the group of "portable" breakwaters, one may place the Leonard-Huspeth membrane wave barrier. The long, sausage-like rubbery membranes, once filled with sea-water can be used as a rigid temporary breakwater. However, itself influencing the wave's impact strenght. Proposed in August 1986, it is still undergoing laboratory testing.

Pontoons and floating breakwaters produce an effect resembling that of a rigid thin barrier plunging to some depth. Some were used on the Normandy coasts during WWII.

The Queens University pontoon relies on reflection along its bottom to reduce wave transmission. A vertical plate supports two cylindrical floats, necessarily 2 wave lengths apart, parallel to the wave crests. Silvester (*op. cit.)* has suggested making a raft of welded together oil drums. Partly and wholly flexible rags have been tested, as have floating water-filled bags, log-jams, U-shaped side-performed wooden booms, fascine mattresses, rectangular and triangular box-type floats, series of submerged horizontal tubes and others.

Somewhat expensive are such transportable breakwaters as caissons, bubble breakwaters and floating mattresses (P.S. Butson, 1967, Transportable breakwaters : *Dock and Harbor Authority*, J. 48, 41-46).

Besides the traditional construction, submerged breakwaters with openings for outflowing surge water, steel pilings, and other configurations exist. Permeable breakwaters offer some advantages but effectiveness may be impaired where they are too porous. Floating breakwaters may not be suitable in areas of high wave energy ; they have the advantage to be able of adjusting their direction with changing winds or currents. Discarded car and truck tires can form an inexpensive floating breakwater when tied together. Scrap tires are commonly used for floating breakwaters construction material. Easily installed, readily adaptable, they have a high energy absorption power.

Units used on the Atlantic coast of the United States frequently consist of twenty tires ; they reach breaking strengths of up to 6237 kg (13750 lbs). A 152.5 by 6.4 m (500x21ft) breakwater, for instance, requires 4260 tires (213 units) (fig. 10). On Narragansett Bay (Rhode Island) such breakwaters successfully withstood the strain of 1.22 m (4 ft) waves, currents reaching 1.3 knots (2.41 km/h ; 1.5 mi/h) and temperatures ranging from 2.2 to 18.9° C (36° to 66° F).

Tying materials can be made of steel, rubber or plastic material and be wires, ropes, chains or belts. They are subject to a large variety of wear and tear factors : besides fatigue, abrasion, corrosion, ultraviolet- and biodegradation will take their toll. The environmentally inert system consisting of rubber conveyor belt edging with nylon fastening is probably the most desirable from all points of view. Still another type of floating breakwater has been tried at the Scripps Institution of Oceanography (La Jolla CA) : rows of spherical balls are anchored to the sea floor but float close to the surface. By moving in the direction opposite to that of the waves, they absorb an estimated 50% of the waves' energy.

The low-cost of floating breakwaters and their suitability where water level differences are great or bottom conditions poor, have made them attractive wave-energy dissipaters when funds are limited. Little energy is reflected[17]. They were used, e.g., along England's southeast coast.

Interestingly a painting by John Constable titled "Floating breakwaters at Brighton" shows open breakwaters installed years ago. It hangs in Brighton's Art Gallery.

The secondary role, as wave energy dissipaters of certain structures whose function is energy production has been praised. At the forefront of these are several wave energy conversion systems : not only do such devices "absorb" wave energy to convert it in electrical or mechanical power, they also fulfill the auxiliary role of *de facto* breakwater.

Segmented breakwaters are discussed later under section 3.

Another device to attenuate the impact of wave energy is the "surgebreaker", a proprietary trademark; it is a modular, previous sill device which apparently favored sand accretion, both seawards and landward. Try-outs were made on the U.S. coast of the Gulf of Mexico along the shore of Basin Bayou State Recreation Area, in Walton County, Florida. Enrichment was also monitored updrift and downdrift of the surgebreaker. However, the steel-fiber reinforced concrete artificial reef, had, as do similar systems, no slowing effect upon bottom and beach erosion in areas at short distances from the reef[17a].

2.1.4 Sea Walls

The sea-wall is a structure separating land and water areas while a bulkhead is placed on a bank or bluff to halt land sliding and protect an inland area against wave action. Venetian *murazzi* belong to this type.

Sea walls are built to prevent the coastline from further regression. *A sea-wall prevents erosion of the land behind ; it does not protect the shore in front of it. They are most appropriate for boating and fishing ports.* Revetments and bulkheads play the same role. Revetment designates a facing placed on a bank or bluff or other stone structure (fig. 22, 23, 24). Revetments are similar, for a sloping coast, to sea walls, while bulkheads are used to prevent backshore property of sliding into the surf zone, as happened, for instance on Long Island (NY). Revetments are layers or blankets of strong, non-erodable material placed on the shore. Stone revetments have been used, for instance, at Johns Pass, Treasure Island (Florida). Revetments built on relatively gentle slopes are stable ; they do not greatly interfere with littoral drift. Nevertheless they may induce local erosion elsewhere ; this is particularly true if they are steep. While themselves either porous or watertight, they are subject to scour at their toe where a rock apron is best placed (fig. 25).

In the Netherlands, paved stones have been traditionally used as revetments although asphalt revetments are not unusual. Studies were recently carried out dealing with the relationships between external wind-wave pressures and pressure gradients underneath the stones in sublayers of granular material. This allows making a quantitative check of the revetment stability.

Asphalt may well become the more common material not only because of a progressive shortage of paved stones but because it permits to complete work in a shorter span of time. Asphalt concrete on sand is an approach that is followed on a large scale along Dutch shores. Sea-walls and bulkheads must be high enough to prevent overtopping, have a toe protected against scour, weep holes to relieve pressure, be securely anchored and tied so as not to tip and defended against percolation and run-off (fig. 25).

Bulkheads resist substantial wave forces by transmitting impact loads into the soil behind the wall, but waves can rapidly destroy a bulkhead if the backfill is lost. Unlike the bulkhead, a free standing wooden breakwater must rely on the strength of its materials to resist wave impacts. Traditional wooden breakwaters used for marina or harbor protection are designed as a total wave barrier with minimal overtopping. To survive waves, they must be heavily constructed and much more expensive than a bulkhead with a similar exposure.

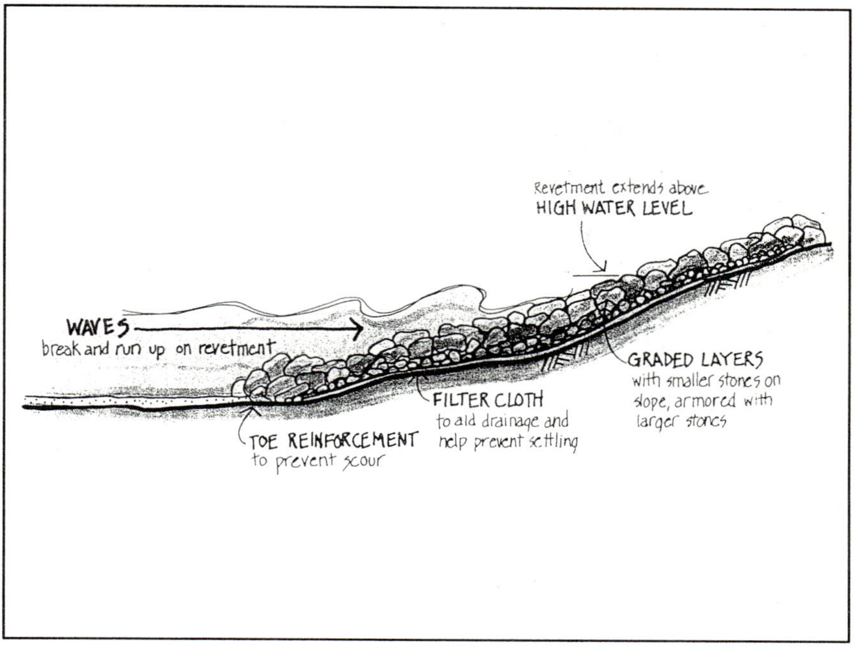

Fig. 22 : Revetment

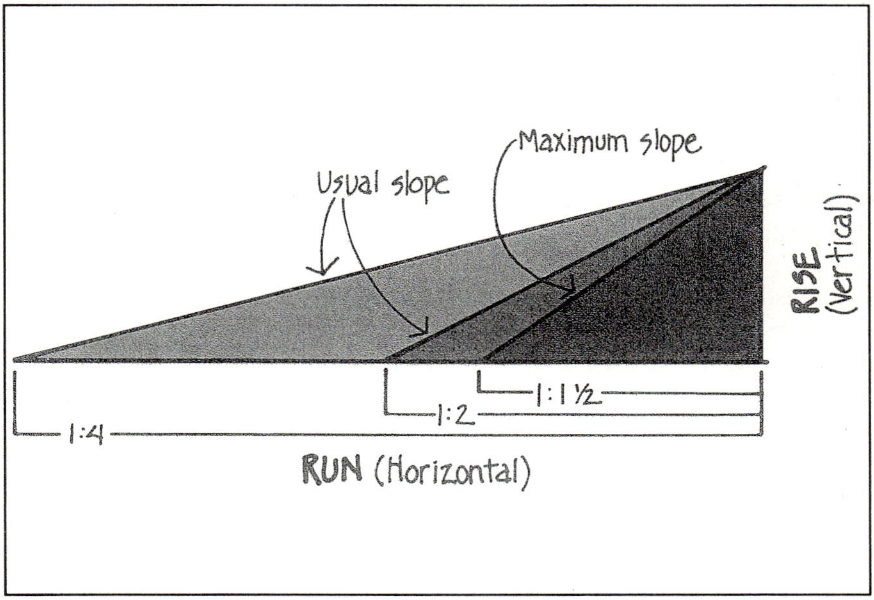

Fig. 23 : Revetment : height effect

Fig. 24 : Gabions as revetments. Rock fill

Low cost sea-walls and bulkheads can be made of treated timber, sometimes combined with steel H-ties, though this raises prices. Timber is most useful in sheet-pile or pile-and-planks designs. Lean concrete filled bags held by hogwire fencing are used above ground (fig. 22).

On eroding beaches with predominant longshore transport, such defenses will not alter the longshore transport and beach erosion in front of the sea-wall will continue. The latter increases wave attack on the sea-wall and the offshore transport. As a consequence the stability of the structure is undermined and it may fall over, a not exceptional occurrence. The beach may even disappear as wave force is reflected seaward instead of being spent across the beach. Sand, when disturbed, may be carried off seawards or along the shore, but away from the wall protected area.

Sea-walls were already being built in the Venice region in the eighteenth century; these *murazzi* were but one of the anthropic inferences into natural processes which can be traced back to the thirteenth century. Additionally stone jetties, protecting harbor entrances acted as groins. Suva peninsula, on Fiji, has an artificial coastline of sea-walls.

Occasionally natural formations may act as temporary or long term sea-walls. This is for instance the case with kurkar ridges (more or less carbonate cemented sands, loams and sandstones) on the shallow shelf, north of the Nile delta (Gaza, Israel, Lebanon), which parallel the coastline. Their effects are the same as human-made sea-walls.

2.1.5 Wave-Return Walls

In cases of flood protection wave-return walls were constructed for example in Great Britain (Ilfracombe, Teignmouth,..) and Zeeland (Flushing). This drastic solution only cf.ms adequate if the sea-ward limits of the coastal barrier cannot be changed any further.

2.1.6 Gabions And Mattresses

Attempts have been made to counteract beach erosion, or to encourage accretion, by using flexible concrete mattresses, or by synthetic seaweed grids (Netherlands) (cf. 2.25). Gabions offer a related approach.

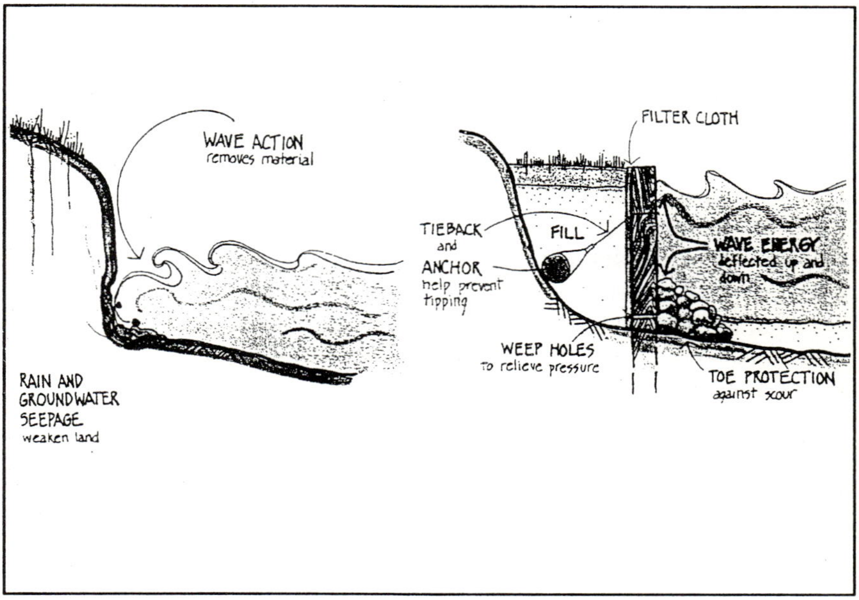

Fig. 25 : Scour of bulkheads and seawalls

They are rectangular shaped metallic cages filled with varied sizes pebbles; spread out one gabion type resembles a mattress[18], usually one half meter (1.64 ft) high and as long as 8 m (26.25 ft) with mesh from 8 to 12 cm (3.2 to 4.8 in) usually. The mattress is laid upon a slope of appropriate gradient. However, the materials must be present in situ and one has to cope with corrosion. Gabions can be used for revetments. Rocks filling the gabions must be at least 10 cm (4 in) in diameter lest they slip through the mesh.

This revetment is permeable, takes on the soft terrain's shape and absorbs wave energy while successfully resisting the strongest currents. Several layers can be placed on top of one another and tied together with wire. However, gabions are poor protection for the beach itself as they may reduce or prohibit accretion by sand of terrestrial origin. It has then been suggested to complement the use of gabions by traditional constructions (groins, etc), by by-passing, by dune building or planting, or simply by renourishment (fig. 26). Gabion breakwaters have been constructed along East Coast Park on the southeastern coast of Singapore.

2.1.7 Artificial Reefs

Denmark and England put both claim for the earliest development of erosion control through use of artificial seaweed. Field attempts were made in Denmark and New Jersey. In the latter location bundles were placed along 275 m (900 ft) of shoreline 245 m (800 ft) offshore in 4.5 m (15 ft) deep water making a 27.5 m (90 ft) wide barrier. No conclusions could be drawn on the effectiveness. Later experiments were conducted in the same countries, The Netherlands, Germany, Norway and France[18a]. Recently, Japan has used them as protective devices.

In The Netherlands and France positive results were achieved in reducing scouring near man-made structures and improvements could be made by lengthening the fronds and increasing their density in the bundles. In the United States emphasis was placed on reducing shoreline erosion. Experiments at Fort Belvoir, VA, at Cape Hatteras Lighthouse, NC, Stone Harbor Point, NJ, Naples, FL, and in about 50 other locations have been negative, or are manufacturers' claims. A rather positive report by Hall has been seriously challenged[18b].

Seaweeds effects, whether natural or artificial, have the greatest impact at and close to the surface hence the most attractive approach by which they can be effective is by reduction of incoming wave height. A seaweed barrier would act as a floating porous breakwater. Where the artificial seaweeds are concerned the European manufactured types appear to be of higher quality than their American counterparts.

But artificial seaweeds fields present two disadvantages : an effective field, even if placed so that it would not be a navigational or recreational hazard, would be more expensive than other beach protection methods, and it appears to have unfavorable effects upon beaches and shorelines further downdrift.

Although artificially, *viz.* by man, put in place, some "artificial reefs" are not made of man made materials but use "natural" material.

The idea has been also put forth by Beavis and Charlier (Int. Seaweed Symp., São Paulo, Brazil, July-Aug. 1986), and others, to use artificial reefs as "natural breakwaters" (fig. 14). Such algae as *Macrocystis pyrifera* (kelp) and *Laminaria* ssp. appear particularly suited for the purpose.

Wave attenuation could be achieved by large kelp beds. Over 0.8 km (0.5 mi) wide and several kilometers long, in water 15.25 m (50 ft) deep such beds could theoretically reduce wave height by half as far as 0.8 km (0.5 mi) into the field. They play a role comparable to that of porous breakwaters[18c].

A natural-weed screen has been proposed as a means of stabilizing sand losses of artificially nourished beaches. A short line of cages, containing e.g. *Laminaria sp.*, could break up rip currents that develop at the head of groins where longshore drift may carry a bed load of sediments seaward (fig. 14).

As artificial nourishment buries the bottom end of groins, back- beach erosion is increased. A weed screen could break up currents confluence and favor landward sand accumulation ; additionally it contains subsequent storm-induced erosion : grains carried seawards strand on the seaweed barrier leading to formation of a sandbank in front of the screen. In short the beach nourishment result could be naturally maintained.

As regards artificial seaweed[18d], the following remarks are of interest :

The hydrodynamic response of artificial seaweed to surface gravity waves is modeled analytically in a perturbation expansion. Because commercially available seaweed is only weakly buoyant, the perturbation expansion involves the density ratio, Dr/r, as the small parameter. The solution indicates that the compliant fronds of the artificial seaweed are excited to oscillations which closely approximate those of the oscillatory water motion. Because of this lack of relative motion between the seaweed fronds and the water, dissipation of wave energy is relatively small.

Relative motion between fronds and water is small, as their oscillations' amplitude nears that of close-to-bottom water motion. The transport capacity remaining about the same, sand accretion and/or erosion reduction is doubtful.

One could consider lengthening the fronds of artificial seaweed but this would compound current problems with the mooring system.

However, because sediment transport is a highly non-linear function of the water velocity, a small retardation of the bottom oscillations induced by artificial seaweed can lead to a large reduction in sediment transport. The formation of bars and shoals over the seaweed bed would consequently be expected.

Design and construction of artificial seaweeds varies depending upon the projected use : either for an artificial reef for anti-erosion coastal protection, or to reduce scouring of footings of at-sea structures and pipelines. Conclusions reached by workers at Scripps Institution of Oceanography (California) are that at fundamental frequency the reef has no significant effect on wave energy, though it absorbs significantly such energy at higher harmonies. Besides inadequate anchoring units are fast buried by scouring action when mass transport is important, and when they can resist it, they suffer from scour around the anchor, bury themselves into the sand and cause no accretion. Observed beach profile changes were not due to reef implantation.

The Nicolon gas injected polypropylene weed was used as early as 1973 to counter scour in The Netherlands. In 5 out of 11 sites positive results were registered. Experiments were conducted around drilling platforms ; claims of reduced scour were made.

Erosion reduction experiments were already tried in 1963 in the Danish North Sea Coast, followed by others in the United States (Ocean City NJ, Wallops Island VA), Great Britain and France (Saintes Marie de la Mer) ; all failed due to anchoring problems. In France, an apparent success in calmer waters was eventually disputed and considered anti-scour action. Of the numerous later tests two were hailed as successes but challenged afterwards : at Cape Hatteras SC the results were attributed to shoreline modification and at Long Beach CA profiling showed erosion over the reef.

Significant changes in elongation, strength weight and flotation showed among seven to thirteen month old samples. Seascape completely failed within eight to 20 months of installation, depending on the type of wave energy, environment and local water quality conditions. If it at first appears cost effective, annual replacement costs render the technology a cost-ineffective solution to erosion in Barbados. Seascape (R) Synthetic Seaweed (SSS) has been promoted in the United States since 1981 as a measure of controlling shoreline erosion. Made of spun bound polypropylene, the units consist of a weighting anchor tube, a mass of vertically floating fronds to capture sand, and foam attachments to improve buoyancy of fronds and to facilitate recovery after flexing. Fronds reduce current flow so as to allow sand to drop from the water column to form an offshore sand bar which reduces wave energy and prevents loss of sand to the offshore. Therefore sand must be available, stirred up, and in suspension.

Reports by Hall, Leadon, Spencer and others concluded from projects conducted up to 1985 that, neither inside or outside seaweed "sheltered" areas, is there any meaningful beach profile modification and state that there is "not a single documented case where artificial seaweed has been successful in controlling shoreline erosion."

Researchers at Scripps Institution of Oceanography at La Jolla CA placed 300 units of *seascape* in service in November 1986, with the cooperation of a member of the staff of *Seascape*, an artificial weed manufacturing company. Three B rows were placed 2 m (6 ft) apart. Previous experience had already led to a substantiated report.

The experiment disclosed the possibility of an additional "handicap to an already long list" : fouling of the artificial seaweed by the natural variety where it is present ! It showed no measurable effect of the seaweed on seasonal erosion losses, and burial of the fronds within a few months due to anchor-induced scour around their base.

In vitro experiments have not been encouraging : at Wallingford (U.K.) onshore transport was retarded by artificial seaweed and the U.S. Army Corps of Engineers, at Fort Belvoir VA, found 12% wave height attenuation for short period waves only which account for only a minor part of ocean energy.

Commercial brands of artificial seaweed include *Seascape, Linear Composites, Sea Leaves* and *Sea Grid.* The Ruhr Universität (Bochum, Fed. Rep. Germany) developed the IMRA system in 1981.

Conclusions reached for artificial reefs cannot be necessarily extrapolated for natural seaweed formations : these have in their stalks gas filled floatation nodules that make them more buoyant and have much longer fronds. Natural seaweeds do retard water motion and move more as compared to wave oscillation making them more effective.

Besides artificial and natural reefs schemes, proposals have also been made to place structures or "sparks" to promote aquaculture or myticulture. These arrangements can play a simultaneous role in coastal protection.

Marsh grasses have been effectively used to control erosion along estuarine shorelines. At sites with relatively low wave activity, marsh and low cost breakwater can offer protection at significantly lower cost than several erosion control alternatives. Rogers based his conclusion on over 25 years of observations. [M. Rogers Jr, 1989, Erosion Control : marsh and low cost breakwaters : *Coastal Zone* '89, I, 781-764]. Successes are numerous but failures have also been common.

Success has been attained along shorelines that have fringed marshes but were denuded (*viz.* stripped).

The recolonization by planting appropriately can re-establish a marsh in two or three years, rather than decades.

Exposed marshes have been eroding in North Carolina at over 1 m/year (3.28 ft/year). Planting success is unlikely if the fetch is greater than 4,8 km. Along with fetch, the shoreline configuration (cove, straight or headland) and sediment grain size are the best indicators of successful plantings.

A fully established marsh is reasonably stable and can provide the divided shoreline protection but the outer edge of the root mat begins to erode, gradually resulting in the destruction of the marsh and the loss of any erosion protection for upland property.

A 3 to 7 year functional lifetime is relatively short compared to the 20 to 30 year expected lifetime of other erosion control options. Several types of structures have been used it conjunction with marsh grass plantings. Temporary structures can reduce the time it takes the planting to become fully established. Permanent structures have been used to maintain marshes in higher wave climates.

Hay bales as temporary protection bales were washed out by storms 5 weeks after placement. Two tiers of tires permitted planting survival and growth was considered successful after two growing seasons. Half the length of the breakwater was removed at that time. Approximately one quarter of the developing marsh width was rapidly eroded along the seaward edge. The remaining marsh appeared to stabilize during the second year after removal.

A fixed breakwater of wooden posts and metal planks supporting tires began to deteriorate soon after construction and the planting failed, but a floating tire breakwater provided sufficient protection for the site to be fully vegetated though partially sank due to sediment accumulation in the tires. Growth was slower were less protected.

Breakwaters provide temporary protection for marsh establishment, but are not suitable for permanent installation. Tires, an inexpensive raw material, are difficult to connect and anchor as functional wave barriers. A tire breakwater requires high maintenance and for marsh protection is not low cost.

Grasses cannot stabilize areas outside a breakwater and unprotected shorelines will continue to erode. When a breakwater protecting a marsh was removed the shoreline response was rapid : marsh seaward of the bulkhead washed out in less than six months. Two years after removal of breakwater, only a few remnants of peat remained and the beach had almost disappeared. The combination of a low inexpensive breakwater and marsh grasses clearly can control erosion for an extended period of time, but even a well established marsh cannot survive the wave climate after breakwater removal. Erosion control with vegetation alone will not be effective unless a small but permanent wave barrier is maintained.

Marshes appear to prevent erosion of the upland shoreline because the array of stems substantially reduces wave height as a wave passes through a fringing marsh and the plants eventually establish a dense mat of roots and partially decayed plant material. marshes appear to provide reasonable erosion protection for upland banks but in higher wave climates evolve to be erosion sensitive themselves.

Protected with a reliable, low cost structure, the combination of marsh grasses and breakwater should provide upland erosion protection for a wide range of wave exposures and for a long functional life.

An experiment near Duck N.C. using a combination of a small wooden breakwater and planting marsh grasses showed substantial potential for controlling shoreline erosion. In a time of rising sea level and concern over shrinking marshes, the combination of low breakwater and planted marsh appears to offer substantial increases over the reliability of the marsh planting zone and still cut the cost of shoreline protection by half of that bulkhead. Along with unprotected marsh planting, the breakwater appears to be one of the few erosion control options which is a clear environmental asset.

2.1.8 Tetrapods and Similar

Quarried rock is not readily available at all sites and its transport being onerous, lumps of "false rocks" have been used. In Japan 50-ton concrete tetrapods have been used to protect the east coast of Honshu. Tetrapod barriers have been built at Silt (Fed. Rep. Germany). Similar are tribars, tripods, dolosses and hanbars[14a]. The hexagonally shaped Australian seabee, a rather recent development, can be manufactured from a variety of materials, among others concrete and vitrified clay. They can be used to build sea-walls or breakwaters. The honey-combed structure was selected for projects near Sydney (Cronulla Beach) and Papua New Guinea (Wewak Causeway in the northwest).

Tetrapods, a double layer even, have been used to armor a breakwater at Rota in the Bay of Cadiz (Spain), more than 1000 of them, each weighing 25 metric tons (27.6 US t). The toe of the structure is protected by concrete blocks of 250 m.t. Nevertheless considerable destruction took place during the 1958 major storm. Burger and his collaborators made full scale tests on tetrapods' resistance (W.W. Burger, H. Oumeraci and H.W. Partensky, 1990, Stresses in tetrapods : results of large scale model tests : *Abstr. 22 I.C.C.E.* (Delft, The Netherlands) 513-514).

It has been suggested (J. De Rouck, *op. cit.*) to use HARO blocks, mentioned earlier, instead of tetrapods, a.o. reasons because the quantity of concrete is then reduced by approximately one fourth. This armoring element has a considerably greater hydraulic stability than cubes and tetrapods, roughly the same as an accropode (Anonymous, 1990, Accropode : *The ideal block for efficient and economic protection of breakwaters* : Grenoble, Soc. Gren. A. Hydr. - SOGREAH), and has in many case a performance better than or equal to dolos.

Quadripods have been used as armor to reinforce breakwaters. The tetrapod is an unreinforced nonblock concrete unit with four truncated conical legs projecting radially from a center. Quadripods are a U.S. newer development of the French product. The tribar is an assembly of three cylinders connected by three radial arms. A dolo resembles a ship anchor ; its straight flukes shaped central shank ends are replaced by triangular heads in a toskane. Modified cubes and hexapods have also been tested. It should be recalled here that concrete armor units have breakage problems which are not encountered with quarrystone.

The above listing of units is not comprehensive : about 40 types are in existence (cf. Hudson, R.Y., 1974, *Concrete armor units for protection against wave attack. Report of the ad hoc Committee on Artificial Armor Units for Coastal Structures* : Vicksburg MS, U.S. Army Engineers Waterways Experiment Station (Document H-74-2)). They were developed in ten different countries, principally the United States, Great Britain, Japan and The Netherlands.

2.1.9 Others

Several other methods and designs have been proposed ; some are proprietary. Beach nourishment (cf. section 3) and vegetation are among the most effective. Timber cribs filled with brush constitute breakwaters that can allow vegetation to become established. Vegetation stabilizes dunes and protects marshes, enhancing simultaneously the environmental aesthetic attractiveness. Its roots and stems traps fine sand and soil, and in marshes absorb some wave and current energy. Types of plants must be carefully selected, and since they are usually less resistant than the natives ones, sand fences should be used during their colonisation period of the area.

Low retaining sills can also trap sand. Being submerged, they have no unaesthetic effect. They create a beach at a height higher than the original level ; such *perched beaches* are generally built where nourishment at the original level would be very costly due to rapid removal of fill material. They do offer some shore protection. Materials and designs are similar to those of breakwaters but must be of interlocking tongue-and-groove elements or have filter material. The beach itself can be artificially nourished (cf. 3) or be built from inshore wave carried sand but then with risk of depletion on adjoining beaches.

It has but rarely been suggested to construct artificial islands or reefs to counter erosion of a shoreline (cf. 2.1.7). Nevertheless the protection role played by a proposed "environmental" island offshore the Belgian coast has been mentioned as a side "benefit". This island would be the site of a waste processing plant (R.H. Charlier and C.P. De Meyer. "An artificial, 'environmental' island offshore the Belgian coast" *Proc. Oceans* '87 MTS/IEEE Int. Conf. (Halifax, N.S.) (1987).

New designs for "hard" protective structures suggested include a variety of groins : stacked bag filled (with sand or grout), timber crib and timber plank groins. Longard tubes can be used in groins, breakwaters and bulkheads. Sheet pile groins and bulkheads - made out of timber, steel or aluminium - are in use, and gabions can be groins or breakwaters. Sandgrabbers, surgebreakers and Z-walls are patented breakwaters.

2.2 Environmental Aspects

Groins and breakwaters will only affect the wave regime and sediment transport in the immediate area. They do not provide solutions for the adjacent coastal areas where the sediment transport capacities are unchanged. To the contrary, they may cause a spectacular increase in sediment transport on the lee-side of the structure causing renewed coastal erosion. In fact, the problem has merely been moved from one area to another.

Although reports have overwhelmingly faulted sea-walls for causing beach erosion, some authors maintain that in several cases this has not been proven[19].

If recreational aspects are important, any coastal structure (e.g. groin) on the beach should be avoided. Furthermore, considerable sums of money have to be repeatedly spent on maintenance and repair due to storm damage to the structures.

In "Impact de la Construction de la digue portuaire de Mohammedia sur les plages avoisinantes", a poster at Bordomer '95, J. Viguier points out that prior to the construction of the seawall, the beach was in virtual equilibrium with a slight accretion at the western end and some erosion at the east end. Erosion was in progress at Monika beach, but it had been "mined" considerably to build a "development", in 1970, in the central area of the high beach. That development is threatened by the sea today.

Construction of the seawall has brought about a widening of the beach, in the western part, of 100 to 120 m (328 to 390 ft), and a loss of 20 m (65.6 ft) in the east of Mohammedia Beach. Monika Beach remains unaffected. But wave effects, under the new wave climate, are posing problems for petroleum facilities. Another extension of the 2,700 m (1,684 mi) long seawall is proposed : some 600 m (1,854 ft) are to be added. Modelling shows that this will result in severe beach modifications : near the old port (west) a widening of 100 m (328 ft), shortening by 200 to 300 m (656 to 927 ft) at the eastern end of Mohammedia Beach, accretion on Monika beach (west) and a loss of 10 m (32.8 ft) width on the right of the development which will be placed in extreme jeopardy.

As a remedy, a "mixed-approach" has been suggested : building of from one to three groins on Mohammedia Beach and artificial nourishment, using sand or crushed limestone, (with a groin field and/or breakwaters) at Monika.

3 Beach Nourishment as Coastal Protection

Appropriate coastal protection solutions are provided by a durable defense system and modifying the beach material itself.

While it is generally thought that the oldest beach nourishment operation took place on the U.S. Pacific coast, and the first one in Europe near Cannes (France), a claim to that fame is laid by Turkey where a white beach on Sedir Island (Bay of Gökova) would have been built up on the orders of Marc-Anthony as a gift to Cleopatra, Queen of Egypt, who had oolithic sand brought from Egypt. The beach is eroding because ... collectors are gathering the oolithes !

3.1 Principles of Beach Nourishment

With the more extensive knowledge of coastal dynamics and the availability of physical and complex mathematical models morphological processes and modifications can be predicted to a considerable extent. Additionally, technical advances in dredging equipment (trailing splithull type, hopper-suction dredgers and sea-going cutter dredgers) provide new and more economical possibilities for beach restoration.

Beach nourishment has been considered a potentially desirable form of shore protection. It is beneficial as wave action spreads the material along the shoreline. Costs have been lowered by use of booster pumps allowing long distance material piping, and the use of hopper dredges which can dredge, deposit pump and transport. Navigation channel maintenance dredged material can be used for beach restoration. Hence both deposition of sand and beach fill are suitable techniques. Beach nourishment is constant with natural coastal processes, augments the natural transport of coastal sediments and use of dredged material keeps it into the littoral system. No substantial biotic effects have been observed.

3.1.1 Hard and Soft Approaches

The earliest beach nourishment project may well have been the dredging, in 1920, of Santa Barbara, CA harbor and deposition of the material on the eroded beach downcoast of the breakwater[19a].

White reported that with beach nourishment involving over 2.500.000 m^3 (3.280.000 cu.yd) of material deposited over three years the beach original fauna was destroyed but a new fauna has developed[19]. Assessments of artificial nourishment of beaches have been made for over thirty years and are still carried out on a frequent basis[20].

Segmented offshore breakwaters were built based on the premise that direct wave pounding on the beach would be attenuated, cause a transformation of the incoming waves and induce deposition of drifting sediment in the lee of the structures (fig. 31). The ensuing beach salient may even become connected to the breakwater segment and form a tombolo.

Wave energy is reduced in the spaces between the successive segments as well. A beach is thus constructed but a certain amount of sand is removed thereby from the longshore transport system.

Beach nourishment has been eyed by the U.S. Army Corps of Engineers as a method to dispose of the 265.5 million m^3 (300 million cu.yd) of dredged material it picks up annually. This material is usually dumped at sea damaging benthic life. Such disposal can also help portuary development and marsh creation. Under the proper circumstances the use of dredged material for beach nourishment not only contributes to the development and maintenance of waterway transportation, but also provides hazard reduction and recreational benefits consistent with environmental mandates and the material laws governing coastal systems.

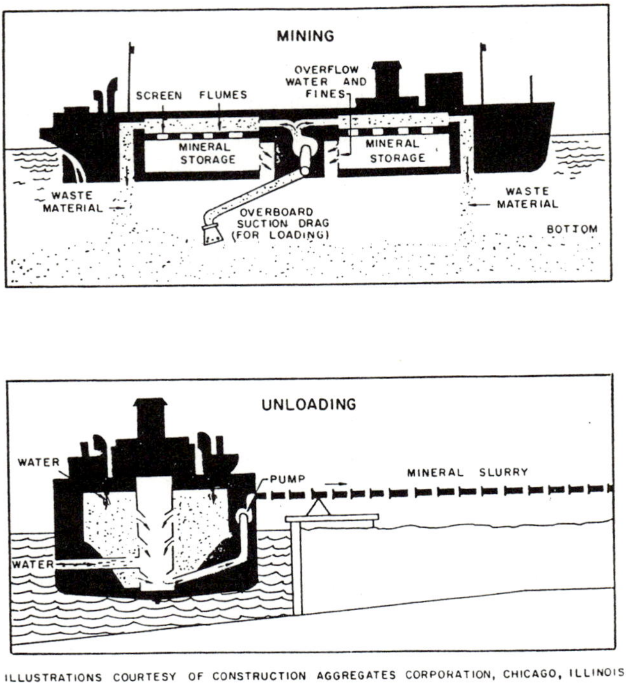

Fig. 26 : Hopper dredge for marine mining use

Beach nourishment is the least-cost environmentally acceptable disposal alternative. Despite its current prevalence, beach nourishment cannot be considered a panacea for dredged material disposal ills. For three out of every four navigation projects, the dredged material is not a practical, cost-effective, and permissible source for beach nourishment supply. Most frequently, the material is too fine-grained to be effective for shore protection. Other constraints are excessive cost, impermissible environmental effects, and technical limitations.

Dredgings have also been dumped at shallow depths off the beaches of Agadir (Morocco) and Anglet (France). They have contributed to the "rebuilding" of the starved beaches.

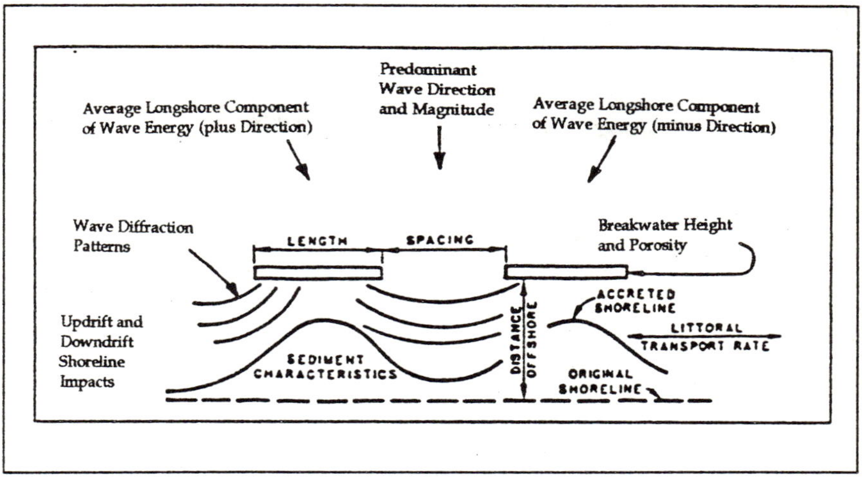

Fig. 27 : Segmented offshore breakwater considerations

In Chile (Chañaral), mine tailings automatically feed as much as 4,400,000 m^3 (155.4 million ft^3) to beaches but the copper content is high, creating a pollution problem affecting fisheries.

The United States hold the record, at least by the number, of artificially nourished beaches. Results have varied widely. In Ocean City MD, thirty replenishments have yielded all but meager results. But several new sites have benefited from such intervention. In Rosslare (Ireland), it was called upon to redress damage done by hard structures. In Seaford (England) £ 9 million (about US$ 15 million) were allocated for shingle replenishment eight years ago.

Perhaps, one of the most successful acitons are the works at Rio de Janeiro's Copacabana beach where the 2 million m^3 (70.6 million ft^3) were placed in shallow water and 1.5 million m^3 (53 million ft^3) were pushed into the beach. The beach gained 90 m (295 ft) and no new fill was needed subsequently.

Artificially nourished beaches loose the material they were fed at various rates. Losses are faster after the first supply has been furnished while a new foreshore profile develops. Evidently, an artificially nourished beach is not as stable as one being naturally recharged. The material will be better retained in protected bays. Apparently, hydraulic sluicing yields better results than truck dumping of "nourishment".

The Sochi beach (Georgia, Caucasus) requires a yearly addition of 70,000 m³ (2,472,000 ft³) after half a million m³ (17,667,000 ft³) were added from 1989 to 1986. The Croisette beach (Cannes, France) remains stable. The Prado beach (Marseille, France) was entirely created by man. Chatelaillon (France) needs yearly 3,000 m³ of sand (105,944 ft³).

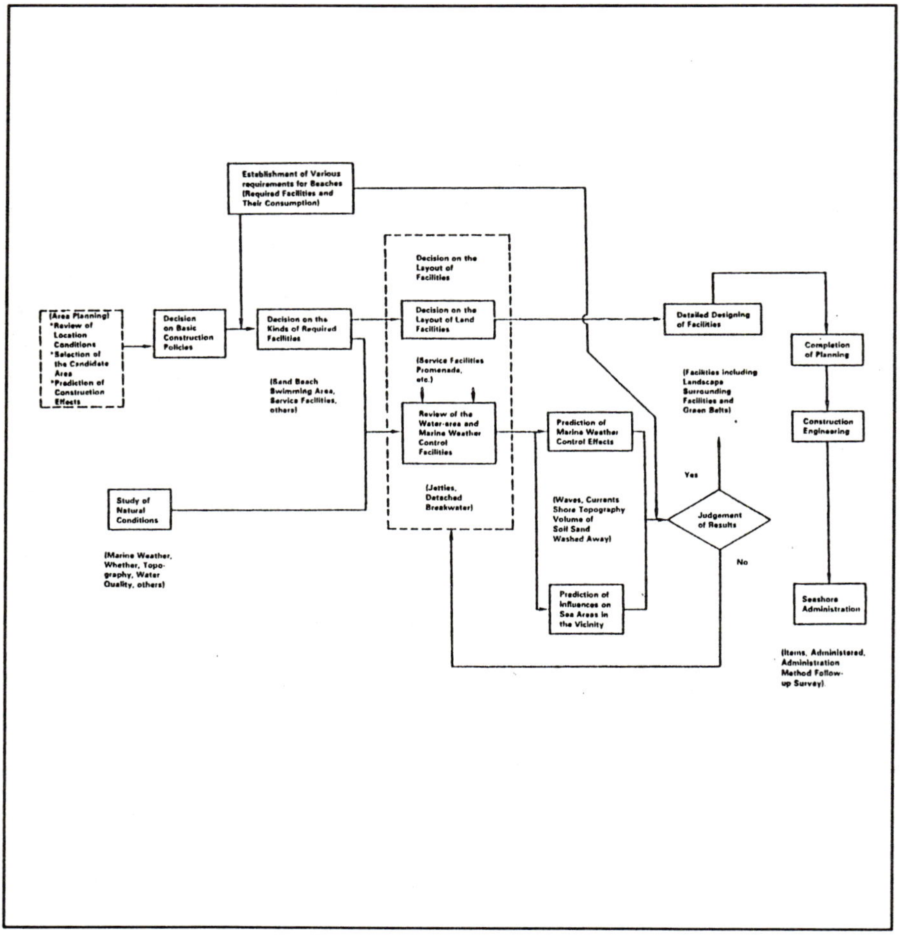

Fig. 28 : Gold Coast alternative nourishment placement zones. (Smith & Jackson 1990)

To balance this loss, an artificial nourishment is desirable equaling the amount removed ; this approach counteracts the negative impact the structures would have outside the zone where they have been placed. However, local conditions may be such that artificial nourishment does not eradicate adverse impact.

Breakwaters are however expensive structures and the lack of examples of segmented systems act as a deterrent to funding. Yet they have been in use in Japan for more than thirty years, generally built rather close to shore20. Others are currenlty found in Denmark, France, Italy, Israel and Singaporte. In the United States, several segmented breakwaters have been constructed in locations spreading from Boston to Virginia, and on the Great Lakes.

Many projects combine sand suppletion with construction of hard defence structures. This was done in Monaco, Israel and Singapore (fig. 7, 8 and 9, section 1.3). This is a commonly recognized approach especially where suitable material is available close-by.

It has often been suggested to counter recession of weak outcrop cliffs by building and maintaining a broad beach, through the construction of groins that intercept longshore drifting, and supplemented by beach nourishment on the foreshore or updrift[21]. When such groins intercept longshore drifting erosion may, however, be accentuated because sediment is withheld from the downdrift area, as has happened in the Newhaven area (Sussex, Great Britain). With sea walls, fronting beaches are rapidly depleted by scour due to the reflection of storm waves from that wall.

However, in some cases, as for instance in Belgium, depending on predominant wave direction governing the littoral drift, simple sand suppletion alone can provide adequate coastal protection (fig. 26). In Westhampton, NY, when sand addition was stopped but more groins added storms stripped the sand beyond the groins field in downdrift direction.

3.1.2 Placement of Material

By adding new sand to a beach, but not to the immediate offshore area, the beach slope may become out of equilibrium with such adjacent underwater area and steepens shoreface profile. So as to flatten the profile, waves, during storms, remove sand and the "shoreline" is again rolled back landward. This must be carefully reckoned with when undertaking an artificial accretion project. The slope of the "filled beach" should match as far as possible the slope of the original beach.

Evaluation of the optimum position and method of placement of the nourishment, sand is required and the number of choices is limited, but their consequences can vary significantly. Basic choices consist in placing the nourishment material on the natural dune line or on the dry back zone of the visible beach and into the swach zone ; or most of the material in the swash zone and the rest into the water out to the fine weather breaking line ; or still place all the nourishment materials on the seabed offshore of the fine weather breaking line.

In the first three options, placement is effected by pumping or truck dumping, allied with some profiling where necessary. The fourth option calls for a decision as to where to place the material on the seabed and how. On a high wave energy coast the work can usually only be done by a floating plant ; i.e. dredges and hopper barges, or vessels that combine both these capabilities. Dredged sand can be pumped from them into shallow water and onto the beach, but for placements further offshore in deeper water, dumping from a hopper vessel is usually the most economic procedure. An offshore placement is thus determined by the draft of the vessels available and the thickness of the dumped material on the seabed. A dredge hopper vessel grounded on its own load is the last thing required. (Fig. 26).

The sand source for offshore nourishment placements is usually from the deeper seabed offshore of the normal active zone, where the sediment mobility is low, its backfilling rate is very slow and its response is largely dominated by only the larger storm wells. The final option that requires consideration (last choice) is whether to spread the nourishment evenly over the offshore shallower seabed (Profile Placement) or concentrate it as heaps along a line nearly parallel to the beach and form a seabed structure that imitates a natural offshore storm bar (Bar Placement). The optimum combinations of options for an offshore placement becomes quite complex for it must consider public relations, politics, nourishment volume, mechanical plant and vessel size and delivery rates, the probably effective "life" of the placement and combine all these factors to attain the maximum economy and public endorsement. The worst enemy of economy is double handling : double the work, double the cost. (Smith, A.W. and Jackson, L.A., 1990, The siting of beach nourishment placements : *Shore and Beach* 58, 1, 17-24). (fig. 27).

Usually, the actual field location is determined before the work is even provisionally designed. Sometimes the reverse applies : the placement location is dominated by the field conditions, so the other design factors have to be modified to suit. Sometimes, both the type of plant and the placement location are determined by the anticipated weather conditions most likely to exist when the work is to be done ; but in the end, the finally selected placement location holds by far the greatest impact upon the behavior of the nourishment material, both during, and after the work is complete.

The number of choices for a nourishment placement is quite small in the dune zone ; on the beach ; in the swash zone and through the fine weather breaking zone ; distributed over the near beach seabed ; on the seabed in the storm bar location.

At Kings Bay, Georgia (U.S.A.), 688,00 m³ (900,000 cu. yd.) were to be excavated from the bay channel section to a depth of 15 m (51 ft). The work has been performed by cutter dredgers and all of the material pumped through pipelines to replenish local beaches. Still at Kings Bay, the excavation of 4.3 million m³ (5.6 million cu. yd.) from the entrance channel section of the project was performed by hopper dredgers and bucket dredgers. The material was then hauled to an open water disposal area located in the Atlantic Ocean.

At Mobile Harbor, cutter dredgers excavated maintenance material, pumping through relatively short pipelines to open-water disposal areas located in the bay near the channel. The bay channel improvement work was done by bucket dredgers and dredged material transported by hopper barges over an average one-way distance of 40 km (25 miles) to an open-water disposal area in the Gulf of Mexico. Hopper dredgers excavated the Bar channel and transported material about 6.2 km (4 miles) to an open-water disposal area in the Gulf. All the material deriving from the improvement work was used to construct a stable underwater berm.

For the Kill van Kull project in New York, bucket dredgers loaded with hopper barges, transported the dredgings to an open water site in the Atlantic Ocean. (Murden, W.R., 1987, The Station of Dredging in the USA, now and in the near future : *47 Mtg. Coast. Eng. Res. Bd* (May 19), 21-24).

As 95 percent of the material, excavated each year in the U.S.A. is not polluted, underwater berms provides a unique and beneficial alternative to the traditional disposal of dredged material.

Of the two viable approaches to the construction of underwater berms, the feeder beach concept involves the placement of beach quality sand in relatively shallow water, with the expectation that the material will move downdrift, but remain in the nearshore coastal region, or area between the surf zone and the 9 m (30 ft) isobath. A stable berm involves placement of dredged material in deeper waters, less subject to waves and currents. The justified expectation is that the material will remain in place. In contrast to the feeder berm, a wide variety of dredged material (silts and fine-grained sand) can be used in the construction of stable berms.

The objective of the feeder placement concept is to add sand, which will result in a less steep underwater slope and contribute to a reduction in the erosion effects on the shoreline. Dredging costs can be reduced, nearshore disposal sites being generally closer to the dredging sites than designated offshore disposal sites.

Already in 1982, dredging equipment routinely used, was capable of constructing significant mounds, even when the material is a mixture of fine sand and silt particles. Mounds, near Norfolk VA, have remained stable in that area despite three hurricanes which have struck the area.

3.1.3 Monitoring, bypassing

Monitoring techniques apply to both the feeder and the stable berms : in the case of the feeder berm, material will move downdrift while a little stable berm is expected to remain stable. An unacceptable movement would lead to a return to conventional open-water disposal (id. *op. cit.).*

Fathometer surveys will be used to measure berm elevations and side-scan sonar will help define the boundaries of the berms.

To track sediment movement, grab samples of surface sediments will be taken. In addition, divers will take core samples to assess to consolidation of the berm material. Waves and currents will be measured and bottom currents tracked, using bed drifters, and wave climate in the vicinity of the berms will also be measured. Aerial photography and satellite imagery will measure any large movement of the berms and detect turbidity plumes that may develop in the stable berm. Effects of the stable berm on fisheries will be evaluated.

An overfill ratio should not necessarily be applied to the entire fill volume when reclamation as well as beach building is involved. The relative proportion will depend on the rate of filling and the extent of the active beach zone.

If the profile formed during material placing is at substantial variance with the natural equilibrium profile for the imported material, initial losses may be very large.

Bypassing is one of the techniques of artificial beach nourishment : sand is "bypassed" from one side of an inlet to the other, reestablishing natural processes.

One advantage of the nourishment approach is that it can easily be adapted to changes in coastal patterns. So-called permanent structures are not flexible and may become harmful in a new situation, necessitating removal. In the long run artificial accretion is less costly.

A beach nourishment program has several options, each with its own merits, some placing an important part of rebuilding the strand to nature. The latter provide far less immediate results. Some options may even be considered more of a maintenance action.

As said earlier, beach nourishment has been eyed by the U.S. Army Corps of Engineers as a method to dispose of the 265.5 million m^3 (300 million cu.yd) of dredged material it picks up annually. This material is usually dumped at sea damaging benthic life. Such disposal can also help portuary development and marsh creation. Under the proper circumstances the use of dredged material for beach nourishment not only contributes to the development and maintenance of waterway transportation, but also provides hazard reduction and recreational benefits consistent with environmental mandates and the material laws governing coastal systems. Beach nourishment is the least-cost environmentally acceptable disposal alternative. Despite its current prevalence, beach nourishment cannot be considered a panacea for dredged material disposal ills. For three out of every four navigation projects, the dredged material is not a practical, cost-effective, and permissible source for beach nourishment supply. Most frequently, the material is too fine-grained to be effective for shore protection. Other constraints are excessive cost, unallowable environmental effects, and technical limitations.

3.1.4 Nature of "Beach Fill"

Material to be used, carefully selected as to nature and source site, may be brought to the beach to be restored and dumped offshore, leaving it to the sea to carry it to the shore and insuring its geographical distribution. It may also be deposited at a single location on the beach; called ***stockpiling***, it allows the artificial supply point to act as a distribution center or ***feeder beach***. Still another approach is to ***spread directly*** the material on the endangered beach, restore it ***stande pede*** and take simultaneously appropriate "soft" measures to protect it. Finally the ***continuous supply method*** involves creation of a fixed, or semi-fixed plant which first intercepts and then passes on the littoral drift to the down-drift side.

In some cases a combined approach has been taken, e.g. offshore dumping and on-shore nourishment. Beach nourishment has even been called upon to restore lake beaches, for instance in the Province of Québec (Lac St. Jean).

The beach nourishment approach has been also used on lakes shorelines ; this has been the case at Lac St. Jean in Canada and on the shores of Lake Michigan, for instance in Evanston, Fort Sheridan, Highland Park, Lake Bluff, Illinois Beach State Park and at several beaches of Chicago (Illinois).

Sand nourishment of Chicago beaches between 1961 and 1964 involved deposition of 381.000 m^3 (498.300 cu.yd).

While the technique is the same for lake and oceanic beaches there are some environmental differences to be reckoned with.

Grain size distribution varies daily due to wave climate, storms, water height, but changes are less pronounced on lakes than on ocean beaches because of water level variations on lakes with ensuing base level readjustments.

Dune nourishment and artificial building may provide additional beach protection. The dune blocks wave attack and provides sand to the beach. Such an approach is being tested at Mantoloking, NY, where a 6m. (19.5 ft) high dune has been erected.

Materials site areas must be examined vertically and horizontally and cores taken several meters below the bottom. Such cores, compared with seismic data, will provide information about sediment nature in each layer and layer extent.

Fill should resemble as much as possible the original material. Coarser fill will erode slowlier but thinner fill will erode laster than the original beach material. Sediment selected for beach nourishment must be at least as coarse texturally as the original beach material. Beach nourishment using sand too fine leads to berm material erosion with offshore redeposition.

The source area must be so selected that no significant deepening of nearshore areas occurs, otherwise wave energy may be intensified and result in or speed up coastal erosion.

When dredging material for beach nourishment, care must be taken to select appropriate depths so as to avoid any beach damage. Somewhat coarser and better sorted sand that the beach material is preferable. Dredging up to 60 m (197 ft) poses no problem and piping it as far as 8 km (4.8 mi) is possible.

Shingle-beach nourishment schemes usually present fewer problems than sand beach nourishment where careful selection of the particle size of material is used is particularly important.

Related to "nourishment" in the broad concept is an effort by Erosion Technology Systems Inc., the developer of "Sea Grid" to counter shoreline erosion by fostering, using artificial reefs, the formation of energy-absorbing sand-bars offshore. No report, so far, has attributed any sand-bar building to any artificial seaweed action[18d].

Unintentional beach nourishment has also yielded good results. Near Workington (Cumbrian Coast, England) coast mining waste and industrial slag dumped there are used as beach nourishment material. In Denmark, at Hoed (Hød), flint gravel and limestone waste products are regularly dumped. Quarry wastes feed the Cornwall beaches at Porthoustock and mine wastes and slag at Porthallow Coal does fulfill the same role (table 2).

The following steps are suggested in the establishment and management of artificial beaches : establishment of various conditions affecting the beach under consideration and choice of the size and methods for the project, assessment of the environmental conditions, detailed facilities design, implementation and, after completion, beach administration (table 3).

The survey program related to a beach nourishment scheme involves bathymetric soundings, beach evolution (eventually by remote sensing) and monthly terrestrial measurements (beach profiling). At least once a year sampling should be carried out on the beach, the nearshore and the seabed. Winds, littoral drift and tides affect sediment transport and effects must be measured. The survey requires hence the gathering of tides, waves, currents and winds data (table 4).

For instance sediment transport study along the Zeebrugge to Zwin area of the Belgian East Coast, showed that tidal currents - irrespective of sedimentation effects - eroded some 250,000 m^3 (327,500 cu. yd) a year initially. With a new outer harbor at Zeebrugge jutting out 3.5 km (2.17 mi) into the sea, this removal reached 300,000 m^3 (393,000 cu.yd) a year, and if one groin were added beach regression would be double and 500,000 m^3 (655,500 cu.yd) a year lost. It should not be overlooked that with the shoreline pushed back 100 m (985 ft) the wider beach offers a large dry sand area for recreational use but also to eolian erosion !

Kerckaert *et al. (op. cit.)* have calculated that the wind removes 80,000 m^3/year (131,000 cu.yd/yr). An attempt to plug up this loss has been made by planting osier hedges and setting up wooden wind screens. Still another 420,000 m^3/year (550,000 cu.yd) is ascribed to wave action, bringing the yearly loss to 800,000 m^3 (1,310,000 cu.yd).

These mathematical predictions, however, proved to be grossly exaggerated but in the early 90s, another 1,000,000 m^3 (1,310,000 cu.yd) of sand were added to the easternmost beach area.

On the Strait of Georgia (Vancouver, Canada), in Monte Carlo (Monaco), at Hook (a.k.a. Hoek) van Holland (The Netherlands) and on Bora Bora, artificial beaches have been built. In Monaco dolomite chips were used instead of sand (fig. 7). In Vancouver large cobbles were used to stem sand cliff erosion, and such nourishment has apparently the desired effect. The use of calcium carbonate sands has been under study for some time[18e].

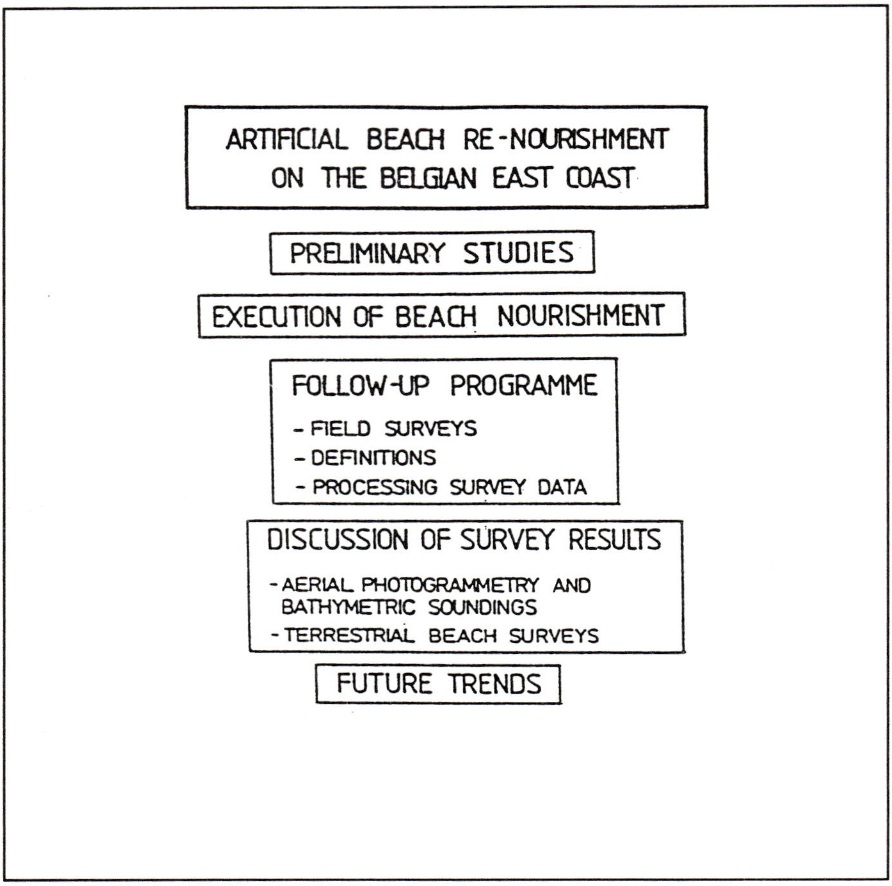

Fig. 29 : Flowchart of artificial beaches construction activities (From Nakayama, Kataoka and Obara, 1982)

3.1.5 Fill Factor Model

In a fill-factor model the compatibility of the native, *viz.* original, and the borrowed, *viz.* nourishment material, sediments is examined ; when they are not entirely compatible a fill is to be taken from a source area to produce one unit volume of sediment equivalent to the original material in grain-size distribution. This factor, also known as **overfill ratio,** is always larger than one.

The largest number of parameters is included in the adjusted SPM (Shore Protection Manual, U.S. Army Corps of Engineers) method fill-factor model. This method assumes that the fill material coarser than the stable grain size will be retained on a beach regardless of whether the grain size is proportional to the original material.

Various processes of the planning, designing, engineering and administration of artificial beaches	Matters to be elucidated Methods to be established
1. Establishment of various requirement conditions of the beach	- Elucidation of water-purifying functions of the beach. - Elucidation of the beach characteristics as the raising environment for various creatures. - Beach characteristics desirable as a recreation space including bathing place, consumption of various facilities, etc.
2. Decision on the kinds and scales of required facilities	- Selection of necessary facilities that match construction objectives of the beach. - Establishment of the decision method of facility scales.
3. Methods of laying out facilities	- Elucidation of mutual relationships among various facility function. - Study of how these facilities are actually used.
4. Estimation of marine weather control effects	- Prediction of wave-deformation in the Surf Zone (refraction, diffraction, reflection). - Prediction of beach current (especially in the vicinity of structures). - Prediction of exchangeability and diffusion of sea-water. - Review of beach stability. - Prediction of bottom quality changes. - Elucidation of functions of marine weather control structures (special weather-breaking facilities, etc.).
5. Detailed designing of facilities	- Safety-related considerations and measures. - Landscape-related considerations, measures, etc.
6. Implementation of construction engineering	- Measures to collect, transport and fill soil and sand. - Measures to improve the yield of soil and sand during construction.
7. Beach administration	- Selection of administration items and review of methods. - Follow-up survey, its analysis and others.

Table 2 : Knowledge and method required in the Process of the planning, designing, construction engineering and administration of artificial beaches. (From Nakayama, Kataoka and Obara, 1982)

Another mathematical model requires calculation of a **renourishment factor**. This factor provides an estimate of the frequency of renourishment with a specific fill required to maintain a beach's dimensions. The frequency is function of the difference in composite grain-size distributions between the original and the fill material.

The model allows an evaluation of different nourishment materials suitability in regards to maintenance costs. As coarser materials have a longer "on-beach" residence time, comparing grain-size distributions enables to predict a fill's long-term performance. A simplified mass-balance equation embodying arriving and departing materials, size and amount is :

$$\mathrm{Log}\,(R) = \Delta\left(\frac{\mu b - \mu n}{\sigma n}\right) - \frac{\Delta^2}{2}\left(\frac{\sigma^2 n}{\sigma^2 b} - 1\right) \qquad (7)$$

with :

R the renourishment factor
μ the mean diameter in φ units
σ the sorting coefficient in φ units
b indicates nourishment material
n indicates original beach material
Δ dimensionless parameter related to winnowing

The renourishment factor, or **relative retreat rate** is the ratio of the respective retreat rates of fill to original materials. **Winnowing** is selective sorting in the environment.

A value of 1 indicates that residence time of nourished and original material is the same ; values larger than 1 indicate the fraction of time the nourished material will be retained as compared to the original material ; inversely, a fractional value of R indicates how many times less often such fill will have to be replaced than the original materials.

This model establishes the frequency of renourishments needed to maintain a stable beach. On the other hand the fill-factor model attempts usually to determine the quantity of fill material needed for a unit volume of grain-size distribution as the natural material. Both aim at providing beach stability, but the latter method considers the original material as optimal for the beach and thus thinks in terms of such size material as fill, in other words to maintain or re-establish the status-quo[18f].

3.1.6 Profile Nourishment

A bottom profile follows usually the equation :

$$y^{3/2} = 0.05\ x \qquad (8)$$

in which y is depth and x is distance from shoreline to point with depth y (Bruun and Schwartz, 1985). The theory has been summarized as follows :

a) The profile is formed by the shear stress due to the wave action and is at right angles to the shoreline. The material detached by the oscillating water is removed by long-shore currents. As the shear stress due to wave action in general - and particularly during storms - is far greater than the shear stress originating from the longshore currents, this assumption cf.ms logical.

b) In the equilibrium profile the shear stress per unit bottom area (τ) may be assumed to be constant, i.e. the "condition" at the bottom is the same ($d\tau/dx = d\tau/dt = 0$). Confirmation of this assumption can only be attained by experiments. One obtains $\tau = k\sigma\ u^2_{ave}$, where σ is the density, k the resistance coefficient and u the bottom water velocity. If k is assumed a constant, then $U_{ave} \cong H\pi/T \sinh 2\pi y/_L$ is also constant where T is the wave period ; H, the wave height ; L, the wave length ; and y, the water depth.

c) dE_l/dx = constant, where E_l is the transported wave energy per unit area of the wave, and x is the distance from the shoreline. The loss of energy is made up of a loss by spilling of the wave and a (very small) loss by internal friction. The correctness of this assumption can only be proven by experiments. Calculations give :

$$x = L_O\sqrt{2\pi y}\left[2\left(\frac{2\pi y}{L_O}\right)+\frac{1}{3}\left(\frac{2\pi y}{L_O}\right)^2+\frac{43}{180}\left(\frac{2\pi y}{L_O}\right)^2\cdots\right] \qquad (9)$$

where y is the water depth and L_O the deep water wave length. The series is convergent for $y < L_O/8$, i.e. for storm waves on the Danish west coast out to depths of about 12 m (37 feet) where $L_O \cong$ 100 m (328 feet). Since $y \ll L_O$, the equation may be reduced to :

$$y^{3/2} = \rho x \qquad (10)$$

where ρ is a constant.
If it now is assumed that the loss of energy is due only to bottom friction and that this loss per unit area is constant, then :

$$\tau = k\sigma u^2_{ave} \tag{11}$$

where k = constant $(a/R)^{3/4}$, a is the length of the ripple marks and R, the half amplitude of the oscillating water motion at the bottom (R>>a). Calculations then give :

$$y^{3/2} = p \quad (y < \text{about } L_O/8) \tag{12}$$

This profile is similar to the one above. Certainly the profile depends on the wave period T, but as the profile is shaped mainly by storm waves and as the variation in T for these is small, the profile in reality will be the same as that given by (10).

Value of p is calibrated to local environmental conditions (waves, materials).

Pier Vellinga (*Beach and Dune Erosion during Storm Surges*, Delft University, Holland, 1986) found the following model law for profiles in the surf zone :

$$n_0 / n_d = \left(n_c / n^2{}_w\right)^{0.26} \tag{13}$$

where n_O is the length scale, n_d is the vertical scale and n_w is the settling velocity scale. Transforming this expression according to Stokes law for settling velocities, one has where D is grain size diameter, thence

$$n_1 \cdot n_D{}^{1/4} \cong n_d{}^{5/4}$$

In the prototype $n_l = n_d = 1$ (14)

one therefore has

$$n_1 n_w{}^{0.5} \cong n_d{}^{5/4} \tag{14'}$$

Putting $n_l = x$ and $n_d = y$ one gets

$$X \cdot n_w{}^{0.5} \cong Y^{5/4} \tag{14''}$$

A profile, however, has different grain sizes varying with depth and actual exposure. The k-value in the above expression [cf. (11)] for shear stresses = $k\sigma u^{max2}$ (U_{max} = max velocity of the orbital flow at the bottom) varies with grain size.

k may be written :

$$k = (D/A)^{3/4} \tag{15}$$

D = grain size diameter taken as D_{50}, A = semi-excursion of water participles right over the bottom.

In order to maintain the same shear stress over the bottom, τ must remain the same, or :

$$\tau_{max} = (D/A)^{3/4}\ \rho.u^2_{max} = \text{constant} \tag{16}$$

u_{max} can be written :

$$u_{max} = \frac{\pi H}{T}\left[\cosh\left(\frac{2\pi d}{L}\right)+\ldots\right] \tag{17}$$

This equation in its first approximation is fairly accurate for the area outside the breaker zone. One therefore has :

$$\tau = (D/A)^{3/4}.\left[\frac{rH}{T}\cosh\left(\frac{2\pi d}{L}\right)+\ldots\right]^2 \tag{18}$$

which shall remain the same over the profile subjected to shear stresses by the oscillating water movement brought about by the wave action. Consider one particular storm of extreme intensity and a narrow range of periods which is normal. Wave heights decrease relatively little moving from the offshore towards the relatively steep bottom. Energy loss will mainly take place by bottom friction and little or less by top-breaking and internal friction. The wave height H, in relation to depth d, may be approximated as :

$$H \cong d^{1/3} \tag{19}$$

or

$$H \cong d^{\,1/4} \tag{20}$$

Eq. (18) may for one particular storm wave period be written as :

$$(D/A)^{3/4}.H^2 \text{ constant} \tag{21}$$

	Iceland	South Coast
Depth = d	H ~ d1/3	h ~ d1/4
meters	D mm	D mm
6	1	1
12	0.7	0.75
18	0.5	0.6
18	0.5	0.6
24	0.4	0.5
	Danish North Sea Coast at Thyboroen	
6	0.22	0.22
12	0.15	0.16
18	0.11	0.13

Table 3 : Examples for Iceland and Denmark

Putting $A \cong H$

one has $D^{3/4}.H^{5/4}$ = constant (21)

and for $H \cong d^{1/3}$

one has $D^{3/4}.\ d^{5/16}$ = constant (22)

Similarly for $H \cong d^{1/4}$, one has $D^{3/4} \cdot d^{5/16} = \text{cons tan t}$ (23)

Using the two assumptions one arrives at the d/D ratios given in table 4 :

$H \cong d^{1/3}$: d -> D	$2d \rightarrow D = 0.7$
	$3d \rightarrow D = 0.5$
	$4d \rightarrow D = 0.4$
$H \cong d^{1/4}$: $d \rightarrow D$	$2d \rightarrow D = 0.75$
	$3d \rightarrow D = 0.6$
	$4d \rightarrow D = 0.5$

Table 4 : Ratios grain size diameter (D) to depth (d)

This is actually similar to what has been found so far, but more details have yet to be worked out. In reality, grain sizes will not become that differentiated because storms are composed of a spectrum of waves, not regular waves, and storms vary in intensity. Each storm tries to produce its own grain size distribution, as indicated in table 3.

The distribution in a less severe and in a severe storm may be as shown below for three different depths (table 4a).

Severe	0.4 mm	0.3 mm	0.2 mm
Less severe	0.3 mm	0.22 mm	0.15 mm

Table 4a. : Distribution of grain sizes in function of storm severity.

In addition, small storms will carry the smaller grain sizes out in less deep water for temporary deposition. This results in that bottom material in values which become somewhat mixed and is what is actually found, e.g. in Iceland. The ultimate result is still a slow migration of all grain sizes away from land until the single grain sizes come to a "rest". Silt and clay sizes will, however, continue their transfer to waters deep enough to make settling possible, if ocean currents do not interfere with the process. Grain sizes, therefore tend to become a little smaller than predicted by the ratio-procedure - and no wonder : samples were all taken during the summer season, when the coarser grains move shoreward, while the finer materials tend to move oceanward to deposit in the calmer offshore waters. This decreases the mean diameter and the sorting of the offshore material. In practical coastal-engineering planning, we must use the winter grain sizes corresponding to maximum exposure as design criteria, because they determine the erosion pattern (Bruun, P., 1988, Profile nourishment : its background and economic advantages : *J. Coast. Res.* 4.2, 218-228).

Dumping material on the beach is probably not the most economic nourishment procedure. Profile-dumping with material of selected grain sizes, depending on the availability of material, will undoubtedly prove to be more practical (and of course more economical) as it provides a higher degree of stability from the beginning[33].

3.1.7 Supplying the nourishment material

Mention has been made earlier of various approaches in securing the nourishment material. Such material may come from land sites but offshore dredging is perhaps currently the most commonly used method.

3.1.7.1 Dredging equipment and methods

Only dredging equipment generally used for beach nourishment, thus no bucket dredger or dustpan dredger, will be discussed here. Nevertheless under some circumstances other equipment may be used. However, the backhoe bucket dredger, a more recent type of dredger, will be mentioned.

3.1.7.1.1 Trailing suction hopper dredger

The trailing suction hopper dredger, or trailer dredger is a sea-going self-propelled vessel equipped with one or two suction pipes designed to trail over the side of the vessel (fig. 29). The suction pipe terminates, at the lower end, in a draghead designed to draw in the maximum amount of bed material. Suction is provided by a dredging pump, situated in the hull of the vessel, which discharges into a hopper in the vessel, where the soil (*viz.* sediment) particles settle and the water is discharged back to the sea over a weir.

Various types of draghead have their own particular advantages and disadvantages, depending on the material to be dredged. Generally, in soft materials the head slowly sinks forward into the sediment and highly concentrated materials are lifted. In harder materials penetration is shallower and a great forward speed is required to maintain output ; disintegration of the soil at the draghead is necessary to obtain a satisfactory output. High pressure water jets directed into the soil and the addition of teeth are used.

The trailer dredger's speed, generally slow, is varied to suit the material being dredged. On the way to the discharge site, speeds of between 9 and 14 knots are usual, depending on the vessel's size.

Maximum dredging depths also vary with vessel's size, usually ranging from 10 to 30 m (33 to 99 ft). To dredge at great depths some modern trailing suction hopper dredgers have their dredging pump located on the suction pipe well below water level, allowing to reduce the height of the suction column.

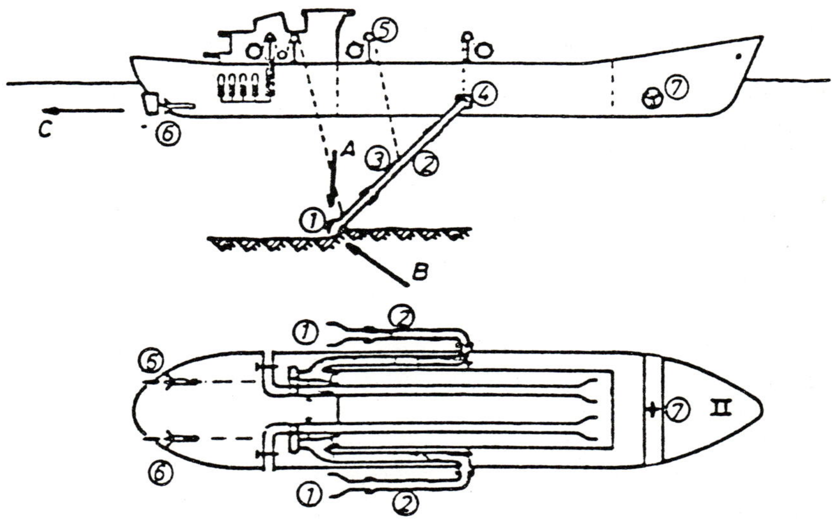

1. DRAGHEAD
2. SUCTION PIPE
3. FLEXIBLE HOSE CONNECTION
4. ELBOW HULL CONNECTION
5. GANTRY AND SWELL COMPEMSATOR
6. PROPELLER
7. BOW PROPELLER

Fig. 30a : Trailing suction hopper dredger

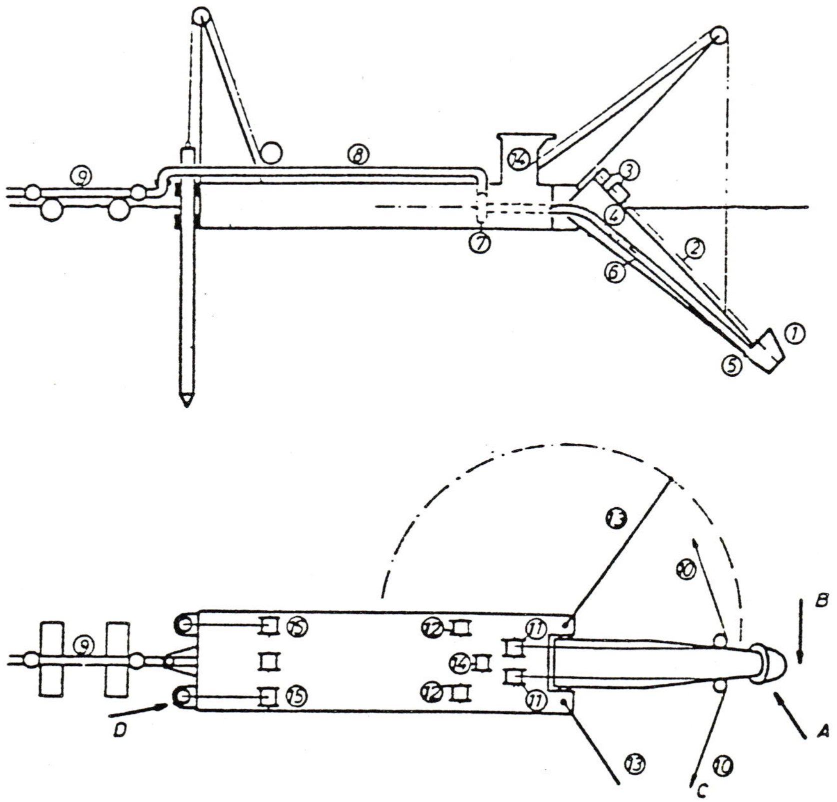

1. CUTTER
2. CUTTER SHAFT
3. CUTTER ENGINE
4. LADDER
5. ENTRANCE SUCTION PIPE
6. SECTION PIPE
7. SANDPUMP
8. RECLAMATION PIPE
9. FLOATING PIPELINE
10. SIDE WIRES
11. SIDE WINCHES
12. WINCHES ANCHOR BOOMS
13. ANCHOR BOOMS
14. LADDER WINCH
15. SPUD WINCHES

Fig. 30b : Cutter section dredger

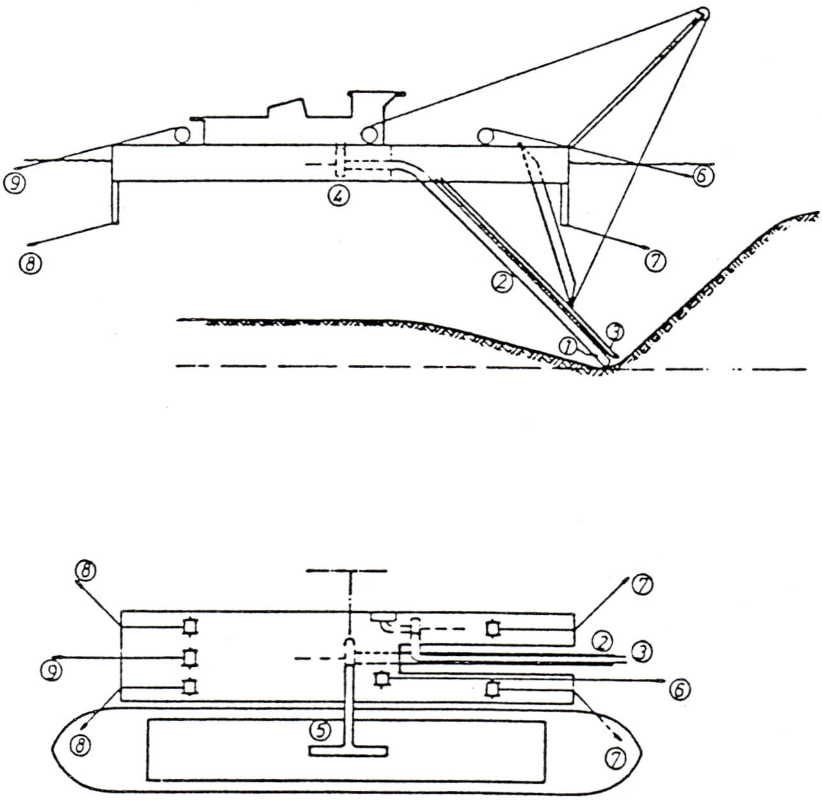

1. ENTRANCE SUCTION PIPE
2. SUCTION PIPE
3. HIGH PRESSURE WATERJET
4. SANDPUMP
5. SPRAYING TUBE
6. HEAD LINE
7/8. SIDE LINE
9. AFTER LINE

Fig. 30c : Stationary suction dredger

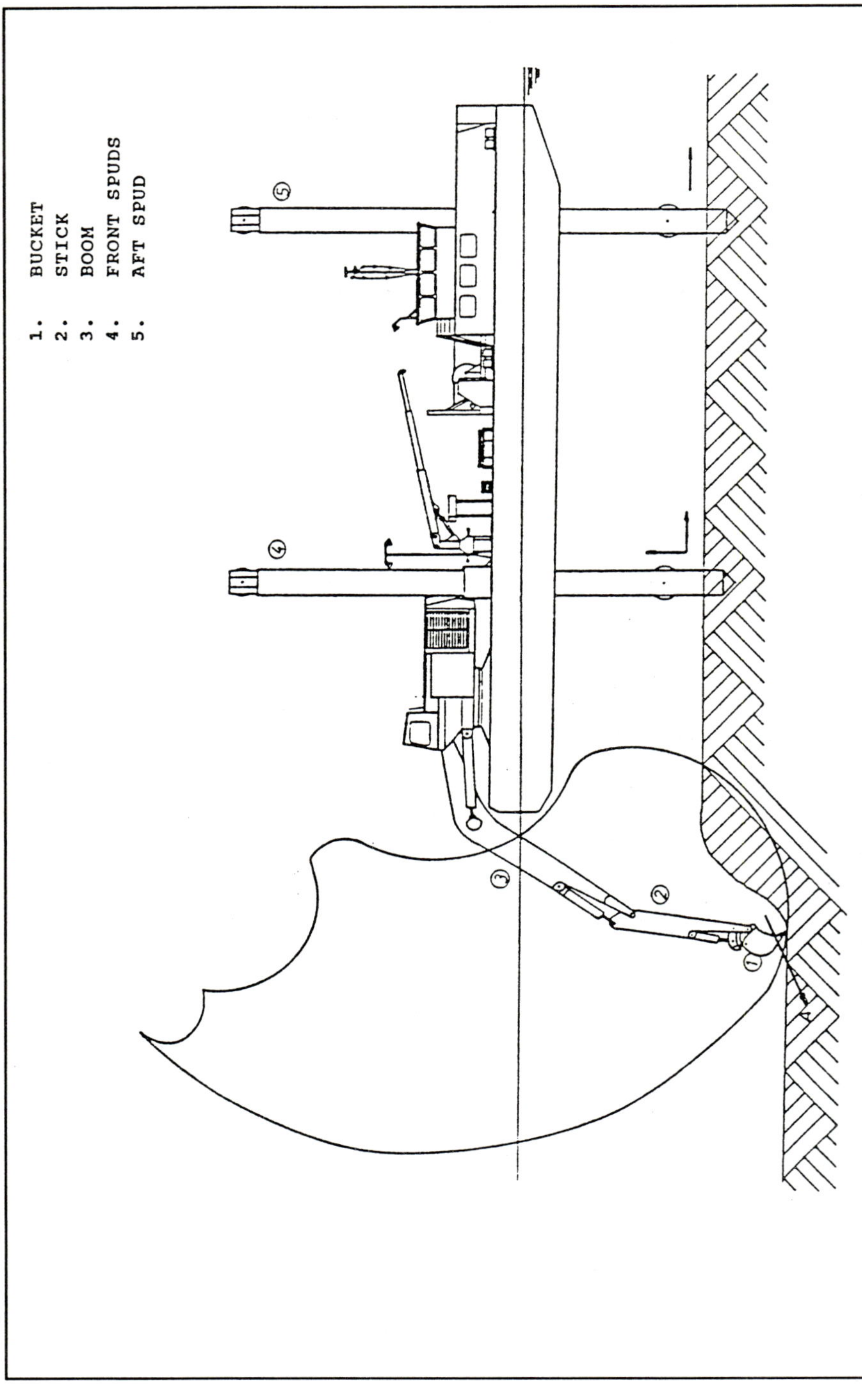

Fig. 30d : Backhoe dredger

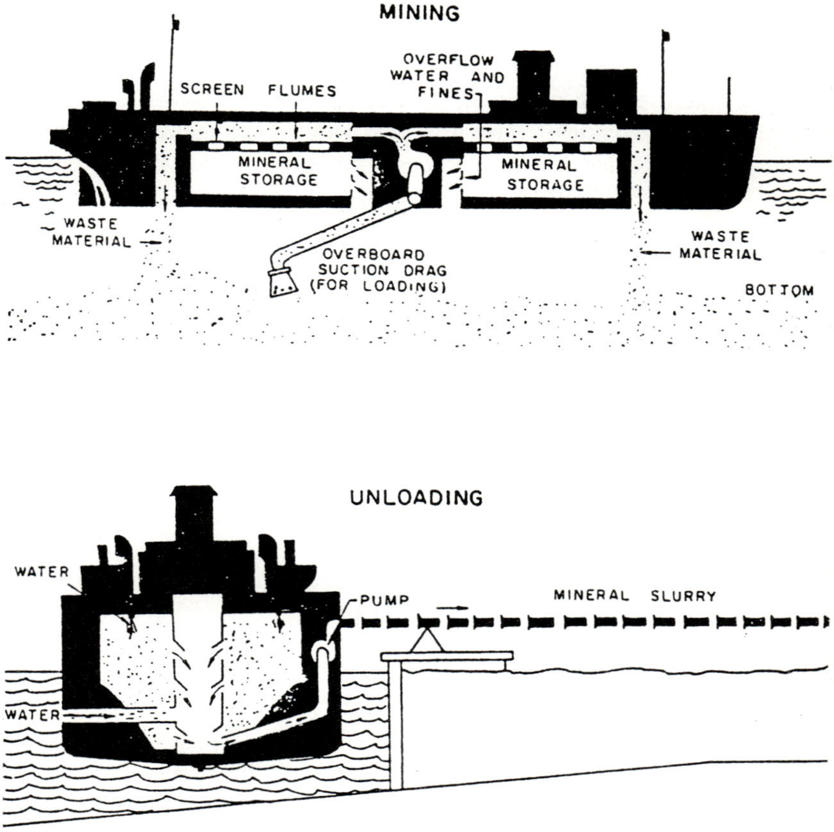

ILLUSTRATIONS COURTESY OF CONSTRUCTION AGGREGATES CORPORATION, CHICAGO, ILLINOIS

Fig. 30e : Hopper dredger for marine mining use

Discharge is normally effected by means of bottom dumping, either through valves or sliding doors or by splitting (splithopper) or where reclamation is desirable by pumping out the hopper into a land pipeline at a specially designed berth.

The shortest operational cycle of the trailing suction hopper dredger is a single run across the dredging area with following sequence : lower draghead ; sail across the site (dredging) ; raise draghead ; turn. The main cycle, for bottom dumping, then becomes : repeat subcycle as necessary ; sail to dump ; dump ; sail to site. The number of subcycles in a main cycle will depend not only on how long it takes to fill the hopper, but also to what extent the hopper is filled. When dredging fine materials, it is often economically desirable to sail to the dump with a load which is less than the possible maximum. Some dredgers are equipped with an adjustable overflow which allows to load a higher volume for lower densities in the hopper.

When the trailing suction hopper dredger is being used for reclamation purposes and is self-discharging ashore, the main cycle, for pumping ashore, is modified to the sequence : repeat the subcycle as necessary ; sail to berth ; moor and connect to pipeline ; pump ashore ; disconnect pipeline ; sail to site.

Trailer dredgers are usually rated by their hopper capacities (in tons) and the power of the sandpumps and propellers. Hopper capacities may vary from some 300 to 11.000 m³ (10,600 to 38,.900 ft³).

3.1.7.1.2 Cutter suction dredger

The cutter suction dredger is the most common of all dredger types. A cutter suction dredger has two main components : the cutterhead and the dredging pump. The cutterhead, located at the entrance of the suction pipe, is used to agitate soft materials or to cut harder materials to bring them to a suitable state for removal by hydraulic means. Cutters are usually rotated at between 10 and 30 rpm, and the rotary motor is located either directly behind the cutter in a submersible drive unit, or at the upper end of the ladder (fig. 29b).

The dredging pump, in the body of the dredger, creates a vacuum in the suction pipe and draws the soil up the pipe and through the pump. The soil is then discharged by being pumped through a pipeline. The raising of the dredged material is due to the pressure of the atmosphere acting on the water column in the suction pipe, thus there is a limit to which dredging depths can be extended before cavitation occurs. This explains why many modern cutter suction dredgers have a dredging pump situated well below water level on the ladder that supports the suction pipe, thereby, effectively reducing the heights of the suction column.

When in operation, the cutter suction dredger makes use of two stern spuds arranged so as to allow the dredger to advance in steps towards the dredging face. In each dredging position, the dredger is swung from side to side by means of side wires anchored either side of the dredging area, the machine being pivoted around one of the stern spuds. The suction tube may be raised or lowered by means of a hoist wire.

The cutter suction dredger provides an almost continuous operation, the dredging pump being stopped only when it becomes necessary to move the pipeline, either due to the advance of the dredger or in order to discharge at a new location.

The basic subcycle is then : cut ; advance on spuds ; cut ; advance on spuds ; (repeat as necessary) ; move side wire anchors. The main cycle consists of repeating the subcycle as many times as necessary and then changing pipeline position. Although the pumping process is continuous, at times the dredger pumps only water, for example, when advancing or moving side wire anchors.

A cutter suction dredger can advance on its spuds, depending on the spud arrangements in various ways. Spud systems work on the principle that one spud is moved while the other is kept firmly in the seabed. Spuds are mounted at the rear of the dredger and forward movement is achieved by swinging on one spud until the other is ahead, then dropping it and lifting the first spud. In recently built large rock cutter suction dredgers, the working spud is mounted in a mobile carriage which allows the dredger to advance over a length of approximately 6 m (± 20 ft) without changing the spud. When the carriage reaches the end of its stroke, the auxiliary spud is dropped, the working spud lifted and the carriage is moved to its advanced position. Hence a new cycle can start after dropping the working spud and lifting the auxiliary spud.

The cutter suction dredger requires side wires used for movements and control of the dredged head. In hard material a considerable force is taken in the side wires and the largest rock cutter suction dredgers have a side pull of about 300 [metric] tons. Anchors must be, when the side wires are at an angle of 30 to 40 degrees to the perpendicular, moved forward to a new position.

The cutter suction dredger is connected to the shore by its floating pipeline and so as to allow the dredger to advance forward as far as possible without having to stop dredging to avoid breaking the pipeline and either extend or move it.

Cutter suction dredgers are rated by the diameter of their discharge pipes and the power on the sandpumps and on the cutterhead.

3.1.7.1.3 Stationary suction dredger

The stationary suction dredger, an older type of dredger, mostly used to mine sand and gravel deposits, can be very useful for beach nourishment in specific circumstances (Fig. 30c).

The deposit is dislodged and transported by a current of water agitated by the dredging pump or sandpump. The suction tube's upper end is connected to the hull by equipment allowing the necessary freedom of movement of the bottom end or suction head, which is suspended by a wire rope, worked by the drum of the suction tube winch.

If the quantity of sediment sucked up is too small, dredging is uneconomical ; if too large, the sand pump may stall when the tube or the pipeline gets clogged. Suction dredgers which raise sediment from the bottom steam ahead, or warp forward and upstream by means of their fore cables and anchors. They aim at making grooves in the bottom as straight as possible. The cutter suction dredger, however, moves in a different way.

The stationary suction dredger with a suction tube hanging in a well, dredges at anchor and discharges sideways into barges alongside moored barges, or in a floating and shore pipeline.

These dredgers are usually equipped with high pressure water jets at the entrance of the suction pipe to desegregate the dense to very dense sand.

Some trailing suction hopper dredgers can be converted to stationary suction dredgers by replacing the trailing suction tube by a suction tube pointing ahead. Stationary suction dredgers need a sandlayer minimum height of approximately 6 m (20 ft) to work economically.

3.1.7.1.4 Backhoe dredger

The backhoe dredger is basically an excavating machine mounted on a pontoon. The backhoe digging bucket is operated in such way that the digging action is performed towards the machine. Thus, when working a face, the bucket is made to penetrate the soil from the top of the face, or, if the machine sits above the face, digging is performed from the bottom of the face upwards. Backhoes are powered by direct hydraulic linkage. The outer rim of the bucket is used as the cutting edge and teeth are fitted to increase the point pressure on the material to be dug. Pontoon and backhoe are often integrated (fig. 29d).

The maximum dredging depth of the backhoe dredger is about 18 m (± 69 ft). The boom, the stick and bucket-capacity have to be adapted to the dredging depth, the bucket-capacity decreasing with the dredging depth. Some backhoe dredgers have a bucket-capacity of maximum 12 m^3 (224 ft^3) which drops to 6 m^3 (212 ft^3) at the maximum depth.

3.1.7.1.5 Barge unloading dredger

This type of dredger raises mixtures of soil [sediment] and water from barges by suction and delivers these through a shore pipeline on areas of land that are to be raised to higher level, filled up or reclaimed. In such cases delivery may take place in large quantities over great distances.

The suction tube with its flattened head is lowered so as to hang lengthwise in the barge. The pump sucks a mixture of soil with water from the barge, delivering it under pressure through a pipeline across the bank of the dumping site. Supply-water is added by means of the waterpump which injects one or two jets of water into the barge through tubes with conical nozzles. The so-called shore-connection between the standpipe of the sandpump and the first part of the shore pipeline must allow the vertical movements of the ship and any small horizontal movement. The barges with sloping hold-sides and closed bottom, which transport the sediment to the dredger, are elevator barges.

3.1.7.2 Production limiting factors

The main characteristics of beach nourishment sand which are of utmost importance to evaluate dredging productions and costs are :

- grain size : essential data to estimate pumping production ;
- relative density or angle of internal friction : to estimate the production of dense sand which can be loosened by the cutterhead or the high pressure water jets. These characteristics have to be estimated from the results of Standard Penetration Tests or Dutch Cone Penetration Tests ;
- angularity of the grain and carbonate content : to ealuate the wear of the dredging equipment which is also dependent of the grain size.

3.1.7.2.1 Trailing suction hopper dredgers

The production is determined by the maximum hopper capacity of the dredger for the soil type to be dredged and by the cycle time. The dredging cycle of a trailing suction hopper dredger consists of loading, turning, turning, sailing and dumping or reclaiming.

Dividing the hopper capacity by the total cycle time gives the hourly production in *hopper* m^3 per hour. To determine the production *in situ*, a bulking factor has to be taken into account.

Idle time for sea and weather conditions, traffic, maintenance and repair and soon has to be calculated to estimate the net weekly production.

3.1.7.2.2 Cutter suction dredger

The relative density of the sand determines the maximum cutting production depending on the cutter power available. This production has to be compared with the suction capacity of the suction pump depending on the characteristics of the pump, diameter of the suction pipe and the dredging depth.

The lower of the two productions will be considered when checking the possibility to pump the sand with a given grain size over the required distance. If the sand cannot be pumped over the required distance, a booster has to be installed in the pipeline or a lower production accepted. Here too, idle time enters into estimation of the net production.

3.1.7.2.3 Stationary suction dredger

Special attention has to be paid to the depth of the groove which can be dredged. If the sediment is not flowing to the entrance of the suction pipe, a stationary suction dredge should not be used.

When loading barges, the same principles as for the trailing suction hopper dredger apply. When pumping through a pipeline those of a cutter suction dredger should be used.

3.1.7.2.4 Backhoe dredger

The production is mainly determined by the capacity of the bucket used which is, as mentioned before, depending of the dredging depth. The latter shall thus mainly determine the production.

Even the cycle time to load and empty a bucket is mainly influenced by the dredging depth and not by the sand characteristics, contrary to the suction dredgers. Knowing the cycle-time (approximately 1 min.) and the bucket capacity, the theoretical output is determined. The main factors influencing the maximum potential output are the delays to advance the dredger and to change the barges. Enough barges are needed to avoid waiting time for the dredger.

3.1.7.3 Reclaiming distance

A very important factor when reclaiming, is the **critical velocity** of the mixture in the pipeline. It is the minimum velocity which is required to avoid deposits in the pipeline and it is also the velocity at which a given production can be reclaimed with a minimum of pump power. This velocity depends upon the mean grain size diameter of the sand and upon the diameter of the pipeline.

Some critical velocity values are given in table 5 below.

GRAIN SIZE d (mm)	DIAMETER PIPELINE D (m)		
	0.65	0.75	0.85
0.2	4.01	4.31	4.59
0.4	4.96	5.33	5.67
0.6	5.38	5.78	6.16
0.8	5.63	6.05	6.44

Table 5 : Critical velocity (m/sec)

The critical velocity increases as does grain-size diameter which also influences the head losses in the pipes and consequently the pump power required.

To demonstrate the impact of the grain-size, the head losses have been calculated for a pipeline with a length of 1,000 m (3,280 ft), having different diameters and for an increasing grain size. For each pipeline diameter (D) a production rate has been assumed corresponding to an apparent concentration of approximately 20% at the critical velocity for the smallest grain size of 0.2 mm (0.008 in).

These productions are :
- 1.000 m^3/h for D = 0.65 m 35,314 ft^3/h for D = 2.13 ft
- 1.500 m^3/h for D = 0.75 m 52.566 ft^3/h for D = 2.42 ft
- 2.000 m^3/h for D = 0.85 m 70,218 ft^3/h for D = 2.74 ft

The values of the apparent concentration have been calculated for increasing grain-size, presuming that the given production has to be reached at the critical velocity corresponding to the given grain size and pipeline diameter.

The results are given in table 6.

PIPELINE D (m)	PRODUC-TION m³/h	GRAIN-SIZE d (mm)			
		0.2	0.4	0.6	0.8
0.65	1,000	20.1	37.9	49.5	59.0
0.75	1,500	20.0	37.5	49.0	58.4
0.85	2,000	19.6	36.3	47.2	56.0

Table 6 : head losses (mWC) for L = 1,000 m

Thus the head loss increases with increasing grain size with each sandpump having a fixed maximum pressure, the maximum pumping distance for a grainsize of 0.8 mm (0.032 in) will be reduced to approximately a third of the distance for pumping the same production having a grainsize of 0.2 mm (0.008 in) (Table 7).

PIPELINE D (m)	PRODUC-TION m³/h	GRAIN SIZE d (mm)			
		0.2	0.4	0.6	0.8
0.65	1,000	405	941	1,334	1,666
0.75	1,500	575	1,334	1,888	2,354
0.85	2,000	771	1,765	2,485	3,087

Table 7 : Required power (kW) to obtain the given production over a distance of 1,000 m

Required power to reclaim a given production over a given distance, is 4 times larger as grainsize increases from 0.2 to 0.8 mm (0.008 to 0.032 in). This illustrates the particle size importance in each project of beach nourishment.

3.1.7.4 Beach nourishment methods

Beach nourishment can thus be done with a variety of dredgers and by various methods. In general, there are two types of beach nourishment methods, *viz.* hydraulic fill placed on a beach or against a bluff, and material dumped by hopper dredgers or barges in the nearshore zone. Each beach nourishment scheme has its own features and local circumstances which lead to the most economical and technically suitable solution for a specific project.

Basically, there are two main features which influence the possibilities available for the execution method : the distance between the borrow area and the reclamation site, and the requirement for nearshore dumping or beach replenishment. The former determines whether direct pumping from the borrow area to the reclamation site is feasible, while the latter determines whether direct pumping with hopper dredgers or barges is needed.

When fill material is required in the nearshore zone but, due to wave conditions, it is not feasible to dump in this zone with hopper dredgers or barges, a rainbow-system can be used which entails jetting the sand-water mixture under approximately 45° from a hopper dredger or from a stationary dredger (equipped if necessary with a flexible floating pipeline and a moored nozzle barge). Distances up to 75 m (274 ft) can be reached. However, it is difficult to control the dumping process.

3.1.7.4.1 Direct dumping in the nearshore zone

α. Conditions
Very good sea and weather conditions are required. The final level depends upon the draught (draft) of the loaded vessels which have to enter the area to dump their load.

β. Equipment and execution :
This method can be performed by trailing suction hopper dredgers capable of shallow dumping. Hopper capacity of the vessels is limited and systems may not have protruding bottom doors such as sliding bottom doors or split hull hopper dredgers.

Dumping by barges loaded by a stationary suction dredger is also a solution if sea conditions make it possible. Seaworthy barges, however, will always be required. By using barges, the minimum required water depth can be decreased.

Monitoring is needed. Survey and dumping vessels should be equipped with radio positioning systems.

3.1.7.4.2 Direct pumping to the beach area

The methods proposed depend upon the location of the borrow area.

α. When the borrow area is at sea, trailing suction hopper dredgers should be used. The vessel loads at sea, sails to a mooring point as near as possible to the beach area and pumps its load ashore.

α.1. Conditions

As the pumping distance range of self-unloading trailing suction hopper dredgers is rather limited, the distance between the mooring point and the beach area should not exceed this maximum range. This can be a restriction for the hopper capacity of the vessel as large trailing suction hopper dredgers require greater water-depth.

The local circumstances should also permit an easy displacement of the mooring point.

α.2. Equipment and execution

The type of mooring point selected and the pipeline connection with the beach depend on the local wave, tide, current and weather conditions. Under calm conditions, it will be possible to use a riser pontoon anchored, or fixed with spuds, on the seaward size of a submerged pipeline. This can be connected to the hopper dredger by means of a flexible floating pipeline and a quick coupling over the bow. When rough weather conditions can be expected, it is not advisable to use these types of pontoons and link to a self-floating flexible pipeline is necessary.

This can then be picked up by the trailing suction hopper dredger under rough weather conditions and connected directly to the submerged pipeline. A greater water depth is required at the mooring point under such circumstances, and the trailing suction hopper dredger must be able to swing around the mooring point to find a good head-on position. When large quantities are required over a long stretch of beach, two or more submerged pipelines should be installed to avoid long pumping distances and to introduce an escape possibility to safeguard a continuous operation.

For rapid or emergency beach nourishment schemes, pumping through a nozzle on the bow of a trailing suction hopper dredger may be the solution. This very expensive method will only be used to replenish areas difficult to be reached by pipelines.

β. Conditions

When the borrow area is situated in a sheltered area, a cutter suction dredger or a stationary suction dredger can be used. The dimensions of the borrow area should be large enough. This method is commonly used.

Fine materials from the borrow site end up in the reclamation area.

3.1.7.4.3 Use of a rehandle pit

This approach can be used when high pumping power is required and when sea conditions are rather rough and a sheltered area available not too far from the beach area.

α. Conditions
The dimensions of the rehandle pit will have to be large enough to accommodate the rehandle pit dredger with sufficient room to achieve optimal production. At the same time, in the other part of the rehandle pit, sufficient room is to be given to the trailing suction hopper dredger to discharge its load safely without any danger of collision. With this working method, the rehandle pit is divided into two parts, thus avoiding a situation in which the hopper dredger dumps just upstream of the stationary dredger. Another advantage of this system is that a certain buffer capacity is introduced so that only the matching of weekly production levels remains necessary.

β. Execution
The layer height dumped by the trailing suction hopper dredgers should be at least 3 to 4 m (± 10 to 13 ft) in order to maintain a continuous high output with cutter dredgers.

During the dredging operation with the cutter dredger a certain amount of spillage will occur, which influences the buffer capacity of the rehandle pit.

Usually, after rehandling operations are finished, these pits are filled by the trailing suction hopper dredgers to the original sea bed level.

An additional advantage of this system is that the fill material is washed out twice, resulting in clean beach sand with almost no particles smaller than 63 μm. However, more pumping horsepower is required and more wear and tear will result.

3.1.7.4.4 Use of barges and a barge unloading dredger

α. Conditions
A barge unloading dredger can only be used when calm weather conditions are envisaged, allowing barges to travel safety from the borrow area to the rehandle pit and permitting the sand winning dredger to work, without any hazards, for a long stretch of time.

β. Equipment and execution

The loading of the barges can be done with a stationary suction dredger, a cutter, or even a trailing suction hopper dredger. The first dredger will, most of the time, be equipped with barge loading facilities, e.g. spray-arms. Cutter dredgers will need a flexible floating pipeline and a separate spider barge, alongside which the barges can be loaded. The last-mentioned loading method, using a trailing suction hopper dredger to fill the barges, is feasible only under very calm weather conditions, while only one barge at a time can be loaded.

For each of these loading systems, a different rate of washing out of the fine particles will result. This is dependent upon the type of loading system, the dimensions of the barge, the discharge flow rate and the concentration.

If no rehandle pit is available or permitted, the unloading is to be done by a barge unloading dredger moored in a sheltered area.

By using this system, the fine fractions from the barges will automatically be deposited on the beach. Therefore, a proper washing out of the "fines" will have to be achieved during barge loading operations. Because only one barge at time can be emptied by the reclamation dredger, an irregular pumping process takes place, which means more water at the reclamation site and thus flatter reclamation slopes. The pumping irregularity makes it impossible to maintain a constant high concentration flow, resulting in a lower weekly production level compared with the rehandle pit method.

3.2 Beach Nourishment Schemes

Several schemes have been undertaken in sites spread over the entire world (table 8). In the United States, it took $ 65 million to restore 25 km m^3 of beach in Miami Beach, FL.; at Redondo Beach 1,100,000 m^3 ($\pm$ 39 million ft^3) of materials were used to rebuild the beach[20, 21].

Other areas in the U.S.A. where beach nourishment was selected as a means of remedying coastal erosion, include the coast of Washington State (Ediz Hook), Coronado Strand (San Diego) and downcoast from the Channel Islands Harbor in California, Sandy Hook (New Jersey), Presque Isle peninsula (on Lake Erie), famed Waikiki Beach (Hawaii) and Cape Hatteras.

YEAR		m3
'19-'78	Southern California coast	(***) 108,000,000
'61-'62	La Croisette, Cannes (France)	100,000
'67-'68	Redondo Beach (California)	1,100,000
'69-'70	Copacabana Beach (Brazil)	3,500,000
'71	Goeree (Holland)	600,000
'72	Silt (W. Germany)	900,000
'74-'75	Bournemouth (U.K.)	650,000
'75-'87	Lagos (Nigeria)	(**) 1,000,000
'77	Rockaway Beach (New York)	48,000,000
'77-'79	Knokke-Heist (Belgium)	8,400,000
'80	Gold Coast, Queensland (Australia)	2,400,000
'80	Miami Beach (Florida)	87,000,000
'82-'83	Isle of Langeoog (W. Germany)	200,000
'83	Pet foreshore (England)	(*)(**) 19,000
-	Walland (England)	(**) 31,000

(*) "shingle"
(**) amount is annual recharge
(***) total amount over a 60 year period and covering 60 projects

Table 8 : Some beach nourishment schemes

Beach nourishment has been used in Florida at Palm Beach and Boynton Beach. The 1963 Atlantic Beach (NJ) beach replenishment was effective for six years. At Long Beach, CA a beach fill project was completed in April 1983 at a cost of $781,000.

3.2.1 Knokke-Heist, Belgium

Offshore transport of beach material resulted in near total beach submergence at high tide east of Zeebrugge (fig. 20). The Appelzak Channel gradually extended itself to within 500 m (1.640 ft) of the harbor's sea-wall favoring beach erosion, while the sea-wall, itself threatened by storm waves, hampered sand transport to and accretion on the beach. Though dune area has expanded during the eighties, fore- and backshore lost ground due to deflection of longshore currents (Duinbergen, Knokke-Zoute) and sand withdrawal (Heist). Beach regression due to the effects of tidal currents affected by the new Zeebrugge harbor was to be expected (Fig. 31).

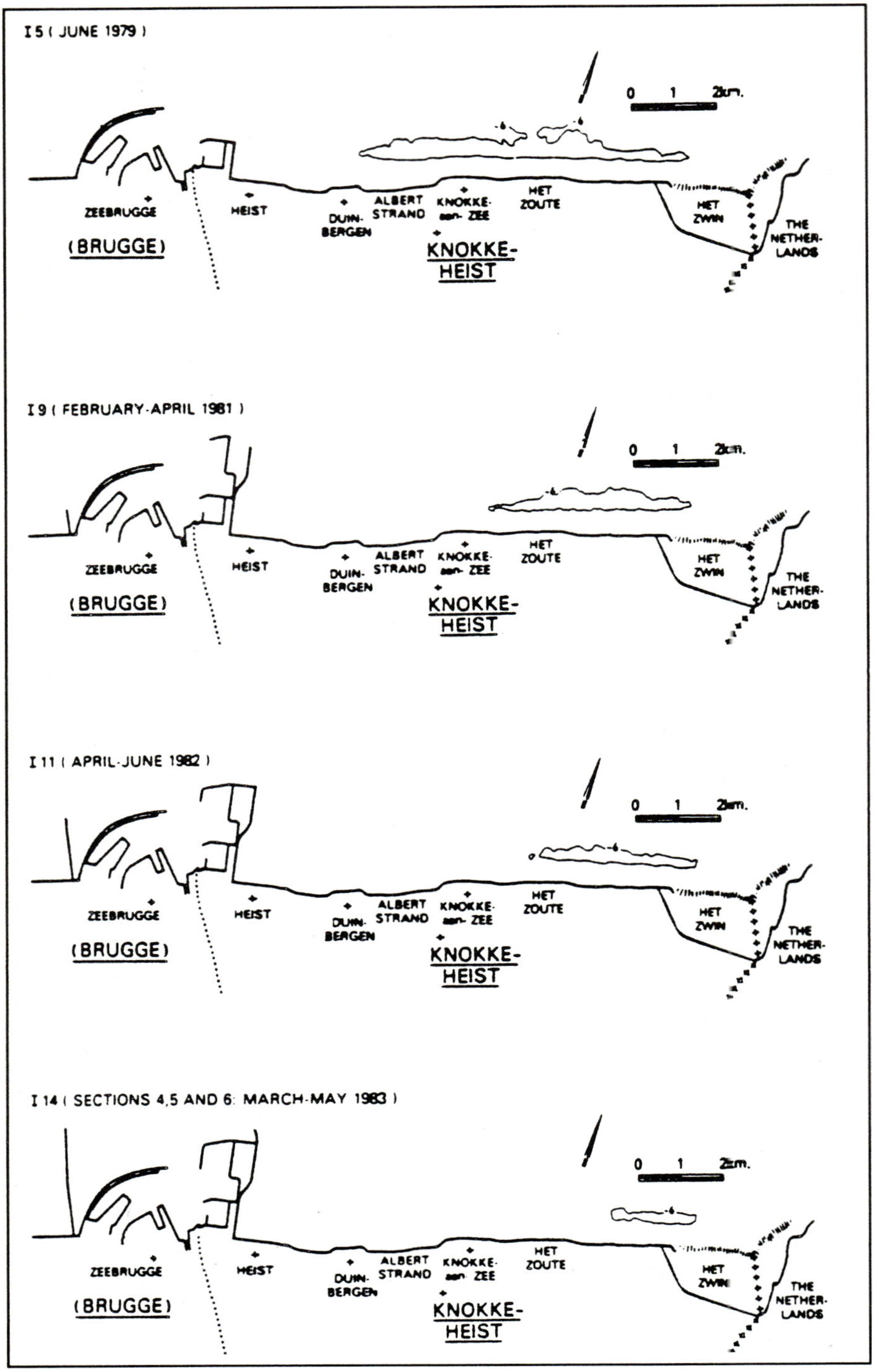

Fig. 31 : Evolution of "Appelzak" gully

In the case of Knokke-Heist predicted sand losses and coastline regressions were compared with solutions in which sand suppletion and groin building were considered (Kerckaert, *et al.*, 1985)[22a]. Total capital and maintenance costs of beach nourishment, with or without groin construction were not significantly different. This, and the fact that beach restoration was at the time a necessity, lead to the choice of beach renourishment as a coastal protection measure (fig. 31 and 32). The entire project was accompanied by a well-defined coastal observation program which is still ongoing. It allows to monitor any coastal changes due to the harbor extension works at Zeebrugge.

A detailed description of the project has been presented in Roovers *et al.*, 1981 and Kerckaert, *et al.*, 1985, and the recent beach evolutions have been the subject of several reports[22b] (fig. 33, 34).

Along this eastern shoreline segment of Belgium, the Appelzak gully, which acted as trap for sediments and materials removed from the beach, has substantially reduced in size, and checks carried out indicate that it is filling up. It is noteworthy that the "filler" material is mainly silt. The westerly segment of the Appelzak is doomed to disappearance (fig. 34).

Due to the siting of a suitable material area at depths exceeding somewhat 10 m (32.8 ft) large sea-going trailing-suction hopper dredgers were used. The dredged material was dumped into a pit wherefrom a cutter-suction dredger, and booster stations pumped them onto the beach. Of the booster stations one was a floating one positioned near the end of the floating pipelines ; two were placed on the beach to-be-restored itself, and two more astern of the cutter dredger. (fig. 36 ; photo 3)

An economic assessment was made based only upon the economic rent derived from beach-based tourism. Income from visitors was estimated in 1984 at $142 million (in 1992 US$) a year. Beach maintenance costs $1,128,000, or about $20 (US) per tourist per day. It is true, however, that the amortization of the investment in the original project is not accounted for in the calculation, and that in this particular instance the dredging can be written off as a no-cost item since it was required to provide a navigation channel.

A technical file card for the Knokke-Heist undertaking would read : funds expended : $ (US) 68 million, sand volume 8,400,000 m³ (10,900,000 cu.yd.), length of beach 8,200 m (5.86 mi), sand volume per length unit/m) 1,000 m³ (1,189 cu.yd/yd), cost per length unit (m) $ (US) 8,333 ($ 7,620 per yd). This compares with $ 1,060,610 for one groin on a normal beach, which requires $ 0,606 of expenditures for repairs after each storm on a yearly basis.

3.2.2 Bournemouth, Dungeness and Portobello, U.K.

In Bournemouth, construction of a sea-wall cut off the sand supply from neighboring cliffs. Beach nourishment was decided upon to remedy the situation.

In this case cost effectiveness of beach renourishment showed that the final cost per running meter of beach amounted to about ± 116/meter (± 35.7/ft) (1975) compared to a conventional sea-wall construction incorporating cut-off steel sheet piling at ± 628/meter (± 193/ft) (1975)[23].

A successful example of beach nourishment in conjunction with groines has been carried out here where indigenous beach sand had a mean diameter of 0.2 mm. Material from a nearby offshore site having a mean size of 0.4 mm was dredged and pumped on the beach. Although some losses of beach occurred over the years, the scheme was cost-effective. An important pleasure beach was restored and the coastal defences were improved.

Using trailing hopper suction dredgers beach rebuilding material was pumped ashore. Some 110,000 m³ (144,100 cu.yd) of sand were placed within 200 m (656 ft) of the shore.

At Dungeness, in Kent, beach material is dredged in surplus areas and spread where the beach is eroding, thereby balancing the sedimentary budget (cf. chapter III).

At Portobello near Edinburgh, Scotland, replenishment of the beach occurs in ten year spans. The results are satisfactory but some question whether this is sufficient.

3.2.3 Silt and Langeoog, F.R. Germany

The projects at Silt and Langeoog (Germany) considered construction of a sand spit (refraction groin) as adequate coastal protection. Under wave action the sand groin moves shoreward feeding the adjacent beach. The orientation of the sand groin has to correspond with predominant wave direction.

Sea floor sand was dredged to rebuild beaches on Silt which had been given earlier a protective system of jetties and tetrapod barriers.

Elsewhere in Germany the century-old tradition of groin construction and maintenance has been set aside in favor of nourishment. The method has been followed on the shores of the Frisian Islands. The first attempt was made on Northerner during the years 1951 and 1952, and followed up by a second, larger scale, intervention in 1957[24]. From then on artificial accretion was used on several beaches : in 1963, Baltrum in 1968, Borkum in 1969 and 1970, Langeoog in 1971 and Westerland-Sylt in 1972, to name the earliest operations. None, however, involved a volume of material as impressive as in Belgium.

3.2.4 France

In the Huttes area of the Médoc, a groin built in 1854 was rapidly broken up and carried to sea. With an average retreat of the beach of 10 m/year (32.8 ft/yr), the beach is artificially renourished and reprofiled each year. One of the earliest beach renourishment schemes was undertaken near Cannes (France), on the fashionable La Croisette Beach where 100,000 m^3 (353/400 cu.ft) were placed on the beach during 1961-1962 (fig. 19).

3.2.5 Scheveningen, Netherlands

Westerly gales reduced the width of Scheveningen, The Netherlands' most celebrated beach, to the extent that at high tide and under bad weather conditions, the sea washed over and against the boulevard. Beach widening became a necessity for both economic and safety reasons.

Here sand is dredged 20 km (12.4 mi) inland. The operating ship berths near the pontoon whereupon a short pipeline ends. This pipeline is connected to the ship; sand is pumped from pontoon to beach. At Goeree, in 1971, beach nourishment works involved 600,000 m3 (787.500 cu.yd).

3.2.6 Former Soviet Union (Baltic and Black seas)

Various beach nourishment schemes have been implemented in the European part of the former Soviet Union, for instance at Sochi, Sukhumi Bay, Gelendzhik and Planerskoye (Black Sea), and on the Sea of Azov. Beaches south of Sochi have narrowed or disappeared altogether due to port construction. Remedial action consisted in building of groins, submarine breakwaters and artificial replenishment. At Odessa, only 52% of nourishment material was lost over a 7-year span[25].

Zenkovich and Kiknadze report the Baltic and Black Sea schemes which are part of a planned coastal protection plant involving hundreds of kilometers. They include coastal processes control and existing projects modifications[26]. The specific case of Georgia is discussed under section 3.1.11.

3.2.7 Australia and New Zealand

Restoration of the Melbourne coast beaches was carried out by piping onshore sand from the floor of Port Phillip Bay. On Australia's Queensland Gold Coast, south of Brisbane, beach nourishment was used to reconstruct beaches[27].

Severe beach erosion at Balaena Beach, Wellington, New Zealand demanded protective action and restoration[27]. The choice was made to artificially rebuild the beach, in 1981-1982, using dredged sandy granular gravel. The beach was monitored between 1982 and 1984. Notwithstanding considerable sediment mobility in response to locally generated wind waves, nearly all the nourished material was retained in the littoral zone.

3.2.8 Rio de Janeiro, Brazil

Famed Copacabana Beach has been subject to erosion for some time. Charlier reported early signs of an accelerating problem in 1967. Artificial nourishment was decided upon after a study on a physical model that took into account the characteristics of wind, waves, tides and sand, and surveyed bathymetric conditions[28]. A mixed approach was decided upon and 2 million m^3 (2,625,000 cu.yd) of sand were dumped offshore while another 1.5 million m^3 (1,967,000 cu.yd) was deposited on Copacabana Beach itself over a distance of 4.2 km. (2.6 mi). This allowed to widen the beach from its remaining 55 m (179 ft) to about 140 m (455 ft).

3.2.9 United States

A project to restore 25 km (15.5 mi) of coast in the Miami Beach area was priced in 1991 at $65 million. At Westhamption Beach, NY an original proposal to install a very limited number of groins and deposit a large amount of sand on the beach met with local resistance. Numerous groins were built and successive storms over a five year period removed the sand from areas downdrift of the groins. And, as in other nearby regions, homes have toppled in the sea as high tide pushes its way under the shoreline buildings.

In Virginia Beach, VA, beach nourishment is required each year and 28,999 m^3 (38,000 cu.yd) are placed on the beach at a cost of $1,5 million. A similar solution has been suggested which would double the width of the beach, but it requires huge expenditures to truck in 2 million m^3 (2,625,000 cu.yd) of sand.

Sea-walls have been prohibited in the State of Maine, and so have their close relative the bulkhead ; North Carolina is following suit. The prohibition is an effort to force better planning in coastal development and to halt interference with the natural systems. In some case however, sea-walls have saved lives, as for instance in Galveston, TX, but they have not solved the coastal erosion problem. As they reflect the waves, they are in fact self-destructing. "The beach is inherently a dynamic system and any attempt to make it stable by ... construction of groins, sea-walls, bulkheads, and other engineering devices ... is ultimately self-defeating.", said Acting Assistant Secretary of the Interior J. Craig Potter at the Cities on the Beach conference, in 1986.

Other areas of the U.S.A. have already been mentioned (cf. 3.2.).

Successful restoration of the Sandy Hook Beach by nourishment has been reported as well as in several locations on the southeast Atlantic and Gulf of Mexico coasts (Huntington Island, SC, Cape Canaveral Beach, FL, Harrison County, MS). Spectacular results were achieved at Key West, FL. where over a 405 mm (16 inch) layer of crushed limestone, a 203 mm (8 inch) blanket of sand was deposited[20]. Virginia Beach, VA fared less well and the problem has been ascribed to the fact that the nourishment ("borrow") material was finer and not as well sorted as was the original ("native") beach material.

In Key West gravel 8.35 mm (3/8 inch) in diameter (so-called "pea-sized") was placed on the beach and gradually intermixed with fine sand during wave storms. Groins existing, the area between them was filled to capacity immediately ; this allows natural sand by-passing and continued deposition processes downdrift. The coarser sand is more likely to resist storm-induced displacement.

In Georgia (USA) a combination of hard and soft counter-erosion measures was taken. Tybee Island has a history of shifting shorelines and severe coastal erosion[22]. Groins, sea-walls and revetments have been built over the years, however, more recently, beach nourishment was undertaken as an alternative and complementary approach.

Beach nourishment was also undertaken further south at Hallandale, Florida. The project was the subject of a thorough ecological impact study which showed that seven years after the operation had been completed no ill effects on fish life could be evidenced.

Close to one hundred sites in the United States have been the beneficiaries of Federal Government beach rebuilding projects with use of dredged material. It has been shown that such use of dredged material has the least environmental impact. Projects were carried out a.o. at Jacksonville Harbor, FL, Sea Girt (N.Y) and Wrightsville Beach (NC) (photo 4)[29a].

The Miami Beach, Jacksonville, Sea Girt and Wrightstown Beach operations, were mentioned earlier; at Redondo Beach 1,100,000 m^3 (1,441,000 cu.yd) of materials were used to rebuild the beach.

On the U.S. side of the Strait of Juan de Fuca (Washington), periodic beach nourishment, and foreshore revetment, are used to combat intensive erosion of Ediz Hook spit. Beaches downcoast of Channel Islands Harbor (California) are nourished with sand removed from a trap built there.

In all these projects, coastal defense was combined with the recreational aims to be attained. The required sand quantities have to contribute to the predominant longshore or offshore sediment transport capacities. Silt and clay particles are easily removed when exposed to wave action. The sand losses (due to longshore or offshore transport) should be considered as future maintenance material for the beach.

Still other nourishment locations in the United States include
Cape Canaveral Beach (2.03 x 10^6 m^3 = 2.3 x 10^6 Cy.yd),
Fort Pierre (0.64 x 10^6 m^3 = 0.72 x 10^6 Cy.yd),
Jupiter Island (2.04 and 0.88 x 10^6 m^3 = 2.4 and 1.0 x 10^6 Cy.yd),
Delray Beach (1.4 x 10^6 m^3 = 1.6 x 10^6 Cy.yd),
Pompano Beach (0.91 x 10^6 m^3 = 1.03 x 10^6 Cy.yd),
Hollandale Beach (0.32 x 10^6 m^3 = 0.36 x 10^6 Cy.yd),
Key Biscayne (0.18 x 10^6 m^3 = 0.2 x 10^6 Cy.yd),
Captiva Island (0.66 x 10^6 m^3 = 0.75 x 10^6 Cy.yd),
Lido Key Beach (0.31 x 10^6 m^3 = 0.35 x 10^6 Cy.yd),
Key West (0.05 x 10^6 m^3 = 0.07 x 10^6 Cy.yd),
and Treasure Island (0.67 x 10^6 m^3 = 0.76 x 10^6 Cy.yd).

The concept of *critical erosion* has found its way in the literature ; areas of *critical erosion* are defined as those where erosion presents a serious problem because the rate of erosion considered in conjunction with economic, industrial, recreational, agricultural, navigation, demographic, ecological and other relevant factors, indicated that action to halt such erosion may be justified.

In the United States only 362 km (226 miles) of coastline have been treated, yet 3,420 km (2,137 miles) are classified as subject to *critical erosion*. Currently 540 km (337 mi) are authorized to receive beach preservation treatment. Plans for the future envision spending $48.8 million annually (in 1990 $). The cost of initial beach restoration averaged between 1950 and 1993 $2,187/m ($663/ft) and periodic renourishment $1,230/m ($375/ft). Of course with labor far less expensive in developing countries these averages would be lower there.

The *Market Study on Beach Preservation* quoted below states 'There were as many [beach restoration/preservation] projects completed in the 1990-1993 period as there were during the entire decade of the 80's".

a. The National Shoreline

Type of Area	Miles of Shoreline	Km of Shoreline	% of Total
Nation's Total Shoreline	84,240	134,784	100
No significant erosion	63,740	101,984	75,7
Critical Erosion, not covered by Federal work	2,137	3,420	2,5
Completed Federal Projects	226	362	0.3
Authorized Federal Projects & Projects in Pre-engineering	151	242	0.2
Authorized Federal Studies	186	298	0.2

b. 1950-1993

Types of Measures	Total Costs in millions	Ave, p/m*	Ave. $/ft.*
Initial Beach Restoration	735,000	2,187	663
Periodic Beach Renourishment	415,800	1,230	375
Structures	153,900		
Emergency Measures	30,200		
TOTALS	1,489,500	4,700	1,433

*Based on Total Cost, assumed applicable to entire 210 miles.

c. Future Plans

Types of Measures	Estimated Expenditure, $*
To complete present programs	
Beach Restoration	18,900,000
Beach Nourishment	734,500,000
Sand bypassing Systems	24,000,000
TOTALS	777,400,000
Planned Future Programs	
Under construction or authorized	714,200,000
Preconstruction Eng. & Design	948,300,000
TOTAL	1,662,500,000**
To Complete & Future Progr. TOTAL	2,439,900,000**
Approx. Future $/yr- (1/50) x TOTAL =	48,800,000

* Assuming Federal funds provide 65% or total cost shown.
** Total life-cycle (50 year) cost

Table 9 : Beach Preservation Activity of the U.S. Army Corps of Engineers*

The above described activity of the U.S. Army Corps of Engineers is the major part, but not a complete record, of beach preservation in the U.S.A.

As far as costs are concerned, the data (in 1993 dollars) in the table 9b is based on 56 large projects covering 210 miles accomplished during the period 1950-1993. Not included is data on 26 smaller projects covering 16 miles during this period.

3.2.10 De Haan, Belgium

Early in 1991, a feeder berm of approx. 2,200 m (7,216 ft) long was put in place for the first time in Europe near De Haan (a.k.a. Le Coq), Belgium. Using two trailing-suction split-hopper dredgers, some 600,000 m³ (21,200,000 ft) of sand were used to build a longshore bar was thereby constructed at 600 m (1,968 ft) from the shoreline on the low water bar.

* Adapted from "*Shoreline Protection and Beach Erosion Control Study,* Phase I : Cost Comparison of Shoreline Protection Projects of the U.S. Army Corps of Engineers" published January, 1994 by the U.S. Army Corps of Engineers.

Mid-1992, a profile nourishment restored the beach profile between the dike and the feeder berm. Due to the distances to cover, a booster station and a double line on the beach were put in place.

Upon completion, an area some 1,000 m (3,280 ft) from the shoreline, about 3,000 m (9,840 ft) long was surveyed.

A regression of 15% has been recorded, certainly due, in part, to some rough weather conditions. A monitoring program studies beach slope evolution as a function of time and weather conditions, and sand budget analysis. The effectiveness of a beach nourishment-feeder berm combination as a new concept of coastal protection will be assessed.

The works cover (in an area from De Haan west to Vlissegem), a distance of 2,150 m (7,052 ft). The fill volume to the new profile of the feeder berm amounted to 245 m^3/m (131 ft) bench. A bench was built with the reclaimed sand on the dry section of the beach. The 40 m (2,653 ft^3). The beach was thus raised an average of one meter (3.28 ft) wide stands above the mean highwater spring tide level. Wind breakers on the dry sand protected again aeolean erosion (Fig. 32).

As part of the deepening of the access to the port of Ostend, sand, with grains of about 200 micron in diameter, was dredged, transported and dumped to build the feeder berm.

The De Haan project avowedly has three aims : keeping navigation channels open in what is labeled the busiest sea-highway in the world and guaranteeing access to the coastal harbors, providing flood protection on the shore and to the adjoining hinterland, and addressing the needs of tourism and recreation.

Fig. 32 shows the relative position of the feeder (underwater) bar, the beach and the upper beach berm (dry benched beach). High and low spring tides are respectively (means) 4.74 and 0.21 m (18 and 0.69 ft).

Fig. 32 : Beach prior to artificial nourishment (De Haan, Belgium)

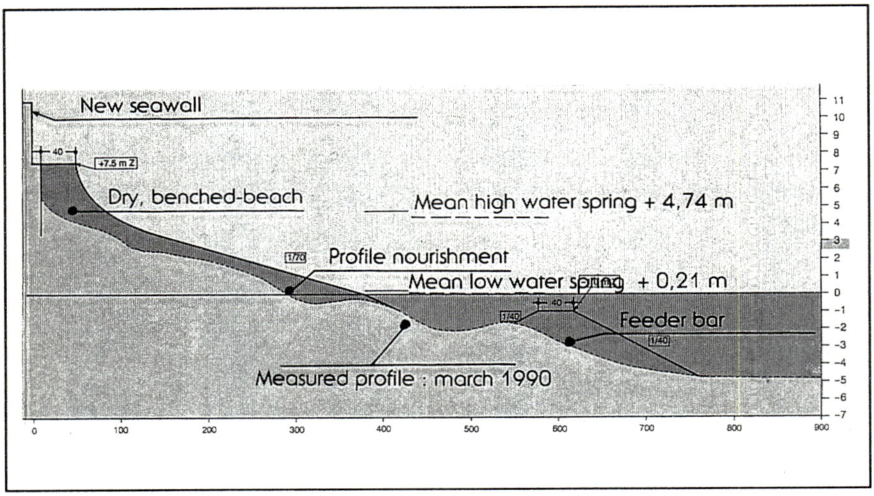

Fig. 33 : Relative position of feeder bar, beach and dry benched beach (De Haan, Belgium)

The undertaking does not exclude the "participation" of the seawall in the protection scheme. The hard structure will act mainly, in harmony with the coastal dunes, to defend the hinterland against flooding.

Approximately 800,000 m³ (2,825/200 cu.ft³) of sand had to be pumped ashore. It was tested for the presence of silt, shells and toxic substances. It had an average grain size of about 300 micron.

Of the 800,000 m³ (2,825/200 cu.ft³) of sand, about 700,000 m³ (247/98,000 ft³) settled on the beach itself, representing a volume of 470 m³ (5,044 cu.ft³) per linear meter was achieved.

The project, using modern technology, combines environmental preservation coastal protection and support of tourist activity (Fig. 34)

Techniques have been improved since the Knokke-Heist project was completed. Another undertaking at Ostend may provide further valuable information. The Winter 1994-1995 has unfortunately by its terrible storms undone some of the work and carried tons of sand seawards.

3.1.11 Other locations

Nourishment works have been undertaken on Belgium's west coast (cf. 3.3.4). Beach nourishment was also called upon south of the Orange River in the Republic of South Africa, at Saldanka, on Singapore, along the Saronic Gulf East Coast (Greece) and Victoria Beach near Lagos (Nigeria). On the latter site over one million metric tons of sand is placed each year (fig. 3,16).

In the Caucasus republic of Georgia, hard defenses have had a disappointing performance and resiliance history.

Of the 314 km (195 mi) Black Sea coastline of Georgia, 155 km (96 mi) were washed out in 1961, 183 km (113 mi) in 1971 and 220 km (136 mi) were in an emergency state in 1981. In 20 years, 1,400 ha (3,460 acres) were lost in spite of attempts at protection by various hydrotechnical constructions at a cost exceeding hundreds of million roubles (official exchange rate then was US $ = R 0,85).

Since 1981, a new approach has been in effect.

A restoration of natural morphodynamics of beaches was started. Artificial wave-suppressing beaches reliably protect dry land against wash-out and restored near-coast landscapes.

Thus, 80 km (31 km) of shoreline were stabilized and 120 ha (296 mi) of dry land restored. Artificial beaches cost half the price tag of hard coast protection (A.G. Kinadze *et al.*, 1995, "Artificial beaches as a method of sea coast zone protection", *Bordomer '95 Abstr.*). The artificial beaches also contribute cleanness of sea water in the coastal zone. Based upon ten years experience on the Georgian Black Sea coast Kinadze concludes that the artificial beaches are the most efficient and economic protective structures.

Fig. 34 : De Haan : beach after completion of works

3.3 Projects Under Study

3.3.1 India

Beach nourishment has been recommended for the 300 km (186 mi) of severely eroding southwest coast of India as far back as nearly twenty years ago, when a short segment was restored at Purakkad.

3.3.2 Lazio Coast, Italy

A feasibility study for coastal protection works on the Lazio Coast in Italy, showed that beach nourishment could be considered as an effective and apparently preferable alternative to the traditional building of detached breakwaters. Appropriate sand qualities are required to contribute to predominant longshore transport. A future maintenance program of the beach based on a 5 year-cycle has been suggested.

The Lido di Ostia is one of the important resorts for Romans going on a shore holiday. Here coastal erosion has considerably reduced the beach area. A feasibility study conducted by N.V. HAECON, a Belgian company, recommended the monitoring of the dredging carried out on the nearby coastal zone, and a beach nourishment scheme which includes the use of 2.6 million m3 (3.4 million cu.yd) of sand. In its Progetto Litorale '83 study, the city of Rome drew-up plans for new developments along the "Roman beach area". The plans consider the establishment of a marina at Isola Sacra. Material removed there, and additional sand from marine sources, could be used for a beach nourishment scheme to restore the damaged Fiumicino beaches.

3.3.3 Ostend, Belgium

True, the soft approach may not be applicable in all instances. A particular case that comes to mind is Ostend, Belgium's most famous coastal resort. Construction of a fort (Fort Napoléon), at the turn of the century, has seriously disturbed, ever since, normal sand transport. Building of groins did not solve the problem. Strong erosion, often the consequence of storm waves caused by refraction-induced wave energy concentration, is endangering beach, dike, and structures beyond it.

Beach nourishment alone is not suited because Ostend is also a cross-channel port, and the navigation channel must be kept open. Big scale beach nourishment would extend beach slopes too far seawards. Yet, the toe of the boardwalk-dike is presently almost unprotected against sapping by the sea ; were it to collapse, it would entrain the weakening of the dike's buildings foundations. Recent storms have evidenced the urgency of adequate protection measures as even boardwalk pavement was uprooted by their fusy (1994, 1995).

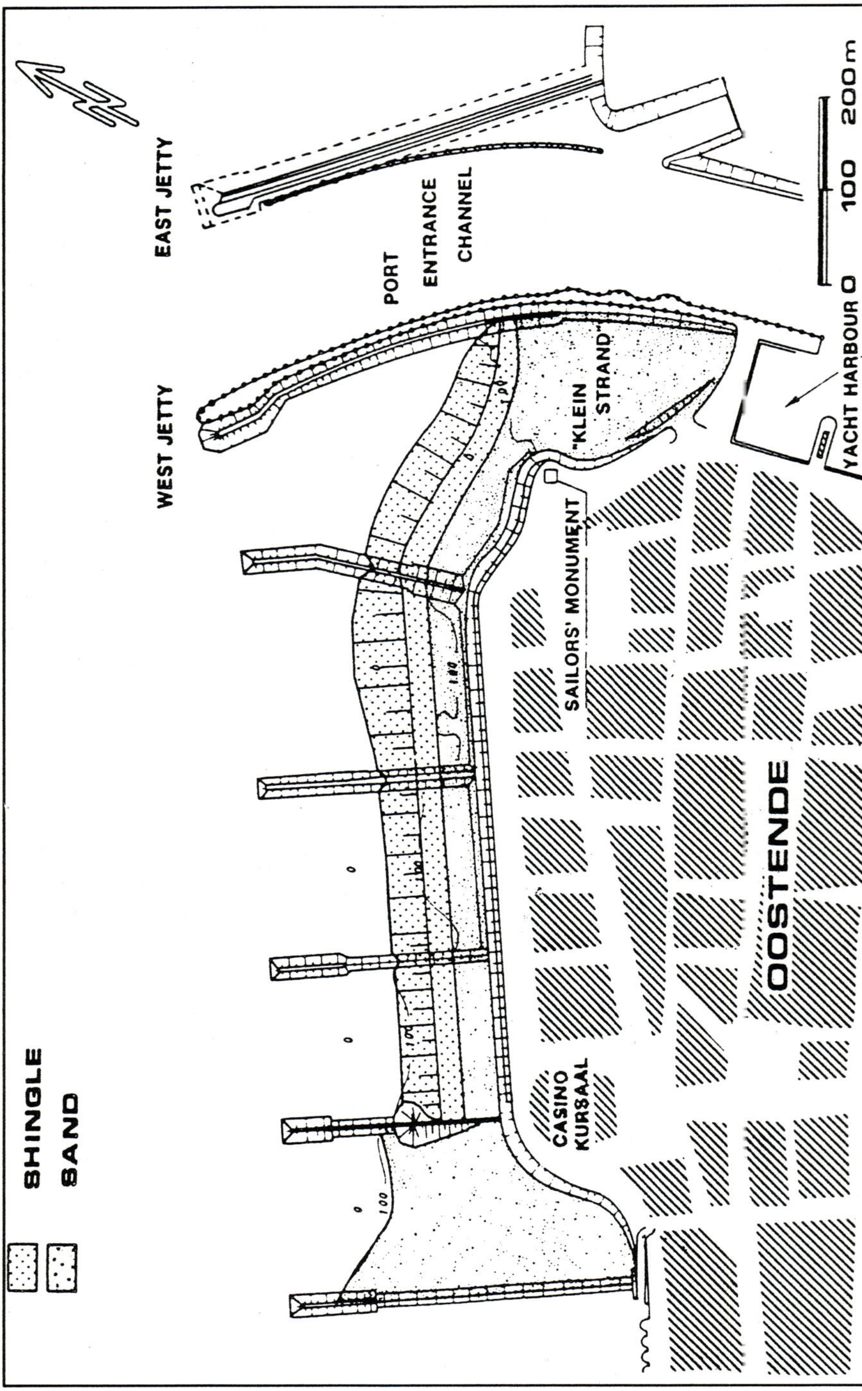

Fig. 35 : Ostend beach rehabilitation project, retained lay-out

Following requirements were considered for the Ostend project : reduction of wave overtopping of the existing sea-wall (flood prevention); prevention of increased sedimentation in existing port access channel, as a result of the coastal defense works; taking into account of future recreational beach activities; and limitation of the project within the existing groin structure extension.

As sand suppletion alone would not effectively contribute to the above-mentioned requirements, the mixed use of sand and gravel is envisioned. This approach allows maximal regard to the transversal coastal dynamics, without any "hard" defense structure. The gravel layer thickness, the equilibrium profile and the overtopping frequencies of the sea-wall were tested in a wave flume. In a further layout, wave attack and deformation of the gravel core were successfully tested in a wave tank at the Hydraulic Research Laboratory of the Belgian Ministry of Public Works at Borgerhout* . The mixed gravel-sand option provides apparently a feasible and dynamically stable coastal protection.

Four alternative solutions were tested based on the idea of confining the sand in a gravel enclosure with steeper slopes. Thereby the volume of materials needed became considerably reduced, and so were expenditures by some 40%. Through continuous monitoring, subsequent shore behavior has been controlled, and, additionally the collected hydro-meteorological data thus gathered has built up a valuable information bank for future projects (Fig. 35).

3.3.4 West Coast, Belgium

The largest beaches are located west of Ostend, in the direction of the French border; heretofore they have not been affected by erosion. However, lately, some concern has been voiced about the area between Coxyde (Koksijde) and Nieuport (Nieuwpoort). A decision has been reached to provide nourishment to the area and preliminary works completed.

3.3.5 Other locations

Still other sites for which beach nourishment has been recommended as a deterrent for coastal erosion are Amanohashidate Beach in Japan[17] and Folly Beach in South Carolina[29a].

---

* Due to "federalisation" of Belgium, this laboratory is no longer "national" but "regional" ; its functions, however, remain the same.
The municipality of Borgerhout has been annexed to the City of Antwerp.

In the latter location beach nourishment over a distance of 5140 m (3,187 mi) involving over 522,000 m³ (685,000 cu.yd) of material is proposed to restore 9 1/2 ha (23 acres) of beach. The feasibility study points to the following impacts : temporary disturbance of the benthos, increase in water turbidity, decline of esthetic values, increase of automobile emissions and in noise levels. All are however temporary.

In Japan 66 beaches have plans for artificial nourishment and 21 others have already adopted this restoration method. Nourishment for sections of the Egyptian coast is being considered[29b].

In Indonesia the severely threatened coast of south Bali is the subject of a proposal for restoration, maintenance and protection of its beaches. Erosion there results from wave attack, airport and hotel construction, and coral removal. While existing protective structures are to be rehabilitated and integrated in the new scheme, the project combines traditional "hard" structures with extensive beach nourishment.

3.4 Beach Dewatering and “Plates” Systems

In Chapter I, the matter of water absorption by the beach was brought forth in connection with erosion. Dewatering has been proposed as an anti-erosion method. Adams (J.W., 1989, The fourth alternative - Beach stabilization by beachfore dewatering : *Coast. Zone ‘89* III, 3958 - 3973) predicts savings ranging from 25 to 50% depending upon original renourishment life and cost.

The method allows to prolong the useful life of a renourished beach, by decreasing the erosion rate, in some instances to extend it indeficultely by stabilizing the beach. Dewatering is, however, not a substitute to hard or soft beach protection, rather it is supplementary. It can, nevertheless, be sometimes applied as a self-sufficient system. The extra cost, however, can be rapidly recuperated, and apparently, even economies realized because over a period of four to five years renourishment expenses can be drastically cut.

The idea, and some systems, have been tested, with excellent results in Denmark (Totsminde, Firsthals beach) and in Florida (Sailfish Point). In The Netherlands polders drainage canals are a *de facto* implementation of the underlying principle, and in South Africa well points were used ; the diamond exploitation company, Consolidated Diamond Mines (R.H. Charlier, 1991, Marine aggregates, Prognosis, *Environmental Conservation,* extended a beach seaward using the dewatering approach.

Patented in the U.S. under the trade-name "*Stabeach*", the system involves buried components : a porous filter drain pipe (inserted in the beachface), a return pipe to the sump, the sump-, and a discharge pipe from sump to a water disposal area ; the sump pump itself is submerged. The principle is simple and has been enunciated from 1948 (U.S. Grant, 1948, Influences of the water table on beach aggradation and degradation : *J. Mar. Res. 7*, 655-660) through 1989 (P. Bruun, 1989, the coastal drain, what it can do and not do : *J. Coast. Res. 5*, 1, 123) : when the water table under a beach is lower than ocean level, the water "flows" (drains) into the beachface depositing sand in the process consolidating it. By draining then the beach, the table is kept below sea level, at low water the seaward flow abandons material and the wave backswash does not take sand from the beach. Furthermore, the water flow into the beach sand condenses it, thereby giving it more erosion resistance.

Monitoring the effects of dewatering in Denmark and Florida has shown that treated areas fare much better than adjoing ones and that no downdrift erosion was provoked.

Drainage has also been used for friable cliff stabilization (clay, chalk, sandstone).

Though unrelated to dewatering, the "Plates system" alternative can be mentioned at this time. This approach has been tested in Barcelona (Spain) about 1993. Little has been heard about the system since then. By attaching at depths of a few fathoms horizontal plates to poles driven into the sea bed, it is expected to reduce wave energy. The large water mass kept in oscillation produces a kinetic energy which creates another wave movement in a phase opposite to that of the incoming waves. The two movements are to cancel one another.

3.5 Feeder Berms

Sand reserves will help natural processes and they have been established on- and off-shore. Berm-feeding consists in establishing such a reserve on the beach-berm so that a supply of material is available for said processes.

Feeder berms consist in the placement of appropriate sediments on the submerged part of the beach slope at waterdepths (below L.W.) ranging from - 1 m to - 8 m (- 3,28 to 26 ft). Feeder berms are a cheap method of placing material close to the beach. Onshore wave action/transport is expected to move the material to the beach according to the equilibrium profile.

Some applications in the U.S.A. have lead to the conclusion that the waterdepth is critical for the location of a feeder berm. Successful applications have been done in Morocco (Agadir) and the U.S.A. (Mobile Harbor, Southern Long Island).

More recently, a feeder berm has been installed at De Haan (Belgium) in conjunction with a profile nourishment (cf. 3.2.10)[36].

3.6 Beachbuilder and Beach Saver

Beachbuilder is a new beach preservation system based on the premise that it is better to utilize erosive wave energy than to dissipate it. The system builds up a beach : a thin flexible flow control sheet of rugged elastomer - coated polyester, or nylon fabric. It is tethered in place over a portion of the surf- zone surface. The surf rushes beneath it while scouring the bottom surface, transporting loosened sand particles towards shore and into the beach.

As the backwash passes above the weighted sheet, resting on the bottom, underlaying trapped sand particles remain undisturbed. Sand built-up during winter provides an improved beach during the tourism season. Costs are supposed to be less than those of artificial nourishment. However, only tanktests have been conducted (Beardsley, 1993) but, perhaps the system could be a supplementary technique in beach nourishment schemes[35].

Besides Beardsley's *Beachbuilder*, another device called *Beachsaver* has been developed by New Jersey based Breakwaters International Inc. It involves hard structures but is taken up here because it claims to be a complementary method to beach nourishment[35].

In this approach to erosion control, a curtain of water is created suspending sand over the beach as offshore current below channels through the backwash flume. Suspended sand is carried black onshore by incoming waves. "Reefs" are placed in the surf zone at depths of approximately 2 m (606 ft) below mean low tide, run roughly 300 m (984 ft). Surf and beach profiles determine the precise siting of ± 100 (21 ton) interlocking reinforced concrete triangular modules 3 by 5 by 2 m (9.84 ft length x 16.4 ft width x 6.6 ft height). Less sand has to be placed on a beach as the beachsaver keeps it nearshore and maintains it in place impeding its return to the sea during replenishment. The reef's slope deflects wave energy, allows sand to wash onshore, while its angled grooves enhance sand collection at the base. On the inland side of the module, a three slotted opening flume projects a water curtain inhibiting sea-ward sand movement.

Along the New Jersey shore (Cape May Point, Belmar and Spring Lake Juntion, Avalon) three installations currently check physical performance against computer projected sand retention.

3.7 Assessments

Numerous studies have been conducted to measure the results obtained with beach nourishment. Many are encouraging. They have been carried out a.o. by the U.S. Army Corps of Engineers, also by HAECON N.V. (Ghent, Belgium). In some cases new techniques have been tried out, for instance with multiple-layered beach fill.

A beach fill composed of a base layer of hydraulically-placed, silty sand covered with a coarser, land-hauled sand layer has stabilized 2.3 kilometers (1.4 mi) of recreation shoreline in Corpus Christi Bay TX for 10 years. It is an example of how to restore low-to-moderate wave energy beaches where suitable deposits are either limited in extent, or expensive to exploit. The overall stability of the beach fill design suggests the concept has merit and may serve as an alternative to homogeneous beach fills when conditions warrant. The beach has experienced erosion since the late 1800's. (James M. Kieslich and Dewey H. Brunt III, 1989, Assessment of a Two-Layer Beach Fill at Corpus Christi Beach, TX : *Coast. Zone '89*, 3975-3983)

The longer the nourishment distance, the smaller the loss of material. Furthermore, structures placed near the ends of a beach nourishment project can reduce materials losses. Submerged shore-parallel structures limit seaward transport of sediment (a.k.a. "perched beach concept") particularly when fed material is finer than the original beach sand.

Most successful schemes have been reported between headlands because removal can take place only transverse to the beach (cf. chapter III.4). Loss to the offshore depends on submarine slopes that must fill before an equilibrium status is reached. Calculation of the beach profile will help to determine amount of fill required. Silvester (*op. cit.*) has commented that sediment coarser than the original will lead to a steeper than original slope, while the opposite occurs with finer than original fill. However, as we mentioned earlier (3.1.) a steeper beach slope may result in wave reflection, which accelerates removal of finer material and exacerbates the existing erosion problem.

D.E. Newman (1976, Beach Replenishment : sea defenses and a review of the artificial beach replenishment : *Proc. Inst. Civ. Engrs.*, I, 60, 445-460) recommends, however, a mean grain diameter size (D_{50}) of at least 1.5 times that of the original sand, in quantities sufficient to establish an ultimate slope compatible with the wave climate. Topping-up should be carried out as need arises.

The following steps are suggested in the establishment and management of artificial beaches : establishment of various conditions affecting the beach under consideration and choice of the size and methods for the project, assessment of the environmental conditions, detailed facilities design, implementation and, after completion, beach administration.

The survey program related to a beach nourishment scheme involves bathymetric soundings, beach evolution (eventually by remote sensing) and monthly terrestrial measurements (beach profiling). At least once a year sampling should be carried out on the beach, the nearshore and the seabed. Winds, littoral drift and tides affect sediment transport and effects must be measured. The survey requires hence the gathering of tides, waves, currents and winds data.

Tables 2 and 3 illustrate a far from comprehensive review of major beach nourishment projects, and the Japanese step-by-step approach to planning, design, construction and care.

In their report on impact T.E. Lanford and B.J. Baca (1989, Comparative environmental impacts of various forms of beach nourishment : *Coast. Zone '89*, II, 2046-2059) showed that impacts resulting from nearshore turbidity, organisms direct burial and severe habitat alterations at deposition site, were of short duration no matter which method was used. More serious consequences are noted for the "source" sites, especially related to hard-bottom communities ; far less impact was observed for near-shore, dune and onshore borrowing sites than for offshore locations. The nourishment methods considered were of short duration no matter which method was used and were upland borrow, offshore pits dredging, inshore dredging (harbor or waterway), intertidal shoal inlet relocation scraping, inlet relocation or -stabilization.

3.8 Sand By-Passing and Compensation Dredging

Whether sand by-passing should be considered a "soft" coastal protection system or a concomitant method with other schemes is open to discussion. What is sure is that it is quite useful in many cases where for one reason or another a beach is starved (cf. 3.1.3).

River mouths but also entrances to lagoons or embayments act as a brake for littoral transport. If subjected to tides, the tidal current may not be strong enough to carry away sediments that accumulate on the downcoast side of the opening. It can be dredged, pumped or one can use the available wave energy to expedite passage ; groin construction is usually necessary. This method has been called by-passing.

However, the term is also used for a mechanical approach whereby dredging units of barge on pipeline disposal are put to work ; or by fixed loading devices feeding a pipeline passing under or around the entrance of the inlet, or of the structure (e.g. groin). By-passing is thus removal updrift before the material accumulates at a site deemed unfavorable. The material can be used for beach nourishment provided it is placed along the beach where wave action can spread it, outside of an effect zone from a downdrift marine structure, yet within the limit of an offshore bar built during the winter.

A low section (weir) in an updrift jetty can provide a sheltered pre-dredged deposition basin acting as a reserve for beach nourishment purposes.

Floating plants can only be used where the impounding zone is subject to periods of light wave action, or by breaking into the landward section of the impoundment and dredging behind the beach berm, leaving a protective barrier for the dredge. Dredging a sand trap in the water of an inlet protected by jetties is a by-passing technique, particularly when tidal currents are strong ; the trap can be periodically emptied. A set-up completed with a shore-connected breakwater is the most efficient and the costliest.

As said elsewhere the three methods of by-passing are land-based dredging, floating dredges and land-based transport (trucks).

Compensation-dredging can be considered as related to sand by-passing in that dredging of material is carried out in a given site and the material used to nourish the beach system elsewhere. The use of dredge spoils, by the US Army Corps of Engineers, mentioned earlier, is in fact “involuntary” compensation-dredging.

3.9 Conclusion

Man's actions at rather considerable distance from a specific site often are additional causes for rapid erosion. Beaches along the Gulf of Bénin have been depleted because of an important barrage built on the Niger River. Indeed river damming and construction of hard structures - piers, harbor jetties, etc - along riverbanks, lead to sediment starvation and as a corollary to an impoverishment in coastal sand supplies.

Existing coastline constructions are thus imperiled. It is necessary that regulations be enacted, governing any further building and setting a safe beach set-back of constructions proportional to the anticipated life of proposed "hard" structures. On the North Carolina coast this is thirty times the annual rate of erosion.

Shoreline retreat cannot be stopped but it can be slowed and loss of beach material can be compensated. The phenomenon is nearly universal, due to natural causes, but compounded by man's intervention and desire to use the coastal zone to its fullest extent. The economic consequences of coastal erosion are considerable. In some areas retreat has been advocated, regardless of loss. Make-shift breakwaters have been installed by private owners in many sites, particularly along the East Coast of the United States. These artisanal structures are in imitation of the usual practice of placing "hard" structures along eroding shorelines.

Hard defenses are either perpendicular or parallel to the coastline, although some structures have been placed at an angle or in a zigzagging pattern. The cost of keeping up the breakwaters, jetties, sea walls, groins, etc. is often high ; such constructions are occasionally severely damaged, even destroyed, during heavy storms. Moreover, in most cases (on the Oman coast rèportedly this did not occur), while a stretch of beach is indeed protected from erosion, the adjoining down-drift segment suffers from increased erosion as it is material starved.

While in some instances a mixed hard-and-soft approach appears adequate or even necessary, artificial sand nourishment offers an alternative which conforms closer to the natural processes. It has been tried for many decades on a small scale. Newer technologies have lowered the relative cost. Some agencies find it a useful and acceptable way to dispose of harmless waste. Many modest projects have been implemented since the end of World War II, in Europe and the Americas. Loss of deposited materials, according to follow-up observation, appears commonly modest so that maintenance is economically bearable. One of the largest beach nourishment schemes, at least as far as volume of material deposited is concerned, has been completed in Belgium simultaneously with the construction of a new harbor at Zeebrugge. The results are satisfactory and similar works are underway for another segment of the coastline. Furthermore the new situation has contributed to the gradual occlusion of a gully to which material removed from the beach was carried.

The "soft medicine" to restore ailing beaches should be considered as an advisable solution for endangered beaches there where economic dictates require their protection and restoration.

4 Social Action

Soft action against coastal erosion can be buttressed by a series of socio-economic measures. They include minimizing damage danger by allowing buildings only back of the foredune crest, denying flood and disaster insurance and utilities service in a delineated area, re-zoning areas which contribute to the erosion problem.

The issue of "disappearing beaches" has been the subject of considerable concern for a long time ; Charlier among many others discussed Long Island shore problems and drew parallels more than three decades ago (1955, *The Professional Geographer* ; 1956, *Long Island Business*), but it is during the last ten years that concern increased even more. A broad panorama of knowledge and vulnerability of U.S. barrier islands and beaches was drawn at an Annapolis workshop[29c]. Increased demand for and use of beaches and dunes triggered a re-appraisal of the *laissez-faire* of "patrons"[30]. And measures once thought in the United States to be in the best interests of both public and property owners were discontinued[31]. This illustrates second thoughts on the ecological impact of counteracting erosion and controlling overwash by "traditional methods" on barrier islands.

Monetary valuations of beach users' recreational experience and preferences towards coastal erosion control (and litter clean-up) for four beaches in Maine and New Hampshire were estimated for various residential, ownership, and environmental characteristics. [*E. Lindsay and C. Tupper, 1989,* Demand for Beach Protection and use in Maine and New Hampshire : A contingent valuation approach, *Coastal zone '89, 79-87,*]. Indeed, natural resources and environmental goods are often viewed as non market goods, not valued in a traditional sense. This particular set of goods is viewed as generating "intangible benefits or Amenities", difficult to quantify in a monetary sense.

There are generally competing economic interests (commercial, industrial, residential and seasonal sectors), that compose the economic activity of a coastal beach area often documented by economic impact studies .

Impact studies should be complemented with an economic valuation of what the actual experience or activity of being on a coastal beach means, a.o. determination of monetary benefits that beach users derive from spending time on a coastal beach.

A recent study commissioned by the International Geographical Union concluded that further research is needed to formulate an appropriate overall plan of action for erosion control. Three quarters of the world's sandy coastlines are in retreat, some at a rate of a meter (3.28 ft) a year. A combination of factors is now destroying beaches that built up in the thousands of years following the end of the latest "Ice Age".

In conclusion, coastal erosion progresses inexorably and nature will have its way, and a sea-level rise ranging from 0.6 to 3.3 m (2 to 11 ft) is predicted for the year 2010. Man, however, can retard the process, alleviate the present situation and help the natural factors preserve, to a considerable extent, the beaches.

Beach nourishment has been eyed by the U.S. Army Corps of Engineers as a method to dispose of the 265.5 million m^3 (300 million cu.yd) of dredged material it picks up annually. This material is usually dumped at sea damaging benthic life (cf. p. 224).

But such disposal can also help portuary development and marsh creation. Under the proper circumstances the use of dredged material for beach nourishment not only contributes to the development and maintenance of waterway transportation, but, as said earlier, also provides hazard reduction and recreational benefits consistent with environmental mandates and the material laws governing coastal systems. Beach nourishment is the least-cost environmentally acceptable disposal alternative.

Be it repeated, despite its current prevalence, beach nourishment cannot be considered a panacea for dredged material disposal ills. For three out of every four navigation projects, the dredged material is not a practical, cost-effective, and permissible source for beach nourishment supply. Most frequently, the material is too fine-grained to be effective for shore protection. Other constraints are excessive cost, not permissible environmental effects, and technical limitations.

5 References

(1) M. Auzel et J. Bourcart, 1950, "Erosion des plages", *Bull. Inf. Comm. Océan et d'Etud. des Côtes* 379-382.
E.C.F. Bird, 1985, *Coastline changes* (Somerset NJ, J. Wiley).
R.H. Charlier and M. Auzel, 1961, "Géomorphologie côtière : migration des sables sur la côte belge, *Zeitschrift fur Geomorphologie* 5,3,181-184.
R.H. Charlier, 1968, "North Sea beach erosion in Belgium", *Proc. XXIII Int. Geol. Congr.* 167-171.
R.H. Charlier, 1967, "Beach erosion and salvage along the Belgian shore", *Proc. 7th. Int. Sediment. Congr.* Paper C-1, 1-3.
S.Y. Chew, *et al.*, 1975, "Beach development between headland breakwaters", *14th Coast. Eng. Conf. Proc.*, 2, 1309-1418.
H. Howa, 1985, "L'érosion du littoral du Nord-Médoc (Gironde)", *Bull. Inst. Géol. du Bassin d'Aquit.* 38, 57-68.
J. Gribbin, 1984, "The world's beaches are vanishing", *New Scientist*, 102, (May 10), 30-32.
F.P. Shepard and H.R. Wanless, 1971, *Our changing coastlines* (McGraw-Hill Book Co., New York).
J.E.L. Verschave, 1961, "La défense et le maintien des plages belges entre Zeebrugge et la frontière Néerlandaise" *Bull. Techn. Assoc. Ing. Univ. Louv.*, 89,1,19-29.
P.P. Wong, 1981, "Beach evolution between headland breakwaters", *Shore and Beach"*, 49, 3-12.

(2) G. De Moor, 1979, "Recent beach erosion along the Belgian North Sea Coast", *Soc. Géol. Belg. Ann.* 88, 143-157.
G. De Moor, 1980, "Erosie aan de belgische kust", *De Aardrijkskunde* NS, 4, 279-294.
G. De Moor, 1979, "Premiers effets du rehaussement artificiel d'une plage sableuse le long de la côte belge", In : Guilcher, A. (ed.), *Les côtes atlantiques d'Europe, évolution, aménagement, protection* CNEXO, (Brest), 9, 97-114.

(3) M. Zunica, 1976, "Human influence on the evolution of the Italian coastal areas", *Proc. Int. Geogr. Congr.* XXIII, 87-93.
M.C. Barth and G.J. Titus, 1984, (ed.) *Greenhouse effect and sea level rise : a challenge for this generation*, (New York, Van Nostrand Reinhold).

(4) P. Bruun, 1962, "Sea level rise as cause of beach erosion" *J. Waterways, Ports and Harbors Div. (ASCE)* 88, 117-130.
J.M. Darling, 1964, "Seasonal changes in beaches of the North Atlantic coast of the United States", *Conf. Coast. Eng. Proc.* IX, 236-248.
W. Kaufman and O.M. Pickley, jr., 1983, *The beaches are moving : the drowning of America's shoreline. With a new epilogue* (Duke Univ. Press., Durham NC).
Khelila, 1985, "Nos plages vont-elles disparaître ?", *La Gazette Touristique de Tunisie*, 14, 23-25.
M.L. Schwartz, 1968, "The scale of shore erosion", *J. Geol.*, 16, 508-517.
J.E. Webb, 1960, *The erosion of Victoria Beach : its causes and cure* (Ibadan, University Press).

(5) R.H. Charlier and M. Vigneaux, 1986, "Study of management and conflicts in the coastal zone", Part 2, *Int. J. Envir. St.*, 26, 206-215.

(6) V.K. Tippie, 1976 "Coastal canals can damage the environment" *Maritimes*, 20,1, 12-14.

(7) R.H. Charlier and A. Haulot, 1975, "Coastal belt touristic occupancy and ecological impact" *Proc. Int. Conf. Ocean Mangmt.* (Tokyo).

(8) The Nature Environmental Research Council, 1973, *Marine Wildlife Conservation* (Publications Series "B" no. 5 Jan).
R.H. Charlier, A. Haulot, *et al.*, 1978, "Coastal belt tourism, economic development and environmental impact" *Int. J. Envir. St.* 10,4,161-172.
R.H. Charlier, A. Haulot, and L. Verheyden, 1978, "Coastal environmental dilemma : economic development versus tourism" *Int. Ocean. Dev. Conf.* (Tokyo) V, E-3,72-78.
J. Clark, 1974, Coastal ecosystems. *Ecological considerations for the management of the coastal zone* (Washington, Conservation Foundation)
C.N. Ehler and D.J. Basta, 1984, "Strategic assessment of multiple use conflict in the Exclusive Economic Zone" *Oceans '84 Proc.* Jt. Ann. Mtg. MTS/IEEE (Washington D.C.).
R. Healy and J. Zinn, 1985, "Environment and development conflicts in coastal zone management", Symposium on coastal management: planning the edge. *J. Amer. Planning Assn.* 51, (Summer) 263-336.
I.P. Joliffe, 1976, "Man's impact on the coastal environment", *Geographica Polonica* 34, 73-90.

A.R. Orme, *et al.*, 1980, (ed.) *Coasts under stress* : (Berlin, Börnträger,).
R. Thompson, 1984, "America's threatened shorelines", *Congress. Quart.* - Edit. Res. Rep. 2,17,819-836.

(9) J.M. Bryden, 1973, *Tourism and development* (Cambridge, Cambridge Univ. Press).
R.H. Charlier and M. Vigneaux, 1986, "Study of management and economic conflict in the coastal zone (Part I)" *Int. J. Envir.* St. 26, 3, 177-189.

(10) W. Harrison, 1967, "Environmental effects of dredging and spoil deposition" *World Dredging Conf. Techn. Papers* 535-559.
G. Govatos and I. Sandi, 1969, "Beach nourishment from off-shore sources" *Shore and Beach* XXXVII, 40-49.
S.J. Williams and D.B. Duane, 1975, *Construction in the coastal zone : a potential use of waste materials.* (Washington D.C., U.S. Army Corps of Engineers (Reprint 2-75).

(11) Anonymous, 1986, "Strategy for beach preservation" : *Geotimes* 30,12, 15-19.
P. Bruun and J.S. Purpura; 1963, *Emergency measures to combat beach erosion* (Gainesville, FL, Florida Engng. and Experiment Station).
A. Führböter, 1972, "A refraction groyne built by sand", *Proc. Int. Conf. Coast. Engng* (Copenhagen), 1451-1469.
M. Petersen, 1963, "German experience on coastal protection by groins : a review" *Bull. Beach Erosion Bd.* XVII, 38-54.
R. Silvester, 1985, "Natural headland control of beaches" *Cont. Shelf Res.* 4,5, 581-596.

(12) R.H. Charlier, 1956, Shifting sands : *Long Island Bus.*, IV, 3, 1-16.
J.K. Mitchell, 1974; *Community response to coastal erosion*, (Chicago, Univ. of Chicago, Dept. of Geogr.) Res. Paper, no. 156.

(13) P.P.L. Roovers, *et al.*, 1981, "Beach protection as part of the harbour extension at Zeebrugge" *Papers XXV Int. Navig. Congr.*, 5,2,755-769.
G. De Moor, 1979, "Premiers effects du rehaussement artificiel d'une plage sableuse le long de la côte belge". In : Guilcher, A. (ed.), *Les* côtes *atlantiques d'Europe : évolution, aménagement, protection.* CNEXO, (Brest), 9, 97-114.
F. Depuydt, 1980, "Het Belgisch Strand", *Verhandl. Kon. Akad. Wet.*, 122.

(14) M. Akle, 1969, "Problèmes d'érosion côtière dans le Golfe du Bénin", *The Siren* 29 Sept., 20-31 (1985).
Hydraulic Laboratory, *Erosion of Victoria Beach at Lagos, sand - by-passing and coastal development* (Delft, NL, Hydraulic Laboratory) Report 1039-Feb.
C.T. Brown, 1978, "Gabions and reno-mattresses as low cost terminal revetment" *4th Austral. Conf. Coast and Ocean Engng.* (Adelaide, Nov. 8-10).
R.H. Charlier, C.F. De Meyer, D. Decroo, 1988, "Soft beach protection and restoration". In : E.M. Borgese and N. Ginsburg, "Ocean Yearbook VII", Chicago, IL, Univ. of Chicago Press.

ONU/DAESI, 1983, *Technologies pour la lutte contre l'érosion* des côtes (New York, Organis. des Nations Unies).
Ministère de la Coopération (France), 1968, *Les ouvrages en gabions* (Paris, Min. Coop.).

(15) C.H. Everts *et al.* 1983, *Sedimentation inversion at Masirah Island* (Oman), (Washington, U.S. Army Corps of Engineers, Engineering Research Center,), Misc. Pap.

(16) M. Yajinia *et al.*, 1983, "Application of sand by-passing to Amanohashidate Beach" *Coast. Eng. Japan*, 26, 151-162.

(17) Adee, P.H., 1976, "A review of developments in using floating breakwaters", *Off-shore Techn. Conf. Proc.*, II, 225-236.
Arafa, R., 1985, Mechanics of wave breaking over floating breakwaters in erodible bed, *7th Miami Int. Conf. on Altern. En. Sources*, 273.
Harms, V.W., 1980, Floating breakwater performance comparison, *17th. Coast. Engng. Conf. (ASCE),* 2137-2158.
Harris, A.J., 1986, Les brise-lames flottants, *Expomat - Actualités*, 74,4, 44-45.
Hayashi, H. Fujisaura, J. Nasuda, K., Uda, T. and Omata, A., 1990, Study of the hydrodynamic characteristics of an offshore breakwater with a compound wave energy dissipation function : *Proc. PACON* 253-262.
Iwata, S. and Natsuishi, N., 1990, Development of new floating - breakwater (strength of composite steel-concrete structural unit) : *Proc. PACON* 243-252.
Ueno, S. Nakatsuka, K., Hoyama, J., Ishino, K., Tanaka, K., Tatsui, H., Minato, R., Hara, K., Uda, T. and Omata, A., 1990, Development of a new submerged breakwater design (Surf.) : *Proc. PACON* 263-274.
Richy, E.P. and Nece, R.E., 1974, Floating breakwaters. State of the Art, Kingston RI, *Floatg. Breakw. Conf.* 1-9.

(17a) Kakuris, P.A., 1983, *Surgebreaker off-shore reef systems* : (Chicago, Great Lakes Environmental Marine Ltd.).

Shabica, C.W., 1985, "Sediment accretion associated with surgebreaker off-shore reef, a low cost shore protection device : an update on the Corps of Engineers section 54 Program" *Oceans '85*, Ocean Engineering and the Environment (MTS/IEEE Joint Annual Mtg) I, 574-580.
Kondo, H. *et al.*, 1983, *"Wave energy dissipating structure"* (Tokyo, Morikita Book, Publ. Co.) [in Jap.].

(18) A.Q. White, 1981, "Impact of beach renourishment on a sandy beach ecosystem" *Estuaries*, 4, 3, 259.

(18a) Benton, S. et al., 1983, *"Cape Hatteras Shoreline Erosion Workshop - Summary Report"* (Columbia NC, North Carolina Dept. of Natur. Res. and Comm. Devel).
Brashears, R.L., and Darnell, J.S., 1967, "Development of the Artificial Seaweed Concept", *Shore and Beach*, 35,2, 35-41.
Geomidi Consultants, 1985, *"Protection against Coastal Erosion (Scour)"* (Saintes-Maries-de-la-Mer (France)), Report.
Linear Composites, 1986, *"Erosion Control Systems"*, [Company release], (North Yorkshire (England), ICI Fibres).
Rogers, S.M. Jr., 1987, "Artificial Seaweed for Erosion Control", *Shore and Beach* 55,1, 19-29.
Seascape Technology Inc., 1984, *"Seascape at Cape Hatteras"* [Company release], (Greenville, DL).
Ten Hoopen, H.G.K., 1976, "Recent applications of artificial seaweeds in The Netherlands", *Coastal Engineering* 166, 2905-2915.

(18a) I. Anonymous, 1986, *Erosion control systems* (North Yorkshire (England), Linear Composites - ICI Fibres).

II. Anonymous, 1986, *Artificial seaweed protects the seabed, estuaries and submerged pipelines* (Amsterdam NL, Nicolon Corp.).

III. Anonymous, 1983, *Non-structural beach erosion protection* (Long Beach CA, Dept. of Publ. Works & Tide-land Agency).

IV. Hall, M.J., 1985, *Final report on the results of monotring a "seascape" installation* (Lawrenceville NJJ, Dept. of Geosciences - Rider College).

V. Jackson, G.A. and Winant, C.D., 1983, Effect of a kelp forest on coastal currents : *Cont. Shelf Res.* 2.1., 15-80.

VI. Jenkins, S.A. and Skelly, D.W., 1987, *Hydrodynamics of artificial seaweed for shoreline protection* (La Jolla CA, Scripps Inst. of Oceanogr.) Center for Coastal Studies Rep. 510-87.16.

VII. London, M.E., 1985, *Monitoring results and analysis, Keewaydin Island, artificial seaweed* (Tallahascf., Fl. Florida Dept. of Natural Resources).

VIII. ten Hoopen, H.G.H., 1976, Recent applications of artificial seaweed in The Netherlands : *Proc. Coast. Engng Conf. XV* (Honolulu HA) 3, 166, 2905-2915.

IX Wang, D.P. and Ditmars, J.D., 1982, *Physical and environmental aspects of ocean kelp farming* (Chicago IL, Gas Res. Inst., Argonne Nat. Lab. Rep. 81/0111).

(18b) Anonymous, 1983, *"Non-structural Beach Erosion Protection"* (Long Beach CA, Dept. of Public Works and Tidal Agency).
Hall, J. *et al.*, 1986, *"Creative Shoreline Management through Community Partnership"* (Santa Cruz, CA, SBPA Annual Meeting).
Rogers, S.M. Jr., *op cit.*

(18c) Dalrymple, R.A., *et al*, 1982, *"Physical and Environmental Aspects of Ocean Kelp Farming"* (Chicago, IL. Gas Research Inst.) (rep. 81-0111).

(18d) Erosion Technology Systems, Inc., 1985, *Sea Grid* (Wilmette IL, Company Release).

(18e) Cunningham, R.T., 1966, "Evaluation of Bahamian Oolithic Aragonite Sand for Florida Beach Nourishment" *Shore and Beach*, 34,1,18-19.
Downie, K.A. and Saaltink, H., "An artificial beach for erosion control", *Coastal Structures '83 Conf. Proc.* (ASCE) 846-859.
M.N. Nichols and C. Cerco, 1987, "Coastal dunes and sand resources". *5th. Symp. Coastal and Ocean Manag., Proc.*
Roos, A., 1977, "Artificial Beach at Hook of Holland", *Shore and Beach* 45,2,19-23.
Tourmen, L., 1968, "The Creation of an Artificial Beach in Larvottao Bay, Monte Carlo, Principality of Monaco", *11th. Conf. Coastal Engineering Proc. (ASCE)* I,35,558-569.

(18f) Hobson, R.D., 1977, *"Review of design elements for beach fill evaluation"* (Vicksburg, MS, US Army Coastal Engineering Res. C.) (Techn. Paper 77.6).
James, W.R., 1975, *"Techniques in evaluating suitability of borrow material for beach nourishment"* (Vicksburg, MS, US Army Coastal Engineering Res. C.) (techn. memo, 60).
James, W.R., 1974, "Borrow material texture and beach ill stability", *Proc. 14th. Int. Coastal Engng. Conf.* 1334.

(19) E.C.F. Bird, 1979, Coastal processes. In : K.S. Gregory, and D.E. Walling (ed.) *Man and environmental processes*, (Canberra, Publ.) pp. 82-101.
O.H. Pilkey and T.D. Clayton, 1987, "Beach replenishment. The National Solution ?", *5th. Symp. Coastal and Ocean Manag., Proc.*
Wiegel, 1987, R.L., Trends in Coastal Erosion Management", *Shore and Beach*, 55,1,3-11.

(19a) Penfield, W.C., 1960, 'The Oldest Periodic Beach Nourishment Project", *Shore and Beach*, 28,1,9-15.

(20) P. Bruun, 1985, "Cost-effectiveness of coastal protection with reference to Florida and the Carolinas U.S.A." *J. Coast. Res.* 1, 1, 47-55.
Berg, D.W., 1965, *"Factors affecting beach nourishment requirements at Presque Isle Peninsula, Erie, Pennsylvania"* (Ann Arbor, Univ. of Michigan, Great Lakes Research Division) (Publ. 13) pp. 214-221.
H.H. Dette, 1977, "Effectiveness of beach deposited nourishment" Proc. Am. Soc. *Civ. Eng. Specialty Conf." Sedimentation '77"* (Charleston, NC) 211-227.
J.V. Hall, Jr. 1952, *"Artificially nourished and constructed beaches"* (Beach Erosion Board), Techn. Mem. 29.
J.R. Lesnik, *"An annotated bibliography on detached breakwaters and artificial headlands"* (Fort Belvoir, VA, Coastal Engng. Res. Ctr.) (MR79-1).
Norby, R.D., 1981, *"Evaluation of Lake Michigan Nearshore sediments for nourishment of Illinois Beaches"* (Champaign, IL., Illinois State Geol. Survey).
Nordstrom, K.F., Allen, J.R., Sherman, D.J. & Psuty, N.P., 1979, "Management considerations for beach nourishment at Sandy Hook, NJ, USA, *"Coastal Engineering* 2, 215-236.
J. Pope, Oct. 3-6, 1986, "Segmented off-shore breakwaters : an alternative for beach erosion control", *Shore and Beach*, 54, 4.
Walton, T.L. and Purpura, 1979, J.A, *"Beach nourishment along the southeast Atlantic and Gulf Coasts"* (Tallahascf., FL. Univ. of Florida, Florida Sea Grant Program), pp. 11-24.

(21) J.C. Boothroyd, 1985, "Geologic processes pose problems for the Rhode Island shore" *Maritimes*, 29,2,1-3.
P. Bruun, 1989, "Coastal engineering and use of the littoral zone" *Ocean & Shoreline Man.* 12, 5/6, 495-516
R.H. Charlier, C.F. De Meyer & D. De Croo, 1989, 'Beach protection and restoration" *Int. J. Env. St.* 33, 3, 33-44 ; *ibid.* 33, 4, 167-191.
Y-H, Chu and E.B. Hands, 1987, "Shoreline erosion protection and beach nourishment" *5th. Symp. Coastal and Ocean Manag., Proc.*
P.J. Godfrey, 1976, "Barrier beaches of the East Coast" *Oceanus*, 19,5,27-40.
L.P. Johnson, 1987, "Natural shore and beach restoration enhancement and preservation systems", *5th. Symp. Coastal and Ocean manag., Proc.*
W. Kaufman and O.H. Pilkey, 1983, *The beaches are moving : the drowning of America's shoreline*, (Durham, NC, Duke Univ. Press).

Laboratoire Central d'Hydraulique de France, 1972, *Plage et littoraux artificiels* (Paris, Ministère de l'Equipement du Logement et de l'Aménagement du Territoire).
M.D. Lemonick, 1987, "Where's the beach ? America's vanishing coastline" *Time* Aug. 10, 40-47
F. Lowenstein, 1985, "Beaches or bedrooms - The choice as sea level rises", *Oceanus*, 28,3,20-29.
M.W. Mugler, 1983, "A problem can be a resource : beach nourishment with dredged material" *Waters spectrum*, Spring, 38-45.
U.S. Department of the Interior, 1985, *Coastal barrier resources systems : Draft report to Congress* (Washington DC, US Dept. of the Int.).

(22) J.F. Oertelm J.E., Fowler, and J. Pope, 1985, *History of erosion and erosion control efforts at Tybee Island, Ga.* (Vicksburg MS, U.S. Army Coastal Engineering Research Center) (Report CERC 85-1).

(22a) P. Kerckaert, 1985, P.P. Roovers, A. Noordam, and P. De Candt, "Artificial beach renourishment on the Belgian Coast" *J. Waterways, Ports, Coastal and Ocean Div. (ASCE)*, Jan.
P. Kerckaert, P.P. Roovers and A. Noordam, 1982, "Artificial beach renourishment on the Belgian East Coast", *18 Int. Conf. Coast. Engng.* (Cape Town, R.S.A.).

(22b) P. Kerckaert, *ibid.*
P.P. Roovers, P. Kerckaert, A. Burgers, A. Noordam, and P. De Candt, 1981, "Beach protection as a part of the harbour extension at Zeebrugge, Belgium", *Proc. PIANC, 25th. Congr.* (Edinburgh) II, 5,755-769.

(23) R.H. Willmington, 1982, "The renourishment of Bournemouth beaches, 1974-1975", *Shoreline Protection Conf.* (Southampton) *Papers*, 19, 157-162.

(24) J. Kramer, 1972, "Artificial beach nourishment on the German North Sea Coast", *Conf. Coast. Engng.* II, 1465-1483.
H. Kunz, 1987, "Shoreline protection of the East Frisian Islands of Norderney and Langeoog". *5th. Symp. Coastal and Ocean Manag., Proc.*

(25) V.V. Dodin and V.V. Ponomarenko, 1972, "Dynamics of artificial beaches for conditions in Odessa" In : *Geologiya poberezhya i dna chernogo morya no. 6* (Kiev University Press), pp. 145-154.

(26) E.K. Grechishchev, L.A. Morozov, and Y.S. Shul'gin, 1972, Artificial sand beaches and organization of their aggravation. In : *Ukreplenie morsk ikh bezegov* (Moskva, Transport Publ.), pp. 60-68).

V.P. Zenkovich, and A.G. Kiknadze, 1981, The sea coast investigations in Georgia. In : *Man and nature in geographical science*, (Tbilisi, Georgian S.S.R. Acad. of Science).
V.P. Zenkovich, 1973, "Geomorphological problems of protecting the Caucasian Coast", *Geogr. J.*, 139, 460-466.

(27) W.S. Andrew, 1981, "Entrance management and beach restoration" *5th Austr. Conf. Coast & Ocean Engng. Shore and Beach*, 53
L. Carter, and J.S. Mitchel, 1985, "Stability of an artificially nourished beach, Balaena Beach, Wellington, N.Z." *N.Z.J. Mar. and Freshwater Res.*, 19, 4, 535-552.

(28) D. Vera-Cruz, 1972, "Artificial nourishment of Copacabana Beach" *Conf. Coast. Engng*, II, 1451-1461

(29) Barrier islands workshop, Annapolis, 1976, *Barrier islands and beaches* (Washington, DC, Conservation Foundation).
R.D. Behn and M.A. Clark, 1979, "The termination of beach erosion control at Cape Hatteras", *Public Policy*, 27, (Winter), 99-127.

(29a) Anonymous, Folly Beach, S.C. 1985, *"Beach erosion control and hurricane protection"* Appendix 2, (Charleston, S.C., Army Engineers District,).
T.W. Kana and M.L. Williams, 1987, "Beach nourishment at Myrtle Beach, South Carolina - An overview". *5th. Symp. Coastal and Ocean Manag., Proc.*
K.F. Nordstrom and J.R. Allen, 1980, "Geomorphologically compatible solutions to beach erosion, *Zeitschr. f. Geomorph.* Suppl. 34, 142-145.

(29b) Nakayama, Y. et al., 1982. "Construction of Artificial Beaches in Japan", *Civ. Engineering in Japan*, 21, 100-113.

(30) E.J. Kahn, III, 1985, "The seashore's secrets", *Boston Magazine* 77, (June), 140-141 + 150-152 + 154).
H.R. Mahoney, 1980, "Dune busting : how much can our beaches bear ?", *Sea Frontiers*, 26 (Nov-Dec), 323-330.
M. Primack, 1982, "Battling the beach buggies", *Sanctuary*, 21 (July-Aug.), 7-10.

(31) Behn and Clark, *op cit.* ref. 29.

(32) Anonymous, 1957, *"Our vanishing shoreline"* (Washington, D.C., U.S. Department of the Interior).

Charlier, R.H., Coastal zone : occupance, management and economic competitiveness, *Ocean and Shoreline Manag.* 12, 5/6, 383-402 (1989). P. Bruun, Coastal engineering and the use of the coastal zone, *Ocean & Shoreline Man.* 12, 5/6, 495-516.
P. Koekkebakker & G. Peet, 1987, Coastal zone planning and management in The Netherlands, *Coast. Z. Man.* J. 15, 121-134
M. Vigneaux, 1987, L'Exploitation de la zone côtière : ses perspectives et ses risques pour l'environnement, *Act. Coll. Mer et Littoral, Couple à risque* (Biarritz) La Documentation Française, Paris, pp. 299-310.

(33) Bruun, P., 1954, *Coast Stability*, (Kobenhavn, Danish Technical Press), 400 p. (cf. also *Technical Memo No.* 44 by the Beach Erosion Board, U.S. Army Corps of Engineers, 1944, 108 p.).
Bruun, P. *et al.*, 1985a. *Design and Construction of Mounds Breakwater and Coastal Protection.* (Amsterdam : Elsevier), 961 p. (cf. review article by Bruun and Losada, printed in *Dredging and Port Construction*, November 1985, 1 pp. 25-32).
Bruun, P. and Schwartz, M.L. 1985b, Analytical Prediction of Beach Profile Change in Response to a Sea Level Rise, *Zeitschrift für Geomorphologie*, Supplement Band 57 : 33-50.
Bruun, P., 1986, Sedimentary Balance (land and sea) with special reference to the Icelandic South Coast from Torlakshöfn to Dyrholaey, River nourishment of shores - practical analogies on artificial nourishment, *Coastal Engineering*, 10, 193-210.
Losada, M. and Desiré, H., 1985, *Incipient Motion of Horizontal Sand Bottom under non-breaking Waves*, Report by Dept. Puertos y Structuras Maritime, Universidad de Santander, Spain (cf. also Losada and Desiré : *Coastal* Engineering, 9, 357-370).

(34) Cur, Rijkswaterstaat, Delft Hydraulics, 1987, *Manual on artificial beach nourishment* (Delft, Rijkswaterstaat).

(35) Beardsley, M.W., 1993, Harnessing steep-wave-energy to build up beaches and prevent erosion : *Proc. Int. Coastal Symposium* Hilton Head Island SC, June 6-9, pp. 174-180.
Creter, R.E., 1994, Offshore erosion control takes on new dimensions : *Sea Technology* 35, 9, 23-26

(36) Charlier, R.H. and De Meyer, C.F., 1995, New developments on coastal protection along Belgian Coast : *J. Coast. Res.* (in press).
Dean, R.G., 1991, Equilibrium beach profiles : characteristics and applications : *J. Coast. Res. J.,* 1
HAECON, Ministerie van Openbare werken, 1990, *Ontwerp van Kustverdedigingswerken te De Haan* (HKVO57/90.3510) : Gent, HAECON.

Warnier, F., Malherbe, B., Lahousse, B. and Rnes, E., 1994, Execution aspects and monitoring of a new beach nourishment concept at De Haan : feeder berm and profile nourishment combination *(in press)*.

Chapter III : Sediment Budget Analysis

1 The Sediment "Bank Account"

Sand available for accretion, erosion or transport makes up a "bank" in which quantities are deposited or from which they can be withdrawn. One can also speak of supply. While we commonly refer to "sand", it often is rather "sediment" which can be of various nature, origin, size, texture and so on. The study of marine sediments and marine sedimentation is a discipline in itself. The elementary traditional view is that sediments are brought to the sea from inland or on-shore coastal sources, and then deposited according to size in a succession of ever diminishing grain "diameters" from shallow to deep waters. Things, however, are not that simple as local phenomena and morphology sometimes give the impression that the pattern has been "inversed" (e.g. tangue).

Sediment transport occurs by dragging over the bottom, or by saltation (bedload transport) or in suspension (suspended-load transport). They occur in general concurrently. Transport zones are the offshore (wave-induced motion over ripples), and the surf (most induced by the passing breaker).

Grain diameter size has been "measured" subjectively and according to several quantitative scales (e.g. Wentworth, Atterberg, AFNOR). For Wentworth, sand has a diameter > 0.074 mm (or 3.75 φ) to < 2 mm (or - 1.0 φ). His "name classes" for sands differ from the Unified Soils Classification in that the USC coarse sands (- 1.0 φ to - 2.25 φ) are considered gravel (- 1 to - 2 φ) and "small pebbles" (- 2 to - 2.25 φ), his very fine sand (3 to 4 φ) is already silt for the USC (upper limit 3 φ). Incidentally $\varphi = \log_2$, a diameter value expressed in mm (4 φ $\cong$ 0.062 mm).

In a preceding chapter the removal agents have been listed, natural and man-made. P.D. Komar calls these the debits. In the credit column appear longshore transport into area, river transport, sea cliff erosion, onshore transport, biogenous deposition, hydrogenous deposition, eolean transport onto beach zone, beach artificial recharge and/or nourishment.

As in any budget reserves may exist. Indeed sand can be withdrawn naturally temporarily from the littoral zone transport system and stored on the backshore and/or dunes. These "reserves" may remain untouched for months or years. They should not be included in a budget calculated shortly after a storm. Eolean transport being slower than water transport, the building up of dune reserves take months or years, those of the backshore may accumulate in days or even hours after a storm. Storms can reintegrate these reserves in the littoral sedimentary transport system. The offshore zone seaward of the beach may be source or sink, but little is known about it.

With worldwide erosion prevalent, it is quite obvious that many "accounts" are in the red. In short, sediment budgets are sediment transport volume balances for a specified coast segment, and many are "negative".

2 Budget Analysis

Analysis requires delimitation of boundaries and examination of the constituent elements. From a sedimentologist's, and engineer's viewpoint, it is thus necessary to identify processes involved, particularly those significant to the specific project or the individual situation, estimate the sediment volume needed, and on the hand of this information balancing the budget.

The area to be studied determines the boundaries, within time limits and according to the aims pursued. It may be useful to gain some knowledge of adjacent areas' sand budgets. The budget of a beach and its nearshore zone has shore-parallel boundaries landward of the line of expected erosion and at, or beyond, the seaward limit of significant transport. Barrier island sand dunes might have a boundary at the island's bay side and the landward edge of the backshore.

The credit lines of a budget are sources ; the debit lines are sinks. The total effect of a line is obtained by multiplying the shown quantity by the length of shoreline over which the "item" operates.

An assessment of the longshore transport rate requires that at least two of the following variables be known : Q_g, Q_n, Q_{rt}, Q_{lt}, γ.

A balanced budget would be expressed by

$$\sum_{i=1}^{4} Qi^{+} + \sum_{i=1}^{4} Q^{*} = \left[\sum_{i=1}^{3} Qi^{-}_{i} + \sum_{i=1}^{3} Q_i^{*} - \right] = 0 \qquad \text{(III.1)}$$

with $Q_i^* = b_i\, q_i$ and $Q_i = k_i\, Q_g$

meaning that the difference of all credits to all debits is nil. Any deviation from 0 indicates an estimated rate of erosion or accretion. Symbols used in (III.1) are identified below.

- Q are credits (+) or debits (-) a.k.a. source and sink points, Q* being a plus (+) or minus total contribution of a budget line calculated over the distance involved (m³/y).
- q are source (+) or sink (-) lines, *viz.* adding or subtracting over an extended segment of a control volume boundary (m³/y/km).
- b is the length of the shoreline segment of activity
- k is the fraction of the gross longshore transport rate (Q_g).
- i (= 1, 2, 3 or 4) identifies the location of source or sink ("where" in littoral zone).

In regards to Q_g, it is the sum of amounts transported to right (Q_{rt}) and left $Q_{(lt)}$ of a given point on the shoreline during a unit of time. The difference between these amounts is the net (Q_n) transport rate. Q_{rt} and Q_{lt} are used in design of jetties and impoundment basins , Q_n for prediction of beach erosion on an open coast. When only a ratio (γ) is known, Q_g and Q_n are related by (III.5). When $\gamma \rightarrow \sim 1$, equation (III.5) $\rightarrow$ and is of no use.
Thus, summarizing.

$$Q_g = Q_{rt} + Q_{lt} \quad \text{(III.2)}$$
$$Q_n = Q_{rt} - Q_{lt} \quad \text{(III.3)}$$
$$\gamma = Q_{lt}/Q_{rt} \quad \text{(III.4)}$$
$$Q_g = \frac{Q_n(1+\gamma)}{(1-\gamma)} \quad \text{(III.5)}$$

Another representation of the longshore transport rate can be given by the immersed weight rate I_l (force unit/time unit, p.ex. . With ρ_s as the mass density of sand (sediment), ρ the mass density of water, a' the volume of solids/total volume (to account for porosity),

$$I_l = (\rho_s - \rho)\, g\, a'\, Q \quad \text{(III.6)}$$

3 The Cell Theory

Mention should probably be made of the position that the sand budget being irregularly distributed, redressing it would provide a viable solution to coastal erosion. The proponent, following into the steps of work published by Collins in 1988, proposed at Bordomer '95, a "Coastal Cell Method".

According to Larcher's paper at that meeting, sand pockets, a.k.a. coastal cells, are sites where beach profiles are healthy and keep wave energy away from bluffs.

Erosion occurs between cells.

The length of coastline within a cell's boundaries determines whether it is a micro-cell (less than 50 m or 164 ft long), a meso-cell (50 to 1,000 m or 164 to 3,280 ft long, or a macro-cell (> 1,000 m).

The irregularity of coastal cells impedes parallel spreading of sediments along coastlines. These cell's morphology determines two kinds of transits : one within the cell, and one between cells. There is hardly any transit between cells under normal wind and sea conditions, with the only noticeable transit within the cell against its limits abutment, with no escape beyond the cell downstream boundary.

Between cells, transit occurs when wind and sea conditions change and reverse abutments. Sediments freed from the cell's attraction, however, by-pass eroded areas only to accumulate in the next cell.

By creating artificial coastal cells, the eroded area can be integrated with the intercell and intra-cell transit. Larcher refers to the approach as adjusting the naturally uneven sand budget. Artificial ripple systems, some made of easily installed hollow modular careens, aim at ondulating the beach and forebeach to stabilize it and create stable sand bumps generating natural sandbank responses.

4 Conclusion

The range of credits on debits due to the various lines on the budget ledger are measured as fractions of the gross longshore transport rate (m^3/y/m or cu.yd/y/ft) of beach front. The relative importance of each is governed by local conditions.

5 References

A.J. Bowen & D.L. Inman, 1966, *Technical Memorandum no. 19* : (Washington DC, Coastal Engng Res. Cent. - US Army Corps of Engineers) ;

Komar, P.D., 1983, Beach processes and erosion - An introduction. In : P.D. Komar (ed.), *CRC Handbook of coastal processes and erosion* : (Boca Raton FL, CRC Press) pp. 1-22.

Chapter IV : Evaluation Of Impacts And Interactions

1 Introduction

In chapter II, environmental impact, *viz.* consequences, of man-made structures upon coastline evolution has been repeatedly stressed. We will therefore limit us here to some additional or concomitant considerations.

Coastal engineering problems may be classified into four general categories : shoreline stabilization, backshore protection (from waves and surge), inlet stabilization, and harbor protection (Fig. 35). A coastal problem may fall into more than one category. Once classified, various solutions are available to the coastal engineer. Some of the solutions are structural ; however, other techniques may be employed such as zoning and land-use management. We deal primarily with structural solutions, but these considerations may also apply to other types of solutions.

Fig. 36 shows the structures or protective works in the four general coastal engineering problem classifications and lists the factors that must be considered in analyzing each problem area. Hydraulic considerations include wind, waves, currents, tides, storm surge or wind set-up, and the basic bathymetry of the area. Sedimentation considerations include the littoral material and processes (i.e., direction of movement ; rate of transport, net and gross ; and sediment classification and characteristics), and changes in shore alignment. Navigation considerations include the design craft or vessel data, traffic lanes, channel depth, width, length and alignment. Control structure considerations include the selection of appropriate protective works (shape, size, site) keeping in mind the potential effects on adjacent areas. Economics must consider maintenance costs, and interest and amortization of initial investment.

As a short term effect scour at structure's toe will form a trough. Scouring can undermine the toe leading to settlement of stone in a rubble mound breakwater. This may be countered by overbuilding the cross section, or by providing excess stone at the toe to fill the anticipated scour trough. The toe of a vertical structure may be protected similarly against scour by the use of stone. Impermeable cutoff walls at the base must be used to protect a gravity wall from undermining by scour. As a general guide, the maximum depth of a scour trough below the natural bed is about equal to the height of the maximum unbroken wave that can be supported by the original depth of water at the toe of the structure.

Placement of a rock blanket with adequate bedding material seaward from the toe of the structure will prevent erosion at the toe and will result in a more stable structure.

For long-term effects, it is preferable to assume that the structure would have no effect on reducing the erosion of the beach seaward of the wall. This erosion would continue as if the wall were not there. Since the determination of scour can only be approximate, general guides are usually adopted.

In fig. 37 the solid line represents an average existing profile ; a structure is to be placed at point A in the fig.. Either the loss of beach width per year or the annual volume loss of material over the beach area, which includes the profile, is known from prior records. In the latter case, the annual volume loss may be converted to an annual loss of beach width by the general rule : loss of 8 m^3 of beach material is equivalent to loss of 1 m^2 of beach area on the berm (loss of 1 cubic yard of beach material is equivalent to loss of 1 square foot of beach area on the berm). This rule is applicable primarily at the ocean front. In shallow, protected bays, the ratio of volume to area is usually much less.

If B is at a waterdepth of 8 m (26 ft), E the point of slope break at about 2 m (6.5 ft), C the berm crest, n being the annual loss of beach width at the berm, then if the structure in A is to last 50 years, in that time span the berm (in the absence of a structure) will retreat 50 n to point D.

Analyses of profiles at eroding beaches allow to assume that the slope seaward of 8 m (25 ft) will remain nearly unchanged, the point of slope break E at about the same elevation, the profile shoreward of the point of break in slope will remain nearly unchanged.

To estimate the final depth at the wall a profile DF parallel to CE should be drawn (D to elevation of E) and B and F connected. DFB will represent the approximate profile 50 years hence if no structure is implanted. A' approximates the elevation of the receded beach at A.

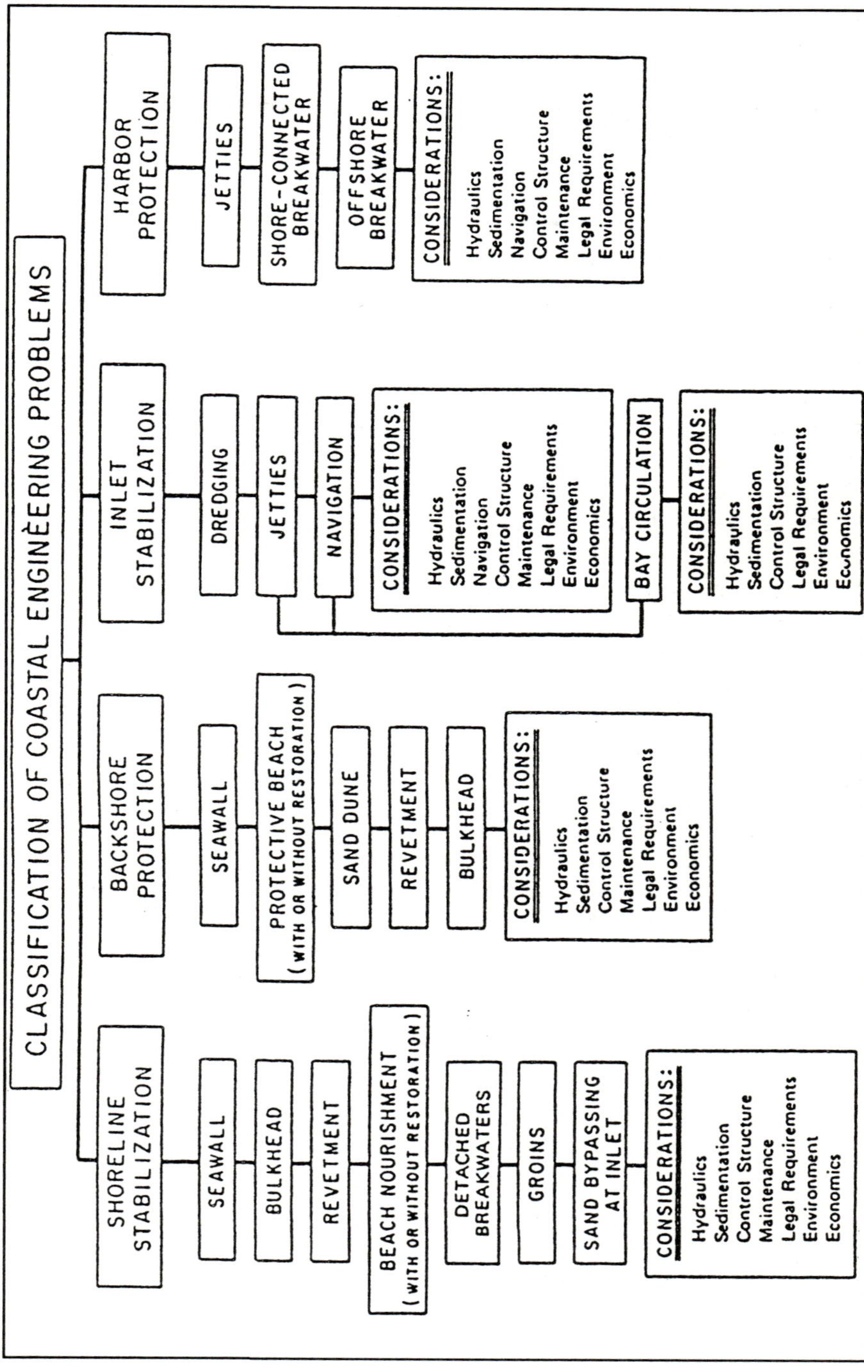

Fig. 36 : General classification of coastal engineering problems

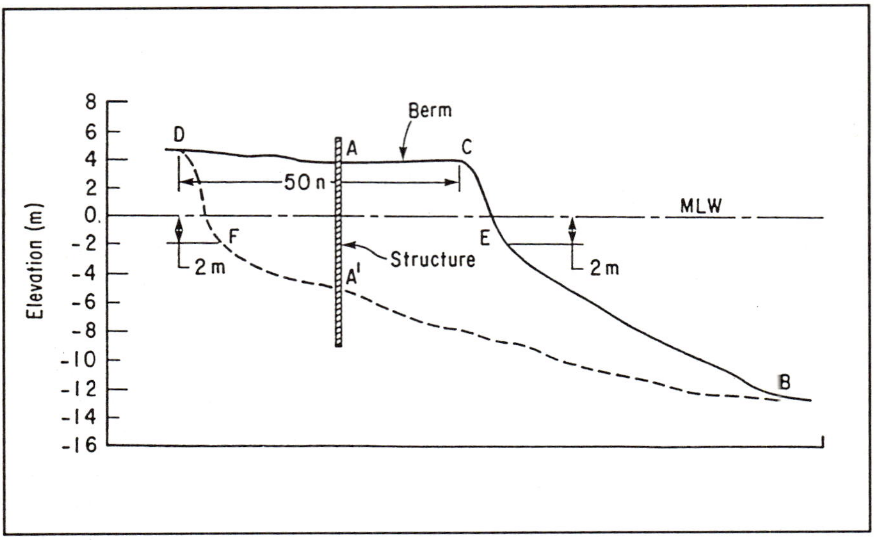

Fig. 37 : Effects of erosion

2 Impact of Existing Structures and Sites

The presence of a promontory, cape, presqu'isle slows down longshore transport, results in deposition of carried material, and a change of direction of the current. Evidently a beach will form on the updrift side, while the downdrift side will be started. Logically thus any hard structure placed in the path of a longshore current will have a similar effect. Groins, sea-walls, breakwaters, jetties are such obstacles. But there are several other constructions not *per se* intended as coastal protection which have similar, or milder effects. Drilling platforms, close to shore, will deflect currents and slow down waves' shoreward progress. Artificial islands will not only have an impact on shoreline morphology, they create erosion problems of their own. An analysis of artificial islands implant and environmental consequences has been recently published (R.H. Charlier and C.F. De Meyer, 1992, Artificial Islands : *Int. J. Env. St.* 40, 4, 244-265).

3 The Beach/Dune System

Beach sand is stored in various coastal indentations, lagoons, off-shore areas, and in massive dunes. But using dune sand may have nefarious effects : exposing areas to flooding.

It requires thus a careful local analysis before using such sand for any purpose. Dune sand, furthermore, being too fine, is seldom suited for beach recharge.

The dune and beach system has gained considerable importance as travel became easier and more readily accessible to persons with modest incomes. If the seasonal occupation of the system exerts a serious stress on that environment, including but not limited to trampling of cf.dlings and plants, dune buggies runs, creation of passages, or even walkways, which facilitate the invasion of storm water and establish wind breaks, *viz.* blowing channels, the building of villas, hotels and other development structures, has led to dune modification, even bulldozing, and outright disappearance.

3.1 Regional Development

Often constructions have been lined up too close to the shoreline and the literature is replete with cases of houses that tumbled in the water. Flood and storm insurance used to compensate owners in the United States ; this has been discontinued and a strict construction line established along all coasts. In some cases the policy of "let us not interfere any longer with nature" is not economically conceivable. The value of property in numerous areas, including hostelries and tourist facilities, is simply too high not to try to protect it.

There are of course various approaches : positioning of hard structures, or favoring nature's work. Dams erected across rivers reduce in many cases delivery of sediment to the beach ; so can electricity generating stations calling upon tidal, wave or current power. The first effect -river dams- has e.g. caused major erosion on the coast of Nigeria. On the other hand construction of the tidal power barrage across the Rance River near St Malo, France, has not had, according to official reports, negative environmental impact.

3.2 Role of Dunes in the Littoral System

A string of dunes is an important protective barrier for coastal properties ; the share of the foredune is evidently primordial in this scheme. After a storm the rebuilding process of beach and dune is far slower than erosion, and recovery may even never occur particularly where waves are fetch-limited.

Other factors than storms can affect the system. So, in some sites beach material can be picked up by winds, carried landward and be entirely removed from the beach system.

Foredunes may occasionally be artificially heightened to avoid overtopping and flooding is unlikely as long as the lagoon's width is less than 8 km (5 mi). In other cases pre-construction investigation is indicated.

Revetments have sometimes been placed in front of dunes to protect the existing slope face. They are a protective armor, not a retaining structure. They are less damaging than sea-walls which cause frequently removal of sand.

Artificial sand dune building has gained much favor to remedy failed natural defenses ; the dune is a flexible last line. Sand dunes can be built artificially by placing permeable screens, 1 m (3 ft) high, half open. Once buried new ones can be placed on top. To anchor the dune marram grass, cf.d grass, shrub buckthorn and trees of the conifers family can be planted or sown. Beach nourishment has also gained acceptability as a dissipater of wave-energy. In contrast to hard structures, compensating the deficiency in the natural sand supply, remedies the erosion cause and benefits adjacent shores.

Where the dune/beach system protects shore constructions, hard protective structures may often be dispensed with. Such massive protection works aim to prevent waves to reach a shore or a harbor (sea-walls, breakwaters, bulkheads and revetments, p. ex.), or to slow down longshore drift (jetties, groins).

3.3 Failure of a dune system

The failure mechanisms of a dune are fairly similar to the mechanisms for dikes (cf. fig. 37) Erosion of the outer slope is however essentially different. The body of the dune contains enough material to take a special shape during storm-surges : “the storm profile”.

During this reshaping process material of the dune is deposited on the near foreshore to flatten the profile. No material is lost. Due to accretion and erosion, the position of the beach and the foot of the dune is changing continually. Different phenomena occur if in this erratical process the underlying trend is accretion, erosion, or dynamic equilibrium.

Dune losses may be reduced by plantations and brushwood fences. Low cliffs may be walled, high cliffs'toes walled or riprap placed in front of it.

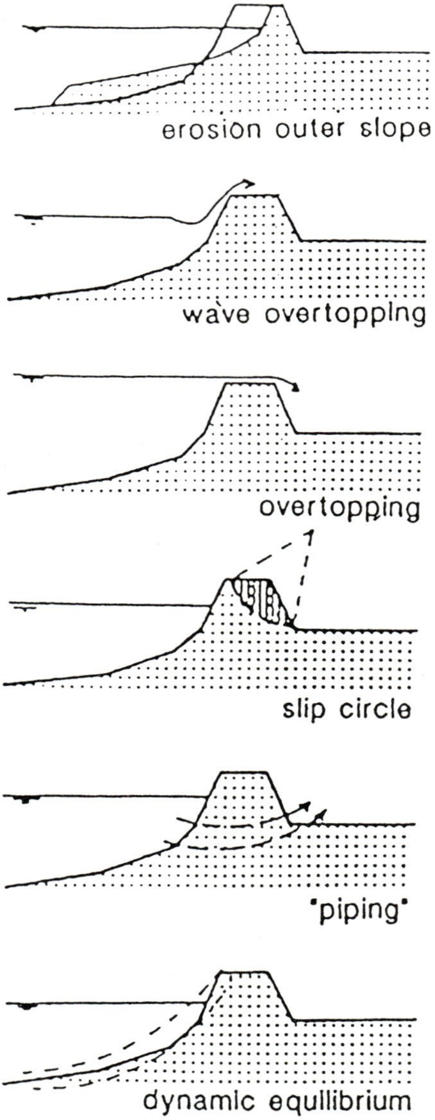

Fig. 38 : Failure mechanisms of a dune system

4 Impacts and Interaction

Man has manifested an irresistible urge to "correct" or "improve" on nature's work. In his relentless efforts to establish ports, to straighten shorelines, to "protect" beaches and cliffs, or to provide unimpeded view, and especially to develop areas, he has erected structures in water and near the shoreline, on barrier islands, beaches and dunes.

Vertical walls have been erected in many shore resorts, often to provide a promenade. This increases wave energy dissipation by beach material fronting the wall and augments erosion. The famed Promenade des Anglais at Nice is at the origin of the loss of the entire sandy portion of the beach of the Baie des Anges.

Engineering structures, whatever their nature or type, have nearly always initiated, or aggravated, erosion in coastal areas. Strangely this has even occurred where there is no longshore littoral transport. One of the earliest and most intensive cases of jetty induced erosion was reported for four India east coast harbors : Paradip, Vishakhapatnam, Kakinada and Madras. Madras is a harbor developed on the open coast, like Zeebrugge in Belgium. No sooner had a breakwater been constructed (1875) that groins had to be implanted to the north side ; a beach built up on the southern side as rapidly as the construction itself.

Completed twenty years after its inception, the "southside" sand soon shoaled the harbor entrance which was relocated northward and an outer quay built. The blockage of the littoral drift resulted in deposition and erosion that extended over many kilometres of shoreline, with maximal effects during the early construction stages ; erosion reached 90 m (275 ft) when exacerbated by storm effects.

Detached breakwaters have equally severe results, witness those near Los Angeles, at Santa Barbara (1927) and Santa Monica. Contrary to expectations they too blocked littoral drift ; if wave energy is decreased by the structure, the longshore transport is at the same time halted though there is no "damming" across the nearshore. Furthermore, accumulation as a spit, on the updrift side eventually reaches the tip of the breakwater, goes still further, and starts shoaling the harbor entrance. Dredging in the only remedy to keep the port's access open. At least, at Santa Barbara, the material is not lost to the littoral system, as it is dumped as replenishment on the downdrift "starved" beach.

Cross-bay and cross-estuary jetties also dam the littoral drift updrift, while downdrift the current carries sand away from the jetty, thus causing shore erosion. A classical case is that of Ocean City MD.

A single groin has the same interaction with the littoral system as a jetty, but a field of parallel groins favors beach formation. Groins being shorter than jetties, longshore transport continues ahead of the toe. Nevertheless a groin field merely transfers downdrift the erosional problem : a classical example is found at Long Beach, NY. Effectiveness of a field with minimized "negative" impact is achieved by concomitant artificial nourishment. By-passing schemes are often a good approach to solve the problem ; they involve dredges, sometimes a dredge protecting breakwater, or a scheme with a deposition basin.

Morphology can play a part in jetty-induced erosion on a coast with no littoral drift. Indeed, erosion factors in this instance encompass the embayment size between pre- and post-jetty shorelines, and shoreline length undergoing sand withdrawal towards a jetty-close area which has to fill to achieve a new shoreline equilibrium. Evidently, equilibrium is eventually reached, which is never the case there where a littoral exists.

In conclusion, any structure that modifies artificially an existing shoreline disturbs an existing equilibrium and its presence will result in erosion and accretion phenomena.

As repeatedly underscored in the preceding chapters, all structures, and remedies, are temporary as only nature could halt coastal erosion. To insure the continued effectiveness of protective measures sustained monitoring and careful maintenance must be carried out. One may include either as "measure" or as "maintenance" the prohibition of practices which result in renewed or intensified erosion, such as mangrove conversion along mud coasts, destruction of coral reefs (an irreversible "damage"), "development" or reclamation of wetlands, poorly planned implantation of constructions at sea or land extensions for whatever purpose(s). On the other hand regular profiling and surveying of regions which have been artificially nourished, dewatered, "replanted" or otherwise protected is necessary both to ascertain *de facto* results of the "intervention" and to identify steps needed to sustain the effects of the engineering work undertaken. Where hard structures are concerned maintenance and repairs must guarantee not only their effectiveness but their very own "survival".

APPENDICES

1 Beach Profiles

Since "beach profile nourishment" is nowadays the recommended approach, the calculation of such profile takes on a special significance. Such profile must be specified because mode of sediment transport differs in near- and off-shore zones. Profile changes in offshore depth-zones are negligible over several decades, but where depths do not reach 20 m (± 70 ft) data must be of recent vintage. Long-term swell stability and short-term transposition from swell to storm beach profile require calculation of beach profiles as well as their correlations with sediments characteristics.

1.1 The offshore profile

Swell waves may, in the long term, be barred from reaching a part of a coast, by creation of a coastal barrier. A new wave climate will ensue. Refraction and/or diffraction may cause wave height decrease. Wave energy dissipated along a longer beach will be concentrated on a shorter distance and the swell is faced more directly. Conversely a beach section may be subjected to lesser waves.

Another scenario occurs with sedimentation when the deposition is hampered where a river debouches, due to natural or anthropic actions e.g. barrage construction upstream. Any change in beach nourishment may bring about a modification in sediment size ; knowing the probable equilibrium profile would help determining needed fill volume to attain stability.

It is therefore useful to attempt establishing a relationship between mean wave conditions and mean sediment characteristics for a stable underwater profile, theoretical or albeit empirical.

A steeper or flatter than so-called "standard" beach profile indicate respectively erosion or accretion. Closely associated, in ocean margins, with sustained swell, the wave climate is even more important in enclosed seas and gulfs as here storm waves are the only energy source.

Some relationship may be calculated using mean energy per ocean area unit and mean sediment characteristics. Several authors have stressed that "in computing such average, some statistical value may have to be used for the energy fluctuations throughout the year, e.g. the average of the highest third of the peak energies". Same energy level or not, data relating to sections of an enclosed sea cannot be used for an ocean coast. Wave length (λ) or period (T) are to be included in the "relationship" as they vary from storm to swell conditions (see below).

1.2 Swell to storm profile

Long-term erosion and/or accretion trends must be taken into account to assess beach encroachment safe limit for construction, and thus establishment of an authorized construction limit, and to advise pipelines positioning depths.

1.3 Theoretical analysis

The Sitarz equations proposed a relationship for surf zone width for swell and storm-waves profiles, plus zonal water cross-sectional area. Average depth can then be derived. With x, the seawards distance from the breaker line, y the depth below SWL, s the quotient sediment density by seawater density [= 2.65], D the median sediment diameter at a given depth (in mm) and H (in m) wave height immediately offshore from the breaker zone :

$$x = ky^2 \tag{106}$$

$$\text{wherein } k = \frac{0.95}{(s-1)^{\frac{1}{2}} \cdot DH^{\frac{3}{2}}} \tag{107}$$

The larger k, the finer the sediment or the lower the waves, hence the milder the slope. In most sand situations, s = 2.65.

The offshore bed curve should be continued up to beach face to the berm. Surf zone is given by :

$$Z = \frac{BH^{\frac{3}{2}}}{(s-1)^{\frac{1}{2}} \cdot D^{\frac{1}{2}}} \qquad (108)$$

With dimensions the same as for (107), D however in (108) is the sediment mean diameter in the beach face near SWL.

For a storm beach surf zone, the cross-sectional water area is given by Sitarz as :

$$A = \frac{3.4g^{\frac{1}{2}} \cdot H^2_{\frac{1}{3}} \; T_{\frac{1}{3}}}{(s-1)^{\frac{1}{2}} \cdot D^{\frac{1}{2}}} \qquad (109)$$

wherein :

$T_{\frac{1}{3}}$ = significant wave period

$g = 9.81 \dfrac{m}{sec^2}$

D = median sediment diameter across surf zone
s as in (107) and $H_{\frac{1}{3}}$ as in (109).

If d_s is t surf zone mean depth :

$$d_s = \frac{A}{Z}$$

Z and d_s can be overlain to predict likely beach degradation. However, recognizing a realistic mean energy input and a stable profile for sediment size remains problematic given that swell waves vary hourly and weekly in height.

When an offshore bar forms, a fully fledged swell profile can only be labeled such after the bar has been entirely swept back to the beach and only then is the bed "in equilibrium". Moral : cross-sections should be made at the end of the period of swell.

2 Littoral Drift

Only transport in the surf zone will be handled under this heading, though it also takes place offshore. Deep water wave characteristics can be related to wave energy - and its longshore component-values. Other factors, e g. energy dissipation, are difficult to assess.

2.1 The current

R. Sylvester (1989, *Ocean Engineering* : Amsterdam, Elsevier Science Publ.) considered it "almost impossible ... to correlate a littoral current with waves breaking on a coast ... oriented almost north-south", in a beach profile example. Temporal variables involved ought thus to be stressed. He quotes researchers stating that the near-shore topography redistributes the breaker influx into different positions along the shore and also undergoes itself displacements and transformations caused by affected waves and currents. Per Bruun studied the beach storm profile.

2.2 Drift

It is virtually impossible to mathematically pass from current to drift without field tests or introduction of empirical values. Setting

$m = \bar{x}_{\text{beach slope}}$

K' = roughness factor

waver power per shore length unit being P

ω = specific weight $_{\text{water}}$

S = specific gravity $_{\text{sediment}}$

θ = sediment internal friction angle

$$P = \frac{2.2\omega H_b^2 \sin\alpha_b \cos\alpha_b}{T} = \frac{wH_0^2 L_0 \sin\alpha_b \cos\alpha_0}{16T} \tag{111}$$

$$\frac{H_b}{L_b} = \frac{H_b}{C_b T} = \frac{H_b}{\sqrt{1.78gd_b}\ T} = \frac{H_b}{\sqrt{1.78g\ 1.3H_b}\ T} \tag{112}$$

$$E_r = \frac{\text{energy dissipated}}{\text{longshore energy component}} = \sin\alpha_b f\left[\frac{mHb}{K'tg\alpha_b L_b}\right] \tag{113}$$

the equation proposed by Castanho for sediment weight transported by second across a plane perpendicular to a beach becomes :

$$G = \frac{E_r P}{\left(1 - \frac{1}{S}\right) tg\theta} \tag{114}$$

For sandy shores he suggests following values :

$$\frac{mH_b}{K'L_b} = 0.12$$

$$\theta = 35°$$

$$I - \frac{I}{S} = 0.623$$

and eq. (114) yields

$$\frac{7GT}{\omega\ H_0^2 L_0} = E_r \sin\alpha_b \cos_0 \tag{115}$$

Taking $E_r = \sin\alpha_b f\left[\frac{0.12}{tg\alpha_b}\right]$ (fig. 38) and $\sin\alpha_b \cos\alpha_0$ (fig. 39) as ordinates and H_0/L_0 as abscissa, since α_b values are a function of α_0 and $\frac{H_0}{L_0}$ (steepness), he proposed following graphs to calculate littoral drift (figs. 38, 39) :

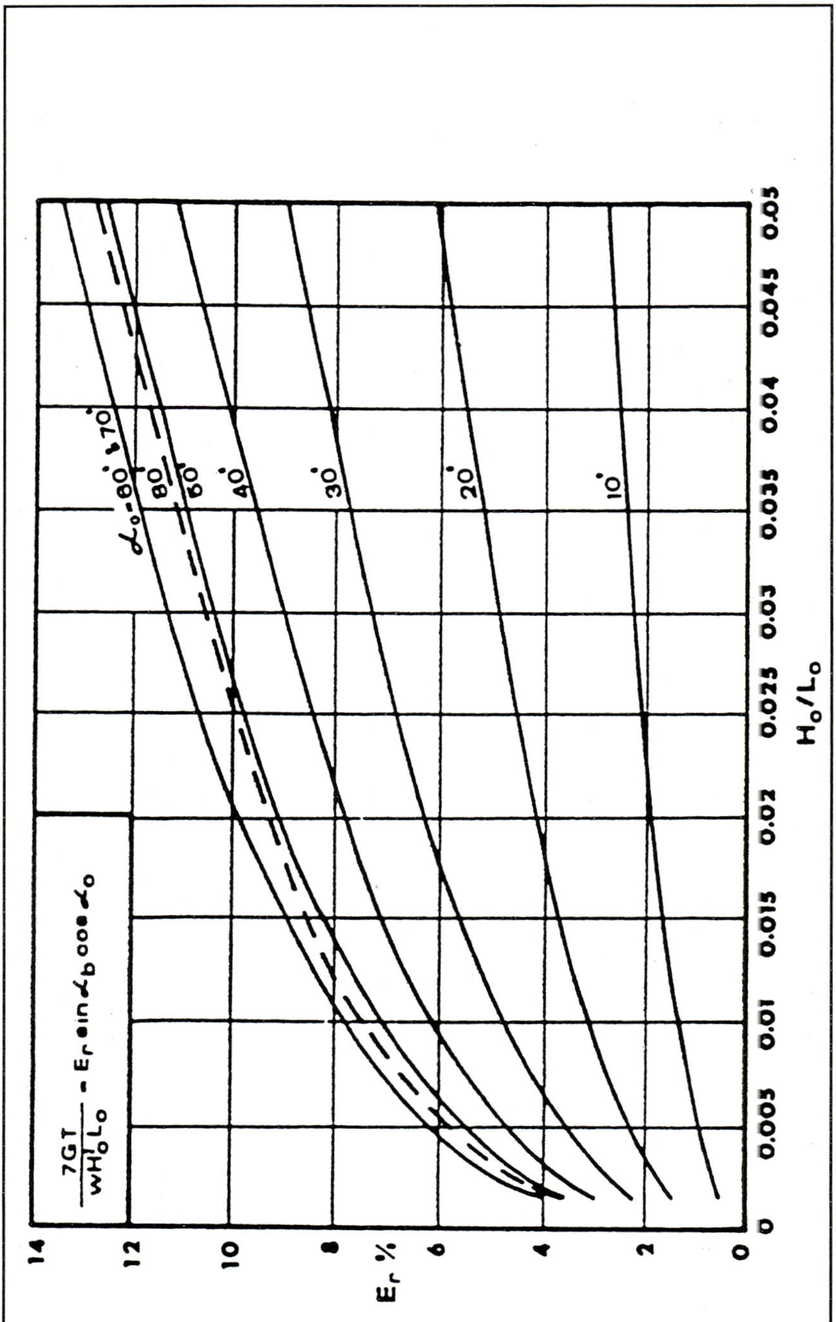

Fig. 39 : Relationship for calculating littoral drift (eq. 115)

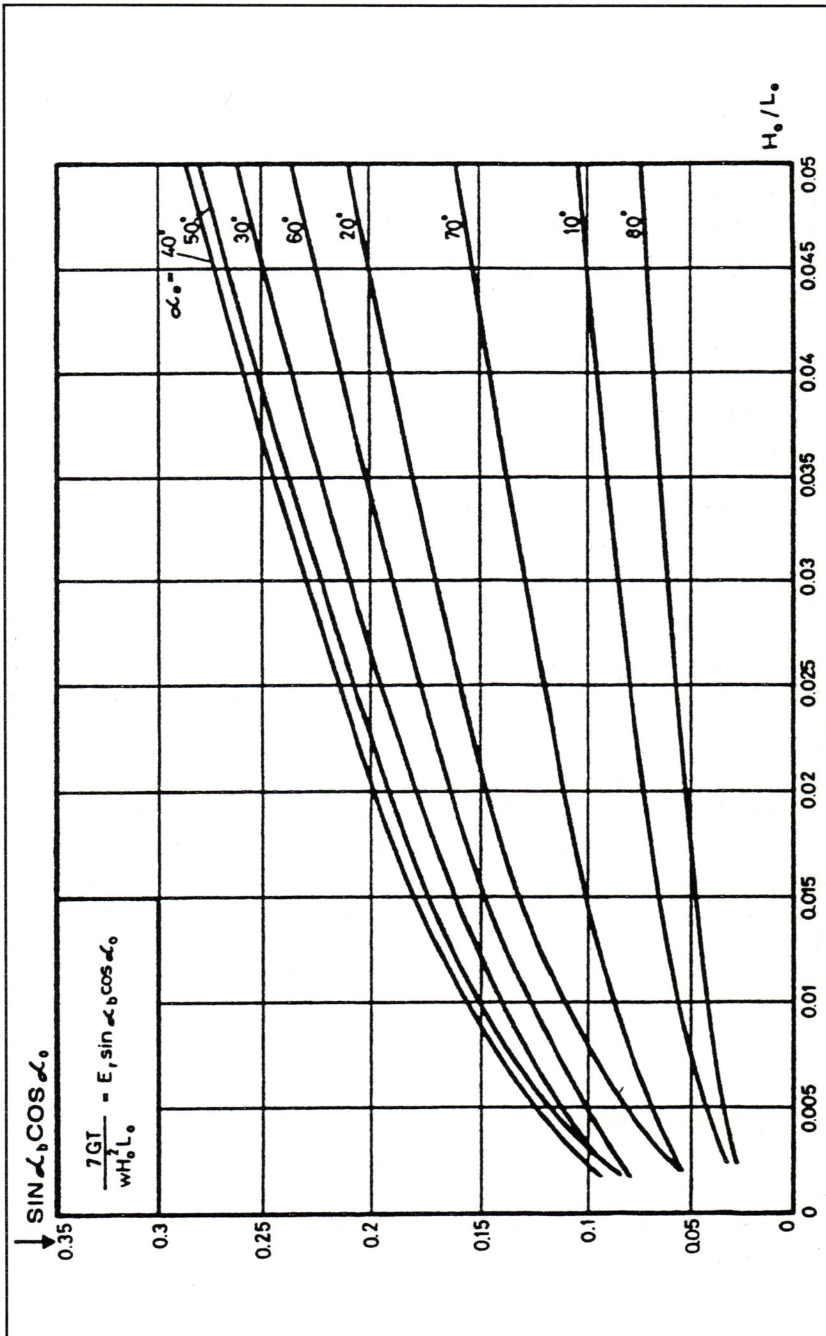

Fig. 40 : Relationship for calculating littoral drift (eq. 115)

An empirical formula has been proposed (Inman and Frautschy)

$$G=\frac{P}{4\left(1-\frac{1}{S}\right)} \qquad (116)$$

Important different frequency waves energy contributions neglect may lead to computations that are too low. Significant wave height in the equation for power computation may cause it to be double of what it should be. Beach planning requires “rigorous evaluation of the relation”.

3 Offshore

Water particle motion differs in a channel from that on an ocean bed as on the latter depths are far less uniform.

The nature of fluid flows leads to parallel ocean bed ripple and dune formations, due to the oscillatory movement, which eventually become a brick pattern and ultimately a rather random distribution. Refraction of a wave-train reaching a coast at an oblique angle, changes its direction while it travels across the shelf, and depth variations modify wave-length and -height, thus lead to changes in amplitude and velocity. Sediment transport values are thus only valid for specific coast sectors.

A discussion of studies at various institutions and by several authors (e.g. Einstein, Kamphuis) goes beyond the scope of this book. Be it simply said that equations worked out are not readily applicable. Some authors had strong reservations about the sediments internal friction values.

4 Transport

Again, because of greater depths offshore measurements of sediment transport prove more difficult there. Methods used include radioactive tracers ; on beach and in surf zone sediment-tracing and -sampling, and silting - or erosion - assessments are used. Sediment transport in the surf zone can be calculated using eq. 115 : G in cubic measure of dry sand. Beyond that zone a volumetric concentration can be estimated by :

$$C = \frac{G}{MT} \tag{117}$$

It appears that the lack of precision in computations and field results is such that they offer little help for design purposes. Even modest height waves are capable to erode or deposit sand around seabed structures.

5 Sources of Beach Fill & Sediment Budget

The US Army Corps of Engineers has utilized sludge as beach nourishment material. It is often a readily -and inexpensive- available material. Its use may resolve, in some cases the increasingly acute disposal problem of what is frequently considered as a waste. However, all sludge is not free of nocive pollutants and caution is advised in the appropriateness of sludge use for that, and evidently also land disposal purposes. The matter is discussed to some extent in the recently (1994) released study on "*Assessment and Remediation of Contaminated Sediments*" or "ARCS Program" (Chicago, IL US EPA, GLNOP).

Not only unintentional artificial nourishment of eroding beaches, but even beach accretion may result from human activities. Besides the shales already mentioned in the south of England, other anthropic handled material may provide beach nourishment ; at Hød, in Denmark, quarry wastes are rejected into the sea leading to a seaward progression of the shoreline, along some ten km (7 mi) approximately, land has thus been gained at the rate of about 1 m (3 ft) per year during the last century. Similarly in Chili, on Chañaral Bay, the beach has widened by some 25 m/yr (about 80 ft/year) because of the near-shore disposal of copper tailings.

Paraphrasing R. Paskoff (1993, *Les littoraux. Impact des aménagements sur leur évolution* : Paris, Masson ; p. 52) studies about transit direction of materials, their origin, nature and volume have identified beaches as independent sedimentary compartments. Each such compartment has its own budget, which is the algebraic sum of the sediments'volume that enters into the system and the volume that exists. When nil, the beach is stable; if positive, aggradation takes place with new beach crests forming. A negative sum, on the other hand, indicates retreat with possible eventual disappearance of the beach.

Chapman, D.M., 1981, Coastal erosion and the sediment budget, with special reference to the Gold Coast, Australia: *Coast. Engng* 4, 207-227.

Clayton, K.M., 1980, Beach sediment budget and coastal modification : *Progr. in Phys. Geogr.* 4,471-486.

Matthews, E.R, 1979, Seasonal variation in volume and morphology of some western Taranaki beaches : N.Z.J. Geol. & Geoph. 4, 465-473.

6 Other Littoral Environments

This book has centered on the coastal erosion affecting beaches. It is self evident that other littoral environments can be subject to erosive processes, resulting from natural processes and human interference. These include lagoons, deltas, estuaries, wetlands and coastal marshes, coral reefs, and cliffs. The consequences of disturbing, or exploiting coral reefs have been touched upon.

Coastal cliffs exist at all latitudes, they are, at least in part, the result of sculptural action of the sea. They retreat in a front that remains parallel to itself and leave at their foot a gently sloping abrasion platform ; sometimes detached blocks are left behind which play the role of temporary natural breakwaters while here and there a highly protruding part of the former cliff maintains itself (stacks, sea-arches and the like). Geomorphologists distinguish live (regularly battered by the waves), stabilized (waves reach it only at times of exceptionally violent storms), and dead cliffs (when the shoreline has moved seawards or aggradation has separated them from the sea).

A cliff is subject to marine and subaereal attack. The sea saps a cliff at its base, destabilizes the scarp, lowers the abrasion platform and "quarries" the cliff. Besides a mechanical effect, the ocean exerts chemical and corrosive effects, while freezing temperatures favor the formation of notches. Haloclastisis, hydroclastisis and biological processes contribute to the progress of erosion. Subaereal processes act upon the scarp itself and on the cliff's summit. The nature of the rock, as with marine erosion, plays an important role. One of the authors has observed cliff destruction due to humidity, including frequent precipitations, on the isle of Wight and close to the Golden Sands north of Varna (Bulgaria).

The rate of retreat of a cliff varies considerably ; it is not unusual for a retreat to occur rapidly and suddenly. The magnitude is often impressive : to cite a frequently given example (Valentin, H., 1971, Land loss at Holderness, 1852-1952. In : Steers, J.A. (ed.), Applied coastal geomorphology : London, Macmillan) over a period of one century at Holderness (Great Britain) cliffs retreated 120 m (nearly 440 ft) along a 60 km (roughly 37 mi) front resulting in a land loss of 720 ha (about 1800 acres) and of 100 million m³ (130 million cu.yd) of clay. During the Winter storms of 1953, the cliffs retreated 10 m (32.4 ft) in two days.

Gone are the days that only forlorn offshore lighthouses dotted the cliff landscape. Human settlements, coastal roads, tourism in most of its manifestations have brought about strong modifications affecting the stability of cliffs : changes in the position of the phreatic layer, weight of constructions, damage to the natural plant cover. Where beaches spread at the foot of the cliff, their shrinking in turn endangers the cliff which no longer remains stabilized. A certain number of steps can be taken to brake the retreat of cliffs ; they include drainage measures, reducing the gradient (scarp slope), creating an artificial beach at the foot of the cliff, placing a barrage of rocks there, consolidating the base into a retaining wall, and even covering the summit with vegetation like Eucalyptus trees.

Keeping in mind that cliff retreat is a natural phenomenon, loss of land must be accepted and a line of construction way back from the edge is a protection which any builder must accept, no matter the song sung by developers. If the drowning of Charleston SC is economically unthinkable, so are some cliff sites and in such cases, protection measures must be undertaken. A classic example followed by one of the authors is the Côte des Basques, near Biarritz, France (Combe, M., 1987, Un exemple de stabilisation : la côte des Basques à Biarritz : *Actes du Coll. Mer & Littoral, Couple à Risque*, Paris, La Documentation Française, pp. 484-495.). At a cost of some US$ 6 million, 1 km (0.62 mi) of coastline was protected by implantation of groins to favor development of a pebble beach, and construction of a rock protected wall.

Bird, E.C.F., 1983 ; Factors influencing beach erosion and accretion : a global review. In : McLachlan, A. & Erasmus, T. (eds), *Sandy beaches as ecosystems* : The Haghe, W. Junk.

Blanc, J.J., 1979, *Protection des littoraux sableux : méthodes d'études, l'exemple de la Camargue* : Marseille, CNEXO 56 pp.

Blanc, J.J., 1993, *Erosions littorales en Camargue : évaluation du recul des plages et analyses prévisionnelles* : Marseilles, Université de Luminy, Fac. des Sci. 59 pp.

Dubois, R.N., 1992, A re-evaluation of Bruun's Rule and supporting evidence : *J. Coast Res.* 3, 618-628.

Healy, T., 1991, Coastal erosion and sea-level rise : *Zit. f. Geomorph.,* Neue R., Suppl. 81, 15-29.

Leonard, L.A., Dixon, K.L. & Pilkey, O.H., 1990, A comparison of beach replenishment on the U.S. Atlantic, Pacific and Gulf coasts : *J. Coast. Res.* (spec. nbr) 6, 127-140.

McLachlan, A. & Erasmus, T. (eds), *Sandy beaches as ecosystems* : The Hague, W. Junk.

National Research Council, 1990, *Managing coastal erosion I :* Washington, DC Nat. Ac. Sci., 182 pp.

Nordstrom, K.F. & Allen, J.R., 1980, Geomorphologically compatible solutions to beach erosion : *Zeit. f. Geomorph.* Suppl. 34, 142-154.

Paskoff, R. 1991, Principes de défense des plages contre l'érosion de la mer : *Act. Coll. "Droit-Littoral-Mer"*, Univ. Nantes, Fac. Droit & Sci. Pol. 35-48.

Pilarczyk, K ;, 1990, *Coastal protection* : Rotterdam, Balkema 500 pp.

Pilkey, O.H. & Wright, H.L. III, 1990, Seawalls versus beaches : *J. Coast. Res.* 6, 1, III-VII.

Rossi G., 1989, L'érosion de littoral dans le Golfe du Bénin : un exemple de perturbation morphodynamique : *Zeit. f. Geomorph.* Suppl. 73, 139-165.

7 Littoral Dunes

In the discussion of the dune-beach system, the authors have made no effort to examine the different types or shapes of littoral dunes. Be it sufficient to mention that generally the following types are identified, not implying, of course, that all types are encountered in all sites : basic or elementary dunes, edge or border dunes, free dunes, and impeded dunes (in humid mesothermal climates). Hydraulic dunes or sub-marine dunes form occasionally underwater nearshore. Dunes can be longitudinal, transversal, parabolic, or are barkhanes. Fixed or anchored dunes remain in place, because they are covered by a pellicule of solidified mud, or have been colonized by vegetation. However if the vegetation dies, the dune will again be subject to deflation. When parabolic dunes get joined, rake-shaped dunes result ; frontless dunes, resembling ramparts may be called drawn-out dunes.

GLOSSARY

Accretion : Accumulation of sand or other beach material at a point due to natural action of waves, currents and wind. A build-up of the beach.

Alongshore : Parallel to and near the shoreline ; same as LONGSHORE.

Backhoe : Excavator similar to a power shovel except that the bucket faces the operator and is pulled toward him.

Backshore : The zone of the shore or beach lying between the foreshore and the coastline and acted upon by waves only during severe storms, especially when combined with exceptionally high water.

Bar : Fully or partly submerged mound of sand, gravel, or other unconsolidated material built on the bottom in shallow water by waves and currents.

Beach : The zone of sedimentary material that extends landward from the low water line to the place where there is marked change in material or form, or to the line of permanent vegetation (usually the effective limit of storm waves). The seaward limit of a beach - unless otherwise specified - is the mean low water line. A beach includes foreshore and backshore.

Beach fill : Sand or gravel placed on a beach by mechanical methods.

Beach, perched : see PERCHED BEACH

Berm : A nearly horizontal part of the beach or backshore formed at the high water line by waves depositing material. Some beaches have no berms, others have one or several.

Bluff : A high, steep bank composed of erodible materials.

Bluff : High, steep bank at the water's edge. In common usage, a bank composed primarily of oil. See CLIFF.

Boulders : Large stones with diameters over 10 inches. Larger than COBBLES.

Breaker : A wave as it spills, plunges, or collapses on a shore, reef, sandbar, rock, natural obstruction, or man-made structure.

Breaking Depth : Stillwater depth where waves break.

Breaker Zone : Area offshore where waves break.

Breakwater : A fixed or floating structure that protects a shore area, harbor, anchorage, or basin by intercepting waves.

Breakwater : Structure aligned parallel to shore, sometimes shore connected, that provides protection from waves.

Bulkhead : A structure or partition placed on a bank or bluff to retain or prevent sliding of the land and protect the inland area against damage from wave action. See also SEAWALL.

Bulkhead : Structure that retains or prevents sliding of land or protects the land from wave damage.

Clay : Extremely fine-grained soil with individual particles less than 0.00038 cm (0.00015 inch) in diameter.

Cliff : A high, steep face of rock ; a precipice.

Cliff : High steep bank at the water's edge. In common usage, a bank composed primarily of rock. See BLUFF.

Coast : The strip of land, of indefinite width (up to several miles), that extends from the shoreline inland to the first major change in terrain features.

Cobbles : Rounded stones with diameters ranging from approximately 7.62 to 25.4 cm (3 to 10 inches). Cobbles are intermediate between GRAVEL and BOULDERS.

Crest : Upper edge or limit of a shore protection structure.

Culm : Single stem of grass.

Current : A flow of water.

Current : Flow of water in a given direction.

Current, Longshore : Current in the breaker zone moving essentially parallel to shore and usually caused by waves breaking at an angle to shore. Also called alongshore current.

Deep water : Area where surface waves are not influenced by the bottom. Generally, a point where the depth is greater than one half the surface wavelength.

Diffraction : Progressive reduction in wave height when a wave spreads into the shadow zone behind a barrier after the wave has passed its end.

Diurnal : Period or cycle lasting approximately one day. A diurnal tide has one high and one low in each cycle.

Downdrift : Direction of predominant alongshore movement of littoral materials.

Dune : A hill, bank, bluff, ridge or mound of loose, wind-blown material, usually sand.

Duration : The length of time the wind blows in nearly the same direction across a FETCH (generating area).

Ebb Tide : Part of the tidal cycle between high water and the next low. The falling tide.

Equilibrium : State of balance or equality of opposing forces.

Erosion : The wearing away of land by the action of natural forces.

Fetch : Area where waves are generated by wind which has a relatively steady direction and speed. Sometimes called FETCH LENGTH. Also, unobstructed distance over water where such generation takes place.

Fetch Length : Horizontal direction (in the wind direction) over which a wind generates waves. In sheltered waters, often the maximum distance that wind can blow across water.

Filter Cloth : Synthetic textile with openings for water to escape, but which prevents passage of soil particles.

Flood Tide : Part of the tidal cycle between low water and the next high. The rising tide.

Foreshore : The part of the shore lying between the crest of the seaward berm (or upper limit of wave wash) and the water's edge at low water. The foreshore is ordinarily traversed by the runup and return of the waves.

Functional life : The period of time during which a structure performs as intended. Performance can be expressed in terms of benefits obtained versus the cost of installation and maintenance.

Gravel : Small, rounded granules of rock with individual diameters ranging from 7.62 to 0.46 cm (3.0 to 0.18 inches). Gravels are intermediate between SAND and COBBLES.

Groin : A fingerlike structure built perpendicular to the shoreline, usually with other groins, to trap littoral drift or retard erosion of the shore.

Groin Field : Series of groins acting together to protect a section of beach. Also called a groin system.

Groundwater : Water within the earth that supplies wells and comes to the surface by seepage or in springs.

Grout : Mixture of Portland cement, fine aggregates (usually sand), and water. Usually used to seal openings or fill bags or other containers.

H-Pile : Straight length of structural steel with an H-shaped cross section designed for driving into the earth.

High Water : See HIGH TIDE.

High Water Line : Intersection of the level of MEAN HIGH WATER with the shore. Shorelines on navigation charts are approximations of the high water line.

High Tide : Maximum elevation reached by each rising tide. See also TIDE.

Hogwire : Short, smooth-wire fencing of the kind normally used to enclose a pig sty.

Impermeable : Not having openings large enough to permit water to freely pass.

Intertidal Zone : Land area alternately inundated and uncovered by tides. Usually considered to extend from MEAN LOW WATER to MEAN HIGH WATER.

Jetting : A method of placing piles by forcing water around and under a pile to displace and lubricate the surrounding soil, allowing the pile to sink to the desired position.

Lee : Sheltered ; part or side facing away from wind or waves.

Leeward : Direction toward which wind is blowing or waves are traveling.

Littoral : Of or pertaining to a shore.

Littoral Drift : The sedimentary material moved along the shoreline under the influence of waves and currents.

Littoral Material : Sediment moved in the LITTORAL ZONE by waves and currents. Also called littoral drift.

Littoral Transport : The movement of littoral drift along the shoreline by waves and currents. Includes movement parallel (longshore transport) and perpendicular (on-offshore transport) to the shore.

Littoral Zone : Indefinite (unprecisely defined) zone extending from the shoreline to just beyond the breaker zone.

Longshore : Parallel to and near the shoreline : same as ALONGSHORE.

Longshore Transport Rate : Rate of transport of littoral material parallel to shore. Usually expressed in cubic yards per year.

Low Tide : Minimum elevation reached by each falling tide.

Low Water Line : Intersection of the low tide level with shore.

Marsh : An area of soft, wet or periodically submerged land, generally treeless and usually characterized by grasses and other low vegetation.

Mean High Water (MHW) : Average height of the daily high waters over a 19-year period. For semidiurnal or mixed tides, the two high waters of each tidal day are included in the mean. For diurnal tides, the single daily high water is used to compute the mean.

Mean Higher High Water (MHHW) : Average height of the daily higher high water over a 19-year period. Only the higher high water of each of the high waters of a tidal day is included in the mean.

Mean Low Water (MLW) : Average height of the low waters over a 19-year period. For semidiurnal and mixed tides, the tow low waters of each tidal day are included in the mean. For a diurnal tide, the one low water of each tidal day is used in the mean. Mean Low Water has been used as datum for many navigation charts published by the U.S. National Ocean Survey, but it is being phased out in favor of Mean Lower Low Water for all areas of the United States.

Mean Lower Low Water (MLLW) : Average height of the daily lower low waters over a 19-year period. Only the lower low water of each pair of low waters of a tidal day is included in the mean. Long used as the datum for Pacific coast navigation charts, it is now gradually being adopted for use across the United States.

Mean Sea Level : Average height of the sea surface over a 19-year period. Not necessarily equal to MEAN TIDE LEVEL.

Mean Tide Level : Plane midway between MEAN HIGH WATER and MEAN LOW WATER. Not necessarily equal to MEAN SEA LEVEL. Also called half-tide level.

Module : A structural component, a number of which are joined to make a whole.

Neap tide : A tide having about 10 to 30 percent less range than the average, occurring about the time of quarter moons.

Nearshore : In beach terminology, an indefinite zone extending seaward from the shoreline well beyond the breaker zone.

Nourishment : Process of replenishing a beach either naturally by longshore transport due accretion, or artificially by delivery of materials dredged or excavated elsewhere.

Offshore : (1) (Noun) In beach terminology, comparatively flat zone of variable width extending from the breaker zone to the seaward edge of the Continental Shelf. (2) (Adjective) Direction seaward from the shore toward a large body of water.

Onshore : The landward direction, away from the water.

Overtopping : The passing of water over the top of a natural or man-made structure as a result of wave runup or surge.

Peat : Residual product produced by partial decomposition of organic matter in marshes and bogs.

Peat Pot (vegetation) : Pot formed from compressed peat and filled either with soil or peat moss in which a plant or plants, grown from seed, are transplanted without being removed from the pot.

Perched beach : A beach or fillet of sand retained above the otherwise normal profile level by a submerged dike or sill.

Permeable : Having openings large enough to permit free passage of appreciable quantities of sand or water.

Permit : A document issued that expresses the assent of a government agency, so far as concerns the public rights and the general public interest, for the accomplishment of certain works (e.g., construction).

Pile : A long, heavy timber or section of concrete or metal that is driven or jetted into the earth or bottom of a water body to serve as a structural support or protection.

Pile, Sheet : Pile with a generally slender, flat cross section driven into the ground or seabed and meshed or interlocked with like members to form a diaphragm, wall, or bulkhead.

Piping : Fluidizing of backfill or an embankment to the extent that large quantities of material are pumped by wave action through holes under or through a bulkhead or revetment.

Plasticity : As applied mainly to clay, the relative ease with which the material yields or deforms under pressure.

Plug : Core containing both plants and underlying soil, usually cut with a cylindrical coring device and transplanted to a hole cut by the same device.

Polyvinyl Chloride (PVC) : Plastic material (usually black) that forms a resilient coating suitable for protecting metal from corrosion.

Profile, Beach : Intersection of the ground surface with a vertical plane that may extend from the top of the dune line to the seaward limit of sand movement.

PVC : (See POLYVINYL CHLORIDE)

Raveling : Progressive deterioration of a revetment under wave action.

Refraction (of water waves) : (1) Process by which direction of a wave moving in shallow water at an angle to the contours is changed. Part of the wave advancing in shallower water moves more slowly than the part still advancing in deeper water, causing the wave crest to bend toward alignment with the underwater contours. (2) Bending of wave crests by currents.

Revetment : A facing of stone, concrete, etc., placed on a bank or bluff of stone to protect a slope, scarp, embankment, or shore structure against erosion by wave action or currents.

Rhizome : Underground stem or root stock. New shoots are usually produced from the tip of the rhizome.

Riparian rights : The rights of a person owning land containing or bordering on a water course or other body of water in or to its banks, bed, or waters.

Riprap : A layer, facing, or protective mound of rubble or stones randomly placed to prevent erosion, scour, or sloughing of a structure or embankment ; also, the stone used for this purpose.

Rubble : (1) Loose, angular, waterworn stones along a beach. (2) Rough, irregular fragments of broken rock or concrete.

Runup : The rush of water up a beach or structure, associated with the breaking of a wave. The amount of runup is measured according to the vertical height above still water level that the rush of water reaches.

Sand : Generally, coarse-grained soils having particle diameters between 0.18 and approximately 0.003 inches. Sands are intermediate between SILT AND GRAVELS.

Sand Fillet : Accretion trapped by a groin or other protrusion in the littoral zone.

Sandbag : Cloth bag filled with sand or grout and used as a module in a shore protection device.

Scour : Removal of underwater material by waves and currents, especially at the base or toe of a shoreline structure.

Screw Anchor : Type of metal anchor screwed into the bottom for holding power.

Seawall : A structure separating land and water areas, primarily designed to prevent erosion and other damage due to wave action. See also BULKHEAD.

Setup, Wind : Vertical rise in the stillwater level on a body of water caused by piling up of water on the shore due to wind action. Synonymous with wind tide and STORM SURGE. STORM SURGE usually pertains to the ocean and large bodies of water. Wind setup usually pertains to reservoirs and smaller bodies of water.

Shallow Water : Commonly, water of such a depth that surface waves are noticeably affected by bottom topography. It is customary to consider water of depths less than one-twentieth the surface wavelength as shallow water.

Sheet Pile : A pile with a generally slender, flat cross-section that is driven into the ground or bottom of a water body and meshed or interlocked with like members to form a wall or bulkhead.

Shoal : (noun) Rise of the sea bottom from an accumulation of sand or other sediments. (verb) - (1) to *become* shallow gradually. (2) To *cause* to become shallow. (3) to *proceed* from a greater to a lesser depth of water.

Shoot : Collective term applied to the STEM and leaves or any growing branch or twig.

Shore : Narrow strip of land in immediate contact with the sea or a lake, including the zone between high and low water lines. A shore of unconsolidated material is usually called a beach. See also BACKSHORE and FORESHORE.

Shoreline : Intersection of a specified plane of water with the shore or beach (e.g., the high water shoreline would be the intersection of the plane of mean high water with the shore or beach). The line delineating the shoreline on National Ocean Survey nautical charts and surveys approximates the mean high water line.

Sill : Low offshore barrier structure whose crest is usually submerged, designed to retain sand on its landward side.

Silt : Generally refers to fine-grained soils having particle diameters between 0.008 and 0.0004 cm (0.003 and 0.00015 inches). Intermediate between CLAY and SAND.

Slope : Degree of inclination to the horizontal. Usually expressed as a ratio, such as 1:25 or 1 on 25, indicating 1 unit vertical rise in 25 units of horizontal distance ; or in degrees from horizontal.

Sloughing : Process where a weakened mass of soil fails and moves downslope.

Sprig : Single plant with its relatively bare roots, as pulled apart from a clump and used for transplanting.

Spring Tide : A tide that rises highest and falls lowest from mean sea level, occurring at new or full moon.

Spring Tides : Increased tidal ranges that occur semimonthly when the moon is new or full and the pull of the sun and moon are nearly in phase. The spring range is larger than the mean range for semidiurnal or mixed tides.

Stem : Main axis of a plant, leaf-bearing and flower-bearing, as distinguished from the root-bearing axis.

Stillwater Level : Elevation that the surface of the water would assume if all wave action were absent.

Storm Surge : Rise above normal water level on the open coast due to action of wind on the water surface. Storm surge resulting from a hurricane also includes the rise in level due to atmospheric pressure reduction as well as that due to wind stress. See SETUP, WIND.

Suspended Load : Material moving in suspension in a turbulent field.

Swell : Wind-generated waves traveling out of their generating area. Swell characteristically exhibits a more regular and longer period and has flatter crests than waves within their fetch.

Tidal Period : Interval between two consecutive like phases of the tide.

Tidal Range : Difference in height between consecutive high and low (or higher and lower low) waters. The mean range is the difference in height between mean high water and mean low water. The diurnal range is the difference in height between mean higher high water and mean lower low water. For diurnal tides, the mean and diurnal range are identical. For semidiurnal and mixed tides, the spring range is the difference in height between the high and low waters during the time of spring tides.

Tide : Periodic rising and falling of water resulting from gravitational attraction of the moon, sun and other astronomical bodies acting upon the rotating earth. Although the accompanying horizontal movement of the water resulting from the same cause is also sometimes called tide, it is preferable to designate the latter as tidal current, reserving the name TIDE for vertical movement.

Tide, Ebb : See EBB TIDE.

Tiller : A plant shoot which springs from the root or bottom of the original plant stalk.

Topography : Configuration of a surface, including relief, position of streams, roads, buildings, etc.

Transplant : SHOOT or CULM removed from one location and replanted in another.

Trough of Wave : Lowest part of a waveform between successive crests. Also, that part of a wave below stillwater level.

Updrift : Direction opposite the predominant movement of littoral materials in longshore transport.

Wake (boat) : Waves generated by the motion of a vessel through water.

Wale : Horizontal beam on a bulkhead used to transfer horizontal loads against the structure laterally along it and hold it in a straight alignment.

Waterline : Juncture of land and sea. This line migrates, changing with the tide or other fluctuation in water level. Where waves are present on the beach, this line is also known as the limit of backrush. (Approximately, the intersection of land with the still water level). See also SHORELINE.

Wave : Ridge, deformation, or undulation of the surface of a liquid.

Wave Climate : Normal seasonal wave regimen along a shoreline.

Wave Crest : Highest part of a wave or that part above the still-water level

Wave Diffraction : see DIFFRACTION

Wave Direction : Direction from which a wave approaches.

Wave Height : The vertical distance between a wave crest and the preceding trough.

Wave Length : The horizontal distance between similar points on two successive waves (for example, crest to crest or trough to trough), measured (perpendicular to the crest) in the direction of wave travel.

Wave Period : The time in which a wave crest travels a distance equal to one wavelength. Can be measured as the time for two successive wave crests to pass a fixed point.

Wave Refraction : See REFRACTION (of water waves).

Wave Steepness : Ratio of wave height to wavelength.

Wave Train : Series of waves from the same direction.

Wave Trough : Lowest part of a wave form between successive crests. Also, that part of a wave below stillwater level.

Weep Hole : Hole through a solid revetment, bulkhead, or seawall for relieving pore pressure.

Wind Setup : See SETUP, WIND.

Wind Waves : (1) Waves being formed and built up by wind. (2) Loosely, any waves generated by wind.

Windward : Direction *from* which wind is blowing.

SOURCES OF ILLUSTRATIONS

PART III

Kursus Pentai (Bandung Indonesia) vol. III. 1-37

PART IV

Akle...17,18
Authors, HAECON, Miscellaneous 1; 2; 7; 8; 9; 21; 26; 27; 31; 32; 33; 34; 35; 36
Claeys...29 (a→b)
Corine...3; 4; 5; 6
Flemming 15
Howa 19
Kawakami *et al*... 6 (a,b)
Kerckaert... 30
Nakayame *et al* 28
Smith and Jackson 27
U.S. Army, Corps of Engineers (Rogers, Golden & Halpern)10, 11, 12, 14, 16, 20, 22, 23, 24, 25

INDEX

A

accretion27, 112; 125; 133, 157; 164; 165; 170; 175; 177; 182; 185; 187; 188; 190; 209; 213; 214; 215; 216; 221; 222; 227; 230; 259; 263; 289; 296; 298; 306; 309
Ackers-White (prediction method)....80; 81
adaptation.... 9; 10; 13, 21; 29; 40; 41; 42; 45
adaptive measure....20; 31
adaptive response....28; 34; 37; 44
allocation criteria....44
anthropic activities....2
artificial nourishment ...13, 173; 198; 215; 222; 223; 226; 270; 276; 278; 294; 309
artificial reef....192; 209; 214; 215; 216; 217; 232
atoll....41

B

β-factor....60
backhoe bucket dredger....242
Bagnold (prediction method)....73
Bakker theory....117; 118
barkhane....321
bar profile....123; 125
barge unloading dredger....257; 311
barrier island....7, 26, 174; 180; 283; 308
beach administration....233; 280
beach nourishment2, 26; 27, 168; 169; 170; 174; 181; 185; 190; 191; 215; 223; 224; 227; 228; 230; 231; 232; 233; 242; 250; 251; 254; 256; 258; 261; 262; 263; 264; 265; 266; 269; 272; 273; 275; 276; 278; 279; 280; 281; 282; 284; 290; 291; 292; 294; 295, 318
beach profile14, 118; 121; 123; 126; 127; 128; 130; 139; 141; 144, 164; 168; 169; 170; 179; 216; 269; 278; 279; 294; 299, 311, 312
beach profiling....233; 280
berm123; 125; 141; 145; 146, 157; 158; 160; 163; 164; 165; 169; 171; 229; 230; 232; 268; 269; 277; 281; 295; 302
Bijker (prediction method)....73; 74
Bijker (transport formula)....114
biological diversity....20
booster.... 223; 252; 261; 269
boundary line....2
box model....137; 138

breaker94, 112; 113; 115; 125; 127; 133; 136; 140; 160; 165; 166; 171; 179; 239; 296
breaker zone 115; 133; 136; 160; 165; 171; 239
breaking85; 90; 91; 92; 93; 94; 95; 96; 113; 114; 136; 144; 145; 160; 166; 187; 208; 227; 228; 235; 239; 249; 281; 288; 294
breaking line 227
breakwater144; 185; 192; 194; 195; 196; 204; 207; 208; 209; 210; 214; 217; 218; 219; 223; 224; 281; 288; 301; 308; 309
bucket dredger 229; 242
bulkhead 209; 210; 218; 219; 265
by-passing 203; 214; 265; 280; 281; 288

C

CERC 113; 114; 117; 119; 133; 149; 292
channelizing 27
Chézy coefficient 74; 76
cliffs 319, 320
climate change 29; 35; 40; 41
coastal ecosystem 25; 26
coastal fringe 1; 5; 10
coastal hazard 20
coastal morphology 117
coastal zone1; 2; 3; 5; 6; 7; 8; 9; 10; 11; 12; 15; 23; 24; 31; 33; 34; 37; 38; 42; 46; 47; 137; 156; 167; 175; 271; 273; 282; 286; 287; 294;
coastal zone management 8; 9; 37; 42; 286
Coastal Zone Management Act : .... see CZMA
Concentration (profile) 62; 68
conflict 5; 12; 35; 286; 287
conservation 5; 8; 9; 10
continuum model 138
convection 62; 69; 97; 98
coral reef 20; 22; 156; 157; 309;
critical erosion 267;
critical velocity 252; 253
cross-shore transport 117; 121
cutter suction dredger 248; 249; 250; 252; 256
CZMA 1; 12

D

dam 206; 308
delta 212
developing countries 3; 6; 10; 24; 28; 29; 30; 34; 38; 40; 42; 43; 185; 189; 267
dewatering 276; 277

diffusion equation ... 62; 97; 98; 99; 129
dike ... 269; 273
dolosses ... 219
draghead ... 242; 248
dredging27; 156; 158; 167; 182; 189; 203; 222; 223; 230; 232; 241; 242; 248; 249; 250; 251; 252; 257; 261; 273; 280; 281; 287
dunal barrier ... 11
dune2; 69; 70; 128; 139; 140; 157; 169; 174; 180; 186; 188; 190; 214; 227; 228; 232; 259; 280; 297; 304; 305; 306; 307; 312, 321
dune restoration ... 2
dunes ... 321

E

economic rent ... 8; 261;
edge waves ... 165; 166
EIA : ... see environmental impact
Einstein (prediction method) ... 72; 73
emergency management ... 43
eminent domain ... 37; 190
energy balance concept ... 54; 73
environmental impact ... 1; 13; 15; 26; 37; 266; 280; 286; 301; 305
equilibrium profile ... 119; 123; 127; 170; 230; 237; 275; 277
eroded profile ... 123; 125

F

Φ-factor ... 60; 61; 62
feeder beach ... 229; 231
feeder berm ... 164; 229; 230; 268; 269; 277; 295
fisheries ... 6; 7; 11; 28; 40; 230;
flood control ... 2

G

geological hazard ... 2
geotextile ... 187
global climate change ... 40; 41
gravel ... 7; 156; 179; 182; 186; 232; 250; 264; 265; 275; 296;
groin168; 169; 179; 197; 198; 199; 202; 203; 221; 222; 233; 261; 262; 263; 275; 281; 309

H

haloclastisis ... 319
harbor6; 8; 109; 112; 182; 185; 190; 192; 195; 198; 210; 212; 223; 233; 259; 261; 280; 281; 282; 301; 306; 308;
headland ... 144; 218; 285; 287
health report ... 14

hinterland........1; 5; 9; 11; 269; 270
history (of protection)........VII
hopper suction dredger........262
human interference........2; 5
hydroclastisis........319

I

Inman formula........317
island nation........29

J

jetties........185; 191; 203; 212; 281; 282; 298; 304; 306; 308; 309

L

legislation........1; 12; 37
littoral drift........314
Longard tube........221
longshore85; 87; 112; 113; 114; 117; 122; 129; 130; 131; 132; 133; 134; 136; 139; 145; 157; 158; 163; 164; 165; 166; 167; 185; 197; 203; 212; 215; 227; 237; 259; 266; 268; 272; 296; 297; 298; 299; 304; 306; 308; 309
longshore current85; 87; 112; 113; 114; 136; 164; 166; 167; 185; 203; 237; 259; 304
longshore transport........113-117

M

management1; 2; 3; 7; 8; 9; 10; 11; 12; 15; 16; 20; 23; 24; 37; 38; 40; 42; 43; 46; 47; 188; 233; 280; 286; 287; 293; 294; 301
mangrove........26; 28; 38; 181; 187; 188; 309;
mariculture........6; 43
marine mining........2; 247
mineral resource........3; 6; 46
mining........2; 5; 7; 8; 23; 156; 158; 181; 232; 247
modeling (math.)........141; 148
models (coastal morphology)........129; 138
moles........112

N

nearshore current........158; 159; 166
notch (in cliff)........319

O

overfill ratio........230; 234

P

planning3; 5; 9; 10; 15; 23; 24; 29; 37; 40; 42; 43; 189; 235; 241; 265; 280; 286; 294
population explosion 1
power station 6; 14; 207
prediction method (see also name of) 81; 82
profiling 13; 216; 228; 233; 280; 309

Q

Quaternary 5

R

ramifications 9
rates (transport) 88
rehandle pit 256; 257; 311
references (and notes)3-6; 18; 19; 45-47; 82-84; 104-106; 111; 148-151; 285-295; 300
relative retreat rate 236
relocation 35; 38; 41; 43; 280
renourishment factor 236
resettlement 21; 22; 29; 38; 43
retreat13; 21; 22; 23; 25; 26; 28; 29; 35; 38; 122; 174; 180; 181; 184; 185; 187; 198; 236; 263; 282; 284; 302
revetment 187; 210; 214; 266; 288
rip current 85; 87; 164; 166; 215
riser pontoon 255
Rivers and Harbors Act 12

S

salinity intrusion 23; 24
salt water intrusion 22; 27; 31; 188
saltation 52; 68; 296
sampling 91; 95; 233; 280
sand2; 7; 20; 27; 85; 88; 90; 95; 99; 113; 114; 115; 128; 129; 130; 133; 134; 139; 140; 156; 157; 158; 159; 160; 164; 165; 166; 167; 168; 169; 171; 179; 181; 182; 184; 185; 186; 189; 190; 197; 198; 202; 203; 209; 210; 214; 215; 216; 220; 221; 222; 223; 227; 228; 229; 230; 232; 233; 235; 250; 251; 252; 254; 257; 259; 261; 262; 263; 264; 265;
sand fence 220
scarp slope 320
seascape 217; 289
seawall 142; 192; 221; 222; 270

sediment budget ... 318
sediment transport 14, 52; 72; 74; 75, 87; 91, 109; 112; 113; 115; 116; 117; 118; 131; 136; 137; 142; 143; 144; 145; 147, 158; 197; 215; 221; 233; 266; 280; 297
setback line ... 20
settling effect ... 63; 65; 66; 68 103; 104;
sheet flow regime ... 90; 96
SIA (Social Impact) ... 33; 34; 35; 36; 283; 307; 308
Sitarz equations ... 312
site study ... 13
spark ... 217
spit ... 262; 266; 308
splithopper ... 248
stable berm ... 229; 230
standing wave ... 164; 165; ; 207
stationary suction dredger ... 250; 255; 256; 257
step profile ... 123; 124; 125
stockpiling ... 231
storm surge ... 29; 112; 127; 128; 139; 140; 141; 158; 166; 167; 169; 186; 301
storm wave ... 157; 179; 186; 227; 237; 238; 239; 259; 273
surf ... 101; 126; 127; 158; 160; 165; 166; 171; 172; 210; 229; 238; 278; 296
surgebreakers ... 221
suspended sediment (detection) ... 107
suspended sediment (fall velocity) ... 101-104
sustainability ... 1; 14
Swart model ... 118-121
swash ... 141; 160; 171; 227; 228
swell waves ... 164

T

tetrapod ... 220
tide ... 107; 110; 169; 190; 255; 259; 263; 264; 269; 278
time scale ... 14; 122; 140
total load transport ... 77
tourism ... 6; 9; 10; 11; 30; 43; 156; 182; 184; 261; 269; 278; 286
trailer dredger ... 242
trailing suction hopper dredger ... 242; 248; 250; 251; 252; 255; 256; 257
transport rats ... 96
tribar ... 220
tripod ... 219
turbidity ... 230; 276; 280
turbidity plumes ... 230
turbulence damping ... 63; 65; 66; 68; 75; 97

V

Van Rijn (prediction method) 74-78

W

wave regime 29; 221
waves14; 24; 30; 85; 90; 91; 92; 93; 94; 95; 96; 98; 112; 113; 114; 117; 121; 124; 133; 136; 137; 140; 144; 157; 158; 159; 160; 162; 163; 164; 165; 166; 170; 172; 173; 179; 186; 187; 189; 198; 202; 204; 207; 208; 210; 215; 217; 223; 227; 229; 233; 238; 241; 264; 265; 278; 280; 301; 304; 305; 306
wetlands 20; 23; 25; 26; 27; 33; 36; 46; 309
wind wave 207; 264

Springer and the environment

At Springer we firmly believe that an international science publisher has a special obligation to the environment, and our corporate policies consistently reflect this conviction.

We also expect our business partners – paper mills, printers, packaging manufacturers, etc. – to commit themselves to using materials and production processes that do not harm the environment. The paper in this book is made from low- or no-chlorine pulp and is acid free, in conformance with international standards for paper permanency.

Springer

Druck: Strauss Offsetdruck, Mörlenbach
Verarbeitung: Schäffer, Grünstadt

Lecture Notes in Earth Sciences

Vol. 1: Sedimentary and Evolutionary Cycles. Edited by U. Bayer and A. Seilacher. VI, 465 pages. 1985. (out of print).

Vol. 2: U. Bayer, Pattern Recognition Problems in Geology and Paleontology. VII, 229 pages. 1985. (out of print).

Vol. 3: Th. Aigner, Storm Depositional Systems. VIII, 174 pages. 1985.

Vol. 4: Aspects of Fluvial Sedimentation in the Lower Triassic Buntsandstein of Europe. Edited by D. Mader. VIII, 626 pages. 1985. (out of print).

Vol. 5: Paleogeothermics. Edited by G. Buntebarth and L. Stegena. II, 234 pages. 1986.

Vol. 6: W. Ricken, Diagenetic Bedding. X, 210 pages. 1986.

Vol. 7: Mathematical and Numerical Techniques in Physical Geodesy. Edited by H. Sünkel. IX, 548 pages. 1986.

Vol. 8: Global Bio-Events. Edited by O. H. Walliser. IX, 442 pages. 1986.

Vol. 9: G. Gerdes, W. E. Krumbein, Biolaminated Deposits. IX, 183 pages. 1987.

Vol. 10: T.M. Peryt (Ed.), The Zechstein Facies in Europe. V, 272 pages. 1987.

Vol. 11: L. Landner (Ed.), Contamination of the Environment. Proceedings, 1986. VII, 190 pages.1987.

Vol. 12: S. Turner (Ed.), Applied Geodesy. VIII, 393 pages. 1987.

Vol. 13: T. M. Peryt (Ed.), Evaporite Basins. V, 188 pages. 1987.

Vol. 14: N. Cristescu, H. I. Ene (Eds.), Rock and Soil Rheology. VIII, 289 pages. 1988.

Vol. 15: V. H. Jacobshagen (Ed.), The Atlas System of Morocco. VI, 499 pages. 1988.

Vol. 16: H. Wanner, U. Siegenthaler (Eds.), Long and Short Term Variability of Climate. VII, 175 pages. 1988.

Vol. 17: H. Bahlburg, Ch. Breitkreuz, P. Giese (Eds.), The Southern Central Andes. VIII, 261 pages. 1988.

Vol. 18: N.M.S. Rock, Numerical Geology. XI, 427 pages. 1988.

Vol. 19: E. Groten, R. Strauß (Eds.), GPS-Techniques Applied to Geodesy and Surveying. XVII, 532 pages. 1988.

Vol. 20: P. Baccini (Ed.), The Landfill. IX, 439 pages. 1989.

Vol. 21: U. Förstner, Contaminated Sediments. V, 157 pages. 1989.

Vol. 22: I. I. Mueller, S. Zerbini (Eds.), The Interdisciplinary Role of Space Geodesy. XV, 300 pages. 1989.

Vol. 23: K. B. Föllmi, Evolution of the Mid-Cretaceous Triad. VII, 153 pages. 1989.

Vol. 24: B. Knipping, Basalt Intrusions in Evaporites. VI, 132 pages. 1989.

Vol. 25: F. Sansò, R. Rummel (Eds.), Theory of Satellite Geodesy and Gravity Field Theory. XII, 491 pages. 1989.

Vol. 26: R. D. Stoll, Sediment Acoustics. V, 155 pages. 1989.

Vol. 27: G.-P. Merkler, H. Militzer, H. Hötzl, H. Armbruster, J. Brauns (Eds.), Detection of Subsurface Flow Phenomena. IX, 514 pages. 1989.

Vol. 28: V. Mosbrugger, The Tree Habit in Land Plants. V, 161 pages. 1990.

Vol. 29: F. K. Brunner, C. Rizos (Eds.), Developments in Four-Dimensional Geodesy. X, 264 pages. 1990.

Vol. 30: E. G. Kauffman, O.H. Walliser (Eds.), Extinction Events in Earth History. VI, 432 pages. 1990.

Vol. 31: K.-R. Koch, Bayesian Inference with Geodetic Applications. IX, 198 pages. 1990.

Vol. 32: B. Lehmann, Metallogeny of Tin. VIII, 211 pages. 1990.

Vol. 33: B. Allard, H. Borén, A. Grimvall (Eds.), Humic Substances in the Aquatic and Terrestrial Environment. VIII, 514 pages. 1991.

Vol. 34: R. Stein, Accumulation of Organic Carbon in Marine Sediments. XIII, 217 pages. 1991.

Vol. 35: L. Håkanson, Ecometric and Dynamic Modelling. VI, 158 pages. 1991.

Vol. 36: D. Shangguan, Cellular Growth of Crystals. XV, 209 pages. 1991.

Vol. 37: A. Armanini, G. Di Silvio (Eds.), Fluvial Hydraulics of Mountain Regions. X, 468 pages. 1991.

Vol. 38: W. Smykatz-Kloss, S. St. J. Warne, Thermal Analysis in the Geosciences. XII, 379 pages. 1991.

Vol. 39: S.-E. Hjelt, Pragmatic Inversion of Geophysical Data. IX, 262 pages. 1992.

Vol. 40: S. W. Petters, Regional Geology of Africa. XXIII, 722 pages. 1991.

Vol. 41: R. Pflug, J. W. Harbaugh (Eds.), Computer Graphics in Geology. XVII, 298 pages. 1992.

Vol. 42: A. Cendrero, G. Lüttig, F. Chr. Wolff (Eds.), Planning the Use of the Earth's Surface. IX, 556 pages. 1992.

Vol. 43: N. Clauer, S. Chaudhuri (Eds.), Isotopic Signatures and Sedimentary Records. VIII, 529 pages. 1992.

Vol. 44: D. A. Edwards, Turbidity Currents: Dynamics, Deposits and Reversals. XIII, 175 pages. 1993.

Vol. 45: A. G. Herrmann, B. Knipping, Waste Disposal and Evaporites. XII, 193 pages. 1993.

Vol. 46: G. Galli, Temporal and Spatial Patterns in Carbonate Platforms. IX, 325 pages. 1993.

Vol. 47: R. L. Littke, Deposition, Diagenesis and Weathering of Organic Matter-Rich Sediments. IX, 216 pages. 1993.

Vol. 48: B. R. Roberts, Water Management in Desert Environments. XVII, 337 pages. 1993.

Vol. 49: J. F. W. Negendank, B. Zolitschka (Eds.), Paleolimnology of European Maar Lakes. IX, 513 pages. 1993.

Vol. 50: R. Rummel, F. Sansò (Eds.), Satellite Altimetry in Geodesy and Oceanography. XII, 479 pages. 1993.

Vol. 51: W. Ricken, Sedimentation as a Three-Component System. XII, 211 pages. 1993.

Vol. 52: P. Ergenzinger, K.-H. Schmidt (Eds.), Dynamics and Geomorphology of Mountain Rivers. VIII, 326 pages. 1994.

Vol. 53: F. Scherbaum, Basic Concepts in Digital Signal Processing for Seismologists. X, 158 pages. 1994.

Vol. 54: J. J. P. Zijlstra, The Sedimentology of Chalk. IX, 194 pages. 1995.

Vol. 55: J. A. Scales, Theory of Seismic Imaging. XV, 291 pages. 1995.

Vol. 56: D. Müller, D. I. Groves, Potassic Igneous Rocks and Associated Gold-Copper Mineralization. 2nd updated and enlarged Edition. XIII, 238 pages. 1997.

Vol. 57: E. Lallier-Vergès, N.-P. Tribovillard, P. Bertrand (Eds.), Organic Matter Accumulation. VIII, 187 pages. 1995.

Vol. 58: G. Sarwar, G. M. Friedman, Post-Devonian Sediment Cover over New York State. VIII, 113 pages. 1995.

Vol. 59: A. C. Kibblewhite, C. Y. Wu, Wave Interactions As a Seismo-acoustic Source. XIX, 313 pages. 1996.

Vol. 60: A. Kleusberg, P. J. G. Teunissen (Eds.), GPS for Geodesy. VII, 407 pages. 1996.

Vol. 61: M. Breunig, Integration of Spatial Information for Geo-Information Systems. XI, 171 pages. 1996.

Vol. 62: H. V. Lyatsky, Continental-Crust Structures on the Continental Margin of Western North America. XIX, 352 pages. 1996.

Vol. 63: B. H. Jacobsen, K. Mosegaard, P. Sibani (Eds.), Inverse Methods. XVI, 341 pages, 1996.

Vol. 64: A. Armanini, M. Michiue (Eds.), Recent Developments on Debris Flows. X, 226 pages. 1997.

Vol. 65: F. Sansò, R. Rummel (Eds.), Geodetic Boundary Value Problems in View of the One Centimeter Geoid. XIX, 592 pages. 1997.

Vol. 66: H. Wilhelm, W. Zürn, H.-G. Wenzel (Eds.), Tidal Phenomena. VII, 398 pages. 1997.

Vol. 67: S. L. Webb, Silicate Melts. VIII. 74 pages. 1997.

Vol. 68: P. Stille, G. Shields, Radiogenetic Isotope Geochemistry of Sedimentary and Aquatic Systems. XI, 217 pages. 1997.

Vol. 70: R. H. Charlier, C. P. De Meyer, Coastal Erosion – Response and Management. XVI, 343 pages. 1998.